# Principles of Agricultural Management

# Principles of Agricultural Management

Edited by Michelle Jacobs

MURPHY & MOORE
www.murphy-moorepublishing.com

Published by Murphy & Moore Publishing,
1 Rockefeller Plaza,
New York City, NY 10020, USA
www.murphy-moorepublishing.com

**Principles of Agricultural Management**
Edited by Michelle Jacobs



International Standard Book Number: 978-1-63987-454-5 (Hardback)



**Cataloging-in-Publication Data**

Principles of agricultural management / edited by Michelle Jacobs.
p. cm.
Includes bibliographical references and index.
ISBN 978-1-63987-454-5
1. Farm management. 2. Agriculture--Management.
3. Agricultural landscape management. 4. Agricultural systems. I. Jacobs, Michelle.
S561 .P75 2022
630.68--dc23

# TABLE OF CONTENTS

# PREFACE

This book aims to highlight the current researches and provides a platform to further the scope of innovations in this area. This book is a product of the combined efforts of many researchers and scientists, after going through thorough studies and analysis from different parts of the world. The objective of this book is to provide the readers with the latest information of the field.

The use of scientific principles for the cultivation of plants and rearing of livestock is termed as agriculture. It aims to produce different types of products for human consumption such as fuels, food, raw materials and fibers. Agricultural practices can be broadly classified into pastoralism, intensive farming, shifting cultivation, and subsistence farming. Agricultural management refers to all those activities and techniques which ensure an effective supervision and implementation at various farming and agricultural sites. It aims to coordinate and plan different types of operations like planting, harvesting, fertilization, etc., to run a farming site in the most effective manner. The topics included in this book on agricultural management are of utmost significance and bound to provide incredible insights to readers. From theories to research to practical applications, case studies related to all contemporary topics of relevance to this field have been included in it. This book will help the readers in keeping pace with the rapid changes in this field.

I would like to express my sincere thanks to the authors for their dedicated efforts in the completion of this book. I acknowledge the efforts of the publisher for providing constant support. Lastly, I would like to thank my family for their support in all academic endeavors.

**Editor**

# Working Effectiveness of Dissemination Channels as Perceived by Potato Growers

*Tahir Munir Butt[1], Muhammad Zakaria Yousuf Hassan[2], Shahbaz Talib Sahi[3], Khalid Mehmood[4], Mashood-ul-Subtain[5] and Naeem-ul Hassan[6]*

Abstract

A research can also reveal the cooperative effectiveness of agencies such as evaluation will lead to discover the weakness and strengths for further improvement of this program. The present study was designed in 2009 especially to see the effectiveness of the working of agricultural extension staff as perceived by farmers in district Okara-Pakistan. The data were collected with the help of a specifically designed and pre-tested interview schedule and total sample was 300 respondents. The results showed that most 45.0% of the respondents belonged to the old age (31-40 years) category and most 30% of the respondents were above illiterate. An overwhelming majority 77.33% of the respondents was in fall radio category. Whereas, only 29.33% of the respondents reported that they had contact with Extension Field Staff. An overwhelming majority 74.33% of the respondents indicated lack of mobility as the major constraints in approaching agricultural extension education services. It was concluded that different mass media were not fully utilized in the area which hindered not only awareness level of the respondents but also adversely effect the adoption level regarding the latest production technology related agriculture sector. On the basis of conclusions it was recommended that for first of all the educational level of the study area should be increased and Government should ensure adequate availability of rural infrastructure facilities to enable more of the dwellers cultivate the habit of utilizing medias channels as source of information in agriculture production.

***Keywords:***
***Agriculture, Potato, Farmers, Information, Communication, Channels***

[1] *Department of Agri. Extension Education, University of Agriculture, Faisalabad.*
[2] *Agricultural Department, Punjab-Pakistan.*
[3] *Department of Plant Pathology, University of Agriculture, Faisalabad.*
[4] *Department of Agricultural Extension in University of Agriculture, Faisalabad-Pakistan.*
[5&6] *University of Agriculture, Faisalabad, Depalpur Campus, Okara-Pakistan.*
*Corresponding author; e-mail: tahirmunir@uaf.edu.pk*

## INTRODUCTION

By 2010 (Government of Pakistan), the population is estimated to have reached 169.90 million. Thus in roughly three generations, While Pakistan has more mouths to feed, more families to house, more children to educate, and more people looking for gainful employment. It is expected that Pakistan will become the fourth largest nation on earth in population term by 2050. Pakistan is a developing country and agriculture is the mainstay of its economy. Its contribution to GDP is nearly 20.9%, and 43.4% of the total employment is generated from agriculture. This sector not only fulfils the daily living requirements of population of the country but is also the main feeder of raw materials to all industries (GOP, 2007). Agriculture remains the dominant source of employment in Pakistan and approximately 66% of the country's population live in rural areas and directly or indirectly relies on the agriculture sector for their livelihood. Pakistan is a federation of four provinces where the provincial governments are primarily responsible for agricultural research and extension activities. Several extension education program/approaches, designed primarily to improve the living standard of rural people through increased agricultural production and improved farm income, have been tried. Crop yields in Pakistan are generally low as compared to the yields of many other developing and developed countries. Moreover the average yields of the major crops are low than their potential agricultural production in Pakistan is affected by a number of factors including lack of continuity in agricultural policies, programs agricultural education, research and extension activities, buyer's and middleman "Mafia", lack of necessary infrastructure for farm exports, deficient management and marketing skills, and unproductive tenancy systems, etc. (NRSP, 1999).

The available technology can be transferred through agricultural extension field staff, because agriculture extension is one of the means available to help alleviate poverty and improve food security and also enhancement the agricultural production and uplift the rural community. It promotes the transfer and exchange of innovation that can be converted into functional knowledge, which is instrumental in helping to develop enterprises that promote productivity and generate income in the present climate of exchange. Agricultural extension is a unique service, which helps to provide small farmers and the rural poor living in remote areas an access to the agriculture innovation. While it can also provide these populations with services to increase their productivity (World Bank, 2003). It is assumed that modern technology available at technology producing centers but not effectively transferred to the ultimately users. So, emphasis should be laid upon the most modern agricultural techniques which were possible by the dissemination of agricultural information among the farmers (Butt, 2002). For the dissemination of these modern technologies extension agencies utilize different methods like individual, group and mass media. Mahmood and Sheikh (2005) stated that creation of awareness is the first step towards the adoption process and enhancement of the agricultural production (Suman, 2003; Yawson et al., 2010).

The present study was designed especially to see the effectiveness of the working of dissemination channel such as radio, TV, agricultural extension staff etc. as perceived by farmers in District Okara, Punjab, Pakistan.

## MATERIALS AND METHODS

The Okara District is famous for its fertile lands, peaceful natural environment and green fields of potato, sugarcane, wheat, rice and maize crops. Okara district comprises three tehsils namely Okara, Renala Khurd and Depalpur. All the three tehsils were selected for the study on the following grounds.

I. The educational level in all the selected tehsils was very low.

II. All the tehsils were similar in socio-economic and agricultural conditions

III. Traditional norms and male dominancy in whole the area prevails.

### *Selection of sample for the study*

As the number of villages in all three tehsils varies too much, therefore, researchers decided to give equal chance to all tehsils and on these bases it was decided that hundred respondents from each tehsil were selected randomly in the figure I. For the selection of villages lottery method was adopted (Thakur, 2003) and names of villages were written on pieces of paper, then tehsil wise ten tickets were selected. Thus making a total of 30 villages, the respondents were selected through simple random sampling method (Ogunjuyigbe, 2005). A well structured interview schedule having open and close-ended questions was prepared (Acharya et al., 2005; Tucker et al., 2005). Out of total these villages 10 farmers were selected randomly in each village at each tehsil. Thus the total sample was 300 respondents. The data were collected with the help of a specifically designed and pre-tested interview schedule. The reliability of data was checked through Cronbach's alpha. The data thus collected were analyzed by using computer program (software), Statistical Package for Social Sciences (SPSS) for reporting results and drawing conclusions.

**1.Tehsil Depalpur:**
10 Village x 10 Farmers = 100 respondents
**2.Tehsil Renala Khurd:**
10 Village x 10 Farmers = 100 respondents
**3.Tehsil Okara:**
10 Village x 10 Farmers = 100 respondents
**Total Sample = 300 Respondents**

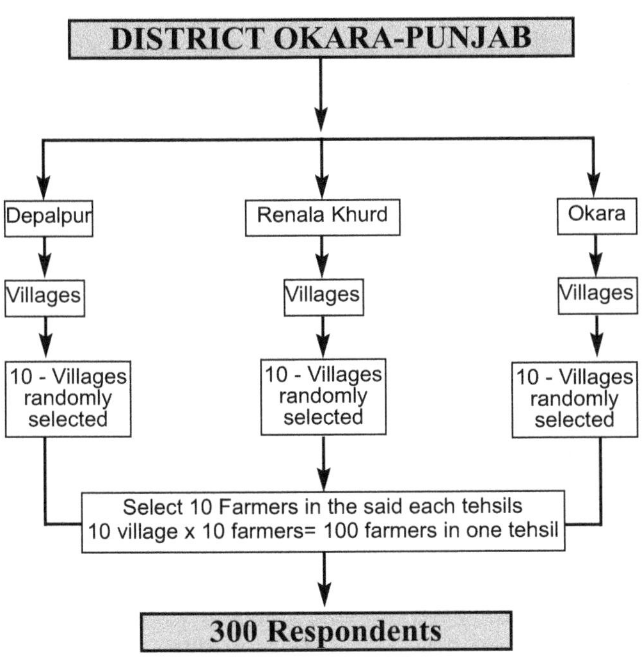


**Figure 1**

## RESULTS AND DISCUSSION

Respondents were interviewed to find out their perceptions about agricultural extension education services in the form of effectiveness of dissemination channels in tabulated, figure and presented in tables. Table 1 reveals that less than half (45.0%) of the respondents were 31-40 years of age. However, 25.0% of the respondents were of upto the age of 30 years. Only 30.0% of the respondents were 41 and above year of age. Table 1 also indicates that a 64.33% of the respondents had an education from primary to matric (ten year education) and above level. However, rest of the respondents (35.67%) was illiterate. The educational level of the study area was very low higher. However, only 12% of the respondents were primary to middle. More or less similar results in the age and education were also observed by Fabiyi et al. (2007). The data presented in table 1 reveals that 55.0% of the respondents had small landholding upto 12.5 acres followed by medium 31.67% up to 12.5-25 acres and large holders were 13.33% owned land above 25 acres. The results of this data confirmed by Irfan (2005) and Hassan et al. (2005) who found that majority 68.3% of the respondents had upto 12.5 acres of land, 20.0% owned 12.5 to 25 acres and 11.7% of the respondents had above 25 acres of land. The sequence of owning is same as of present research study. Table 2 depicts that respondents were very in favor of agricultural dissemination services for them as 93.33% of the respondents were agree, followed by 6.00% who were agree. On the other hand, with the statement that farmer perception about the dissemination channels essentially need agricultural extension services, only 0.67% respondents disagreed. So we can say that majority of the respondents was in favor of the agricultural dissemination services for rural women. In this connection Sailaja and Reddy (2003) reported that for consistent growth of agricultural production and stabilization of income in agriculture, it was necessary to involve women in increased adoption of improved farm practices and Medias. This could be attained through effective training and extension dissemination channels programmes

Table 1: Distribution of the respondents according to the age, education and land holding (n=300 respondents)

| Characteristic | frequency | Percentage |
|---|---|---|
| **Age** | | |
| Upto 30 year | 75 | 25.00 |
| 31- 40 year | 135 | 45.00 |
| 41-above year | 90 | 30.00 |
| **Education** | | |
| Illiterate | 107 | 35.67 |
| Primary | 78 | 26.00 |
| Middle | 67 | 22.33 |
| Matric & above | 48 | 16.00 |
| **Family land holding** | | |
| Small (up to 12. 5 acre) | 165 | 55.00 |
| Medium (12.5-25 acres) | 95 | 31.67 |
| Large (above 25 acres) | 40 | 13.33 |

specially designed and based on technological felt needs of farmer to enhance productivity. Table 3 shows that simple to large majority (65.0-91.67%) of the respondents were aware of potato varieties like Desiri, Santey and Diament respectively, whereas, awareness about other varieties was at low level. On the other hand, adoption level for all varieties was low except Santey which was adopted by 71.67% of the respondents. Table 4 reveals that radio was the most used mass media for the dissemination of potato technology and utilized by 77.33% of the respondents. The other method which was benefited by 46.0% of the respondents was meetings. Whereas, other Medias were utilized by lesser number of respondents. However, internet was used by negligible number (05.0%) of respondents. As radio was less expensive and portable therefore use by both literate and illiterate persons, whereas internet was a new technology and due to its pre-requisites less available in the study area therefore its utilization was very low. Table 5 an overwhelming majority 74.33% of the respondents indicated lack of mobility as the major constraints in approaching agricultural extension education services with mean 0.98 and S.D. 0.27. Mobility means motorcycle, car and any vehicles. There may be less availability of vehicles to access these agricultural extension services in the form of dissemination channels. Moreover a large majority 71.66% of the respondents indicated lack of access to credit as the core constraint in approaching agricultural extension education dissemination services having mean 0.92 and S.D. 0.17. Shelly and Costa (2000) also indicated lack of credit facility and lack of resource availability as the main constraints which are being faced by rural peoples involved in the agricultural development programme in Bangladesh. More or less similar results were also observed by Raju et al. (2001), Sadaf et al. (2005) and FAO (2001).

## CONCLUSIONS AND RECOMMENDATIONS

It was concluded that different dimensional channels were not fully utilized in the area which hindered not only awareness level of the respondents but also adversely effect the adoption level regarding the latest production technology related to potato. Following recommendation has been made: On the basis of conclusions it was also recommended that for first of all the educational level of the study area should be increased so as farming community is able to get benefit from printed material. Govt. also arranges the video vans to create awareness and enhancement the education system each Tehsil level for the betterment in agricultural development. Government should organize seminars and workshops for both public and private sector, for creating awareness about more advanced and sophisticated production technology of agriculture. To increase

Table 2: Distribution of the respondents according to their perception about the dissemination channels. (n=300 respondents)

| Statements | Disagree | | Somewhat agree | | Agree | |
|---|---|---|---|---|---|---|
| | f | % | f | % | f | % |
| Agricultural extension dissemination channels | 2 | 0.67 | 18 | 6.00 | 280 | 93.33 |

Table 3: Distribution of the respondents according to their awareness and adoption of the recommendations regarding different varieties of potato (n=300 respondents)

| Varieties | Awareness | | Adoption | |
|---|---|---|---|---|
| | f | % | f | % |
| Desiri | 195 | 65.00 | 100 | 33.33 |
| Cardinal | 60 | 20.00 | 40 | 13.33 |
| Roberz | 40 | 13.33 | 25 | 8.33 |
| Ustericks | 65 | 21.67 | 35 | 11.67 |
| Berna | 50 | 16.67 | 20 | 6.67 |
| Caroda | 75 | 25.00 | 55 | 18.33 |
| Valja | 80 | 26.67 | 25 | 8.33 |
| Diamint | 275 | 91.67 | 75 | 25.00 |
| Ejaks | 35 | 11.67 | - | - |
| Santey | 250 | 83.33 | 215 | 71.67 |

Table 4: Distribution of the respondents according the use of multidimensional channels in the promotion of latest technology for potato production (n= 300 respondents)

| Mass Media | f | % |
|---|---|---|
| Radio | 232 | 77.33 |
| Meetings | 138 | 46.00 |
| Agri. Deptt.(Extension Field Staff) | 88 | 29.33 |
| Key Informant | 88 | 29.33 |
| TV | 62 | 20.67 |
| Newspaper | 50 | 16.67 |
| Campaign/Focus Group etc | 40 | 13.33 |
| Print Media | 25 | 8.33 |
| Internet | 15 | 05.00 |

Table 5: Constraints faced by them in agricultural dissemination channels services (n=300)

| Constraints | f | % | Mean | S.D. |
|---|---|---|---|---|
| Mobility (any vehicles) | 223 | 74.33 | 0.98 | 0.27 |
| Access to credit | 21.5 | 71.66 | 0.92 | 0.17 |

the use of computer and internet both public and private sector should create awareness and establish telecenters for the dissemination of modern potato technology that ultimately results in the uplift of potato production.

## ACKNOWLEDGEMENTS

We would like to express our thanks to Muhammad Ahmad, Muhammad Yaqoob students of B.Sc (Hons.) Agriculture with class mates for their extensive help during the data collections.

## REFERENCES

1- Acharya, S., E. Yoshino, M. Jimba & S. Wakai. (2005). Empowering rural women through a community development approach in Nepal. Community Development Journal Advance Access publication, 1-13.

2- Butt, S. A. (2002). Role of television in the dissemination of agricultural technologies among the farmers of tehsil Faisalabad. M.Sc (Hons.) Agri. Ext. Thesis, Univ. of Agri., Faisalabad.

3- E. F. Fabiyi, B. B. Danladi, K. E. Akande & Y. Mahmood, (2007). Role of women in agriculture development and their constraints: A case study of Biliri local Government area, Gombe state, Nigeria. Pakistan Journal of Nutrition 6(6). 676-680.

4- FAO, (2001). Research and Extension: A gender perceptive. Women in Development Service (SDWW): Women and Population Division, Rome, Italy.

5- GOP. (2007). Economic Survey of Pakistan. Finance Division, Economic Advisor's Wing, Islamabad, Govt. of Pakistan, 166.

6- GOP. (2010). Economic Survey of Pakistan. Finance Division, Economic Advisor's Wing, Islamabad, Govt. of Pakistan.

7- Hassan, M. Z. Y., A. Majeed, & I. U Rehman. (2005). Correlation of demographic characters of the respondents with usefulness and effectiveness of technical trainings as organized by PRSP in district Muzaffargarh. The Indus Cotton, 2(3), 219-231.

8- Irfan, M. (2005). Comparative effectiveness of mass media in the dissemination of agricultural technologies among farmers of Lahore district. M.Sc. (Hons) Agri. Ext. Thesis, Univ. of Agri., Faisalabad.

9- Mahmood, M. A. & A.D. Sheikh. (2005). Crop yields from new technologies. "DailyDawn" March 28-April 3, 2005, III.

10- NRSP (1999). Training module on production technology of cotton for small farmers, Punjab Agricultural Task Force, Punjab.

11- Ogunjuyigbe. P. O., A. Akinlo & J. A. Ebigbola, (2005). Violence Against Women: An examination of men's attitudes and perceptions about wife beating and contraceptive use. Journal of Asian and African Studies, 40(3), 219-229.

12- Raju, K. A., G. S. Roy, T. S. Kamala, & Rani, M. S. (2001). Constraints and Suggestions for Effective Implementation of Farm Women Development

Programmes. MANAGE Extension Research Review, India.

13- Sadaf, S., Javed, A. & Luqman, M. (2005). Constraints faced by rural women in approaching agriculture extension services: A case study of district Faisalabad, Pakistan. Indus Journal Biology Science (2), 483-88.

14- Sailaja, A. & Reddy, M.N.(2003). Changing needs of Farm Women in Agriculture, MANAGE Extension Research Review. National Institute of Agricultural Extension Management, Rajendranagar, Hyderabad, India, 164–75.

15- Shelly, A.B. & Costa, M.D. (2000). Women in Aquaculture: Initiatives of Caritas Bangladesh. Bangladesh Institute of Development Studies (BIDS), Bangladesh, 77-87.

16- Suman, M. (2003). Role of Communication Channels and Constraints in Adoption of Soil Conservation Technologies in Food- fodder Production. In watershed based rainfed soils of bundelkhand. International Conference on Communication for Development in the Information Age: Extending the Benefits of Technology for All. 07-09 January 2003 Eds. Basavaprabhu Jirli Editor in Chief, Diapk De, K. Ghadei and Kendadmath, G.C., Department of Extension Education, Institute of Agricultural Sciences, Banaras Hindu University, Varanasi, (India).http://agropedia.iitk.ac.in

17- Thakur, D. (2003). Research methodology in Social science. Deep and Deep Publications (Pvt.) LTD. Rajouri Garden, New Dehli,475.

18- Tucker, J.S., Wenzel, S.L., B.Straus, J., Ryan G.W. & Golinelli. D. (2005). Experience interpersonal violence perspectives of sexually active, substance-using women living in shelters and low income housing. Violence against women, 11(10), 1319-1340.

19- World Bank, (2003). Operationalizing Agricultural Extension Reforms in South Asia- A Case of Pakistan. Country Paper: Regional Workshop, Delhi, India.

20- Yawson, D.O., Armah, F. A., Afrifa, E.K.A., & Dadzie, S.K.N. (2010). Ghana's Fertilizer Subsidy Policy: Early Field Lessons from Farmers in the Central Region. Journal of Sustainable Development in Africa Clarion University of Pennsylvania, Clarion, Pennsylvania, 12 (3).

# 2

# Affective Factors in the Wheat Farmers' Adoption of Farming Methods of Soil Management in West Azerbaijan Province, Iran

*Soleiman Rasouliazar * and Saeid Fealy*

Abstract

Soil Management (SM) is critical to human well-being that it is more important now because of meeting the high demands for food production and satisfying the needs of an increasing world population. Therefore, the purpose of this descriptive-correlation study was to investigate the effective factors on wheat farmers' adoption of Farming Methods of Soil Management (FMSM). The research instrument was a structural questionnaire with close-ended questions, which its validity and reliability was confirmed. The target population included all wheat farmers in West Azerbaijan Province (N=24949) that among of them, 371 wheat farmers was chosen by using Krejcie & Morgan's table through multi-stage sampling (n=371). The descriptive results showed the majority of farmers (237 or 63.90%) had moderate adoption of FMSM. These results also indicated farmers used three FMSM namely 1) using crop rotation, 2) using animal fertilizers, and 3) using soil testing more than others did. On the other hands, there were significant relationships between some of personal, farming, social, economic, and extension-education characteristics of farmers and the amount of their adoption of FMSM. Finally, stepwise regression analysis revealed that 35.30% ($R^2$=0.353) of the variances in the amount of farmers' adoption of FMSM could be explained by the five variables namely farm size, knowledge about FMSM, the amount of extension contacts about FMSM, distance between farm and agricultural service centers, and the amount of attitude toward FMSM.

*Keywords:*
***Adoption, Soil erosion, Soil management, Wheat farmer, West Azerbaijan Province***

*Department of Agricultural Management, Mahabad Branch, Islamic Azad University, Mahabad, Iran.*
** Corresponding author's email: rasouli88s@yahoo.com*

## INTRODUCTION

Reports stated that Asia continent takes pains for Soil Erosion (SE) in comparison with the other continents and among Asian countries, Iran has the amount of soil erosion too high (Dregne, 1992). The more accurate investigations showed that the amount of soil destruction in Iran varies from 20 to 30 million tons annually (Chizari *et al.*, 2003). From this amount, more than 10 million tons silt up dams annually that result in the decrease reserve capacity of water and electricity production of dams (Ebrahimi *et al.,* 2003). The report of the Asia Productivity Organization (APO) explains that the amount of the annual SE in Iran is a tree, two, 14, and 16 times as much, on average, of soil erosion in Asia, Africa, America, and Europe continents, respectively (Abedi and Tabatabaei, 2007). Ma'roof (2007) also quoted from FAO that the amount of SE in Iran varied from one billion ton in the year 1973 to two milliard ton in the year 2003.

SE not only reduces soil fertility, crop production, and biodiversity but also changes to water quality and increases risks of global climate change and food insecurity (Blanco and Lal, 2008). This status results in migrating rural people to cities, increasing in agricultural arrears and pollution of surface and subsurface water (Alonge and Martin, 1995). Control and management of SE are important because when the fertile topsoil eroded away the remaining soil is less productive with the same level of input. While we cannot control SE completely, but we must manage excessive erosion to minimize adverse effects on productivity (Blanco and Lal, 2008).

Parvizi (2005) described SM as an optimal use of farm soil resources for improving production management and achieving sustainability goals. Cramb (2004) stated SM includes all activities in farms that prevent from destruction of soil and improve farm crop sustainability. Research's Lal (2003) indicated that SM could play important roles in improving utilization, increasing self-sufficiency of nutritious crop, decreasing poverty level, food security and sustainable agriculture. Totally, SM includes farming and mechanical methods. FMSM consists of using animal fertilizers, soil testing, sprinkler irrigation system, mineral fertilizers, tillage practices, green Fertilizers, crop residues, and crop rotation (Qassim, 2003; Debarry, 2004; Lal, 2003; Maiangwa *et al.*, 2007; Srivastava & Pandey, 1999; Yadav *et al.*, 2006; Davenport, 2003) whereas mechanical methods consists of using dams, gradient breaker, terracing, and windbreaker (Davenport, 2003; Debary, 2004).

The success of programs of SM depends on increasing farmers' knowledge and awareness about SE and SM to improve their attitude toward it and finally to adopt and use it on farm level. On the other hands, the amount of farmers' adoption of such technologies and methods is restricted because of being not considered social, economic, and farming circumstances of farmers in the process of technology development (Balali & Afkhami, 2006). The several researches also showed that personal, economic, social, farming, extension-education factors affected on the amount of farmers' adoption of FMSM (Pezeshki Rad *et al.*, 2010; Onweremadu and Matthews-Njoku, 2007; Bayard *et al.,* 2006; Isife *et al.*, 2006; Shahroudi *et al.*, 2010; Demeke, 2003; Bandara and Thiruchelvam, 2008; Fe'li *et al.*, 2010 and 2011; Ghorbani and Kohansal, 2011; Rezvanfar *et al.*, 2010; Mahboubi *et al.*, 2005; Matata *et al.*, 2008; Lapar & Pandey, 1999; Cramb, 2004; Chomba, 2004). Therefore, the main purpose of this study was to investigate the effective factors on wheat farmers' adoption of FMSM. The specific objectives of this study were:

1- To describe farmers' personal and farming characteristics;

2- To investigate the amount of farmers' adoption of FMSM;

3- To identify the amount of farmers' knowledge about FMSM;

4- To investigate the amount of farmers' attitude toward FMSM;

5- To study the amount of farmers' extension contacts about FMSM;

6- To examine correlation between farmers' personal, farming, social, economic, and extension-education characteristics and the amount of their adoption of FMSM; and

7- To identify the predication equation of the amount of farmers' adoption of FMSM.

## MATERIALS AND METHODS

This study was a quantitative study from a philosophical point, an applied study in terms of goals, and descriptive-correlation in terms of method. All wheat farmers in the West Azerbaijan Province (state) of Iran in the farming year 2012, were the target population for this study (N=24949). The population frames were obtained from the West Azerbaijan's agricultural organization. The sample size was determined and supported by the studies of Krejcie and Morgan (1970). The sample also was obtained through multi-stage sampling (n=371).

The instrument for gathering data and information was questionnaire whose questions was designed in three parts based on the review of literature. The first part was devoted to identifying the amount of respondents' use of farming methods of soil management including 9 items in a four-part Likert scale which are ranked from not at all (0), seldom (1), sometimes (2), to all the time (3). The second part was devoted to test the respondents' knowledge about soil management methods with 13 four-choice questions and investigate their attitude toward soil management methods including 11 items in a five-part Likert scale which are ranked from completely disagree (1), disagree (2), no opinion (3), agree (4), to completely agree (5). The last part was devoted to information gathering about the individual, farming, social, economic, and extension-education characteristics of the respondents. It is mention to state that for investigating social, economic, and extension-education has been used from a six-part Likert scale which are ranked from at all (0), very low (1), low (2), moderate (3), high (4), to very high (5).

Content and face validity were established by a panel of experts consisting of faculty members in the departments of agricultural extension and education at Tarbiat Modares University, agricultural management at Mahabad Branch, Islamic Azad University and agricultural officers of Mahabad Township. A pilot test was conducted with 30 irrigated wheat farmers in the Bookan Township (out of a sample size) of the West Azerbaijan Province three weeks before the study. Minor changes in wording were made because of the pilot test. Questionnaire reliability was estimated by calculating Cronbach's alpha. Reliability for the main sections of instrument was estimated from 0.71 to 0.83.

Data were collected by personal interview with farmers at their farms. The data were coded and analyzed by using the Statistical Package for the Social Science (SPSS 16) for windows. Descriptive statistics (frequencies, means, standard deviations, minimum, and maximum) were used to describe analyze data. Spearman and Pearson correlation coefficients and multiple regressions were employed to analyze the relationships between variables.

## RESULTS

### Objective one - To describe farmers' personal and farming characteristics

The mean of age of farmers in the study was 42 years old (SD=11) that the majority of them (n=165 or 44.50%) ranged from 39 to 51 years old. On average, farmers had 22 years of expe-

Table 1: Farmers' personal and farming characteristics (n=371)

| Variables | M | SD | Min. | Max. |
|---|---|---|---|---|
| Age (Year) | 41.90 | 10.76 | 25 | 80 |
| Agricultural experiences (Year) | 21.35 | 11.48 | 2 | 70 |
| Agricultural experiences in cultivating wheat (Year) | 18.10 | 11.65 | 1 | 60 |
| Farm size (Hectare) | 15.62 | 15.36 | 2 | 70 |
| Land under wheat cultivation (Hectare) | 7.14 | 8.09 | 1 | 39 |
| The amount of wheat produced per hectare (Ton) | 4.59 | 1.54 | 3 | 10 |
| Distance between farm and agricultural service centers (Km) | 7.97 | 6.26 | 1 | 45 |
| Education level (Year) | 7.90 | 5.42 | 0 | 18 |

rience in agriculture (SD=11) that the majority of them (n=180 or 48.50%) ranged in agricultural experience from 2 to 18 years. On average, they also had 18 years of experience in cultivating wheat (SD=12) that the majority of them (n=200 or 53.90%) ranged in experience of cultivating wheat from 1 to 15 years. The mean of farmlands were 16 hectares that farmers had allocated 7 hectares of it to cultivate wheat. Farmer's education level average was 8 years whose 83 persons (22.40%) were illiterate and 74 farmers (19.90%) had a secondary school level education. On the other hand, the average of the distance of farmers' land to the nearest agricultural service centers was 10 Km and its standard deviations was 5 Km (Table 1).

**Objective two- To investigate the amount of farmers' adoption of FMSM**

Farmers were asked to indicate the amount of their adoption of FMSM for nine items. The nine items were measured on a four-point, Likert-type scale, that ranged items from not at all=0, seldom=1, sometimes=2, and all the time=3. Means and standard deviations for the nine adoption items were reported in table 2.

One of the nine items had a mean more than 2.00 indicating "all the time". This item was "using crop rotation". Another seven items had mean score more than one indicating sometimes. The highest mean was for the item of "using animal fertilizer" (Mean=1.99 & SD=0.86). One item also had a mean score less than 1.00 indicating seldom. This item was "using green fertilizers".

As table 3 results were shown, the amount of farmers' adoption of FMSM is divided into 3 levels with equal distance according to scores range. These results showed that Majority of farmers had moderate (n=237 or 93.90%) adoption of FMSM, whereas nearly 22 and 14 percent of them had good and weak adoption of FMSM, respectively.

**Objective three- To identify the amount of farmers' knowledge about FMSM**

For evaluating the amount of farmers' knowledge about FMSM, 13 statements about FMSM concepts, principles and methods are propounded. Farmers were asked to indicate their opinion about being right or false. One score is given to right item and 0 score is given to false and not respondent items. Hence, Farmers' knowledge

Table 2: Ranking the amount of farmers' adoption of FMSM (n=371)

| Item | M* | SD | Rank |
|---|---|---|---|
| Using crop rotation | 2.21 | 0.68 | 1 |
| Using animal fertilizers | 1.99 | 0.86 | 2 |
| Using soil testing | 1.73 | 0.92 | 3 |
| Using chemical fertilizer according to soil testing | 1.66 | 1.00 | 4 |
| Using sprinkler irrigation system | 1.52 | 1.25 | 5 |
| Using crop residues | 1.51 | 1.09 | 6 |
| Using mineral fertilizers | 1.45 | 1.03 | 7 |
| Using tillage practices | 1.32 | 1.11 | 8 |
| Using green Fertilizers | 0.98 | 1.10 | 9 |

Note: •not at all=0, seldom=1, sometimes=2, all the time=3

Table 3: Classification of the amount of farmers' adoption of FMSM (n=371)

| Classification of score knowledge | Category | Frequency | Percent |
|---|---|---|---|
| 0-8 | Weak | 52 | 14 |
| 9-17 | Moderate | 237 | 93.90 |
| 18-27 | Good | 82 | 22.10 |

Table 4: Classification of farmers' knowledge about FMSM (n=371)

| Classification of score knowledge | Category | Frequency | Percent |
|---|---|---|---|
| 0-4 | Weak | 0 | 0 |
| 5-8 | Moderate | 27 | 7.30 |
| 9-13 | Good | 344 | 92.70 |

Table 5: Ranking farmers' attitude toward FMSM (n=371)

| Statement | M* | SD | Rank |
|---|---|---|---|
| Soil as life resource is forming during a long time• | 4.27 | 1.13 | 1 |
| I encourage other farmers to use SMM for improving the quality of their farms• | 4.10 | 0.75 | 2 |
| Soil erosion results from agricultural activities is an important problem in farm• | 4.00 | 0.87 | 3 |
| If our zone farmers use SMM, I also use one• | 4.00 | 1.14 | 4 |
| Farmers are familiar with SMM, so they don't need to extension agents or experts•• | 3.87 | 0.98 | 5 |
| The much use of fertilizer causes water, air and nutrition pollution• | 3.85 | 1.05 | 6 |
| I must use SMM to improve and protect the quality of farm soil•• | 3.83 | 0.85 | 7 |
| I prefer soil testing for fertilizer suggestions against the farm soil observations•• | 3.78 | 0.96 | 8 |
| SM is an agricultural method that must do by government supports• | 3.57 | 1.25 | 9 |
| The use of the amount of chemical fertilizer suggested based on soil testing is not useful•• | 3.38 | 1.11 | 10 |
| The much use of chemical fertilizer results in the low use of animal fertilizer in farm• | 3.28 | 1.19 | 11 |

Note: • completely disagree=1, disagree=2, no opinion=3, agree=4, and completely agree=5
•• Completely disagree=5, disagree=4, no opinion=3, agree=2, and completely agree=1

about FMSM ranged from 0 to 13.

As table 4 results were shown, farmers' knowledge about FMSM is divided into 3 levels with equal distance according to scores range. These results showed that all farmers had "moderate (n=27 or 7.30%) and good (n=344 or 92.70%)" knowledge about SMM.

**Objective four- To investigate the amount of farmers' attitude toward FMSM**

Farmers were asked to indicate their attitude toward FMSM for 11 statements. The 11 statements were measured on a five-point, Likert-type scale, that ranged in 8 positive statements from completely disagree=1, disagree=2, no opinion=3, agree=4, completely agree=5 and in 3 negative statements from completely disagree=5, disagree=4, no opinion=3, agree=2, and completely agree=1. Means and standard deviations for the 11 attitudes statements were reported in Table 5.

Four of the eight positive statements had a mean equal or more than 4.00 indicating "agreement." The highest mean was for the statement of "soil as life resource is formed during a long time" (Mean=4. 27 & SD=1. 13). Another four positive statements had mean score more than 3.28 indicating they had no opinion about SMM. Three negative statements also had a mean score more than 3.38 indicating disagreement.

As a table 6 result was shown, the amount of farmers' attitude toward FMSM is divided into 3 levels with equal distance according to scores range. These results showed that majority farmers attitude (n=243 or 65.50 %) toward FMSM was at favorable level and 34.50%

Table 6: Classification of the amount of farmers' attitude toward FMSM (n=371)

| Classification of score knowledge | Category | Frequency | Percent |
|---|---|---|---|
| 1-15 | Unfavorable | 0 | 0 |
| 16-40 | Neutral | 128 | 34.50 |
| 41-55 | Favorable | 243 | 65.50 |

Table 7: Ranking the amount of farmers' extension contacts about FMSM (n=371)

| Item | M* | SD | Rank |
|---|---|---|---|
| Visiting with extension agents and experts at agricultural service centers and agricultural officers | 2.69 | 1.32 | 1 |
| Attending extension-education courses | 2.50 | 1.66 | 2 |
| The existing amount of extension agents and agricultural experts in fields | 2.08 | 1.36 | 3 |
| Reading extension publications | 2.03 | 1.43 | 4 |
| Visiting sample fields and extension-research projects in that zone | 2.01 | 1.36 | 5 |
| Showing extension films | 1.77 | 1.45 | 6 |

Note: • Very little=1, little=2, moderate=3, high=4, and very high=5

(n=128) of their attitude toward FMSM was neutral.

### Objective five- To study the amount of farmers' extension contacts about FMSM

Farmers were asked to indicate the amount of their extension contacts about FMSM for six items. The six items were measured on a five-point, Likert-type scale, that ranged from Very little=1, little=2, moderate=3, high=4, and very high=5. Means and standard deviations for the six extension contact items are reported in table 7.

One of the six items had a mean less than 2.00 indicating very little. This item was "showing extension films". Three of six items also had a mean close to 2.00 indicating little. Finally, two items had a mean more than 2.50 indicating moderate. These statements were "visiting with extension agents and experts at agricultural service centers and agricultural officers" and "attending extension-education courses", respectively.

As table 8 results are shown, the amount of extension contacts of farmers about FMSM is divided into 3 levels with equal distance according to scores range. These results showed that the amount of extension contacts of majority farmers (n=216 or 58.20 %) about FMSM was at moderate level and nearly 29% (n=107) of their extension contacts was at weak level.

### Objective six- To examine correlation between farmers' personal, farming, social, economic, and extension-education characteristics of respondents and the amount of their adoption of FMSM

Pearson Correlation Coefficients were employed for measurement of the relationships between personal, farming, social, economic, and extension-education characteristics of respondents and the amount of their adoption of FMSM (except for education level). Table 9 showed that there was no significant statistical relationship between the years of agricultural experiences ($r=-0.006$ & $p>0.05$) and the amount of access to agricultural inputs ($r=0.069$ & $p>0.05$) with the amount of farmers' adoption of FMSM. It is noteworthy that a Hinkle et al.'s (1988) model for describing the magnitude of correlation has been used; 0-0.30: negligible association, 0.30-0.50: low association, 0.50-0.70: moderate association, 0.70-0.90: substantial association and 0.90-1: very strong association.

The calculated Pearson Correlation Coefficients showed that there was a negatively significant relationship between age ($r=-0.123$ & $P\leq0.01$) as a "negligible association", and agricultural experiences in cultivating wheat ($r=-0.006$ & $P\leq0.05$) as a "negligible association", and furthermore positively significant relationships between farm size ($r=0.277$ & $P\leq0.01$) as a "low association", land under wheat cultivation

Table 8: Classification of the amount of farmers' extension contacts about FMSM (n=371)

| Classification of score knowledge | Category | Frequency | Percent |
|---|---|---|---|
| 0-9 | Weak | 107 | 28.80 |
| 10-20 | Moderate | 216 | 58.20 |
| 21-30 | Good | 48 | 12.90 |

Table 9: Correlation between farmers' personal, farming, social, economic, and extension-education characteristics and the amount of their adoption of FMSM (n=371)

| Variables | The amount of adoption of FMSM $r_s$ | p | Description |
|---|---|---|---|
| Age | -0.123** | 0.001 | negligible association |
| Agricultural experiences | -0.006 | 0.869 | negligible association |
| Agricultural experiences in cultivating wheat | -0.121* | 0.001 | negligible association |
| Farm size | 0.277** | 0.000 | negligible association |
| Land under wheat cultivation | 0.459** | 0.000 | Low association |
| The amount of wheat produced per hectare | 0.284** | 0.000 | negligible association |
| Distance between farm and agricultural service centers | -0.185* | 0.013 | negligible association |
| Education level | 0.236** | 0.000 | negligible association |
| Knowledge about FMSM | 0.287** | 0.000 | negligible association |
| The amount of attitude toward FMSM | 0.296** | 0.000 | negligible association |
| The amount of extension contacts about FMSM | 0.206** | 0.000 | negligible association |
| The amount of social participation | 0.129* | 0.014 | negligible association |
| The amount of social status | 0.330** | 0.000 | Low association |
| The amount of access to agricultural inputs | 0.069 | 0.087 | negligible association |

$*P \leq 0.05$ $**P \leq 0.01$

(r=0.459 & P≤0.01) as a "low association", amount of wheat produced per hectare (r=0.284 & P≤0.01) as a "low association", distance between farm and agricultural service centers (r=0.284 & P≤0.05) as a "negligible association", knowledge about FMSM (r=0.287 & P≤0.01), the amount of attitude toward FMSM (r=0.296 & P≤0.01) as a "negligible association", the amount of extension contacts about FMSM (r=0.188 & P≤0.01) as a "negligible association", the amount of social participation (r=0.129 & P≤0.05) as a "negligible association", the amount of social status (r=0.284 & P≤0.01) as a "low association", and the amount of farmers' adoption of FMSM. Finally, the calculated Spearman Correlation Coefficient showed that there was a positively significant relationship between education level (rs=0.240 & P≤0.01) as a "negligible association" and the amount of farmers' adoption of FMSM.

**Objective seven- To identify the predication equation of the amount of farmers' adoption of FMSM**

In multivariate regression analysis, stepwise method has been used. A feature of the aforementioned method is that, at the first, the most important variable comparing with all other independent variables is there to be inserted in the equation, and that is as the most important variable that has much more power to explain the dependent variable. This trend will be repeated in other steps until no independent variable has the ability for being inserted in the

Table 10: Multivariate linear regression analysis (the amount of adoption of FMSM as dependent variable) (n=317)

| Independent variables | Unstandardized coefficient | Standardized coefficient | t | Significant level |
|---|---|---|---|---|
| Constant | -6.939 | - | -2.442 | 0.015 |
| Farm size ($X_1$) | 0.151 | 0.416 | 9.186 | 0.000 |
| Knowledge about FMSM ($X_2$) | 0.853 | 0.254 | 5.518 | 0.000 |
| The amount of extension contacts about FMSM ($X_3$) | 0.144 | 0.184 | 4.057 | 0.000 |
| Distance between farm and agricultural service centers ($X_4$) | 0.133 | 0.144 | 3.272 | 0.000 |
| The amount of attitude toward FMSM ($X_5$) | 0.125 | 0.112 | 2.412 | 0.016 |

Note: R=0.594, $R^2$=0.353, F=38.441, Sig= 0.000

Regression Linear equation. The results of table 10 showed five variables namely 1) farm size; 2) knowledge about FMSM; 3) the amount of extension contacts about FMSM; 4) distance between farm and agricultural service centers; and 5) the amount of attitude toward FMSM explained 35.30 percent of the variance of the amount of farmers' adoption of FMSM. According to Unstandardized coefficients, its prediction equation can be written below:

$$Y = \text{Constant} + b_1 (X_1) + b_2 (X_2) + b_3 (X_3) + b_4 (X_4) + b_5 (X_5)$$

$$Y = -6.939 + 0.151 (X_1) + 0.853 (X_2) + 0.144 (X_3) + 0.133 (X_4) + 0.125 (X_5)$$

## DISCUSSION

The results of this research indicated that the majority of farmers' adoption of FMSM were at moderate level. This result supports the previous finding study of Rezvanfar *et al.* (2010). The more accurate investigations showed that the farmers adopted and use three FMSM namely 1) using crop rotation, 2) using animal fertilizers, and 3) using soil testing more than other methods. The finding of researches' Onweremadu and Matthews-Njoku (2007) and Chomba (2004) also implied that "crop rotation" was the main methods of farmers for managing their farm soil in Nigeria and Zambia, respectively. This result also was against result of researches' Rezvanfar *et al.* (2010) and Pezeshki Rad *et al.* (2010) because the results of their researches showed Iranian farmers used FMSM namely "crop residues" and "soil testing" more than the others. This result means the farmers adopt and use some of FMSM according to time and place circumstances such as a farm slop, farm soil quality, etc. Besides, the other results of the study indicated the majority of farmers' knowledge, extension contacts, and attitude about FMSM were at moderate and favorable levels, respectively.

The correlation analysis results also showed that there was a positively significant relationship between education level (consenting to researches' Shahroudi *et al.*, 2010; Onweremadu and Matthews-Njoku, 2007; Lapar and Pandey, 1999; Fe'li *et al.*, 2010; Rezvanfar *et al.*, 2010; Bandara and Thiruchelvam, 2008), farm size (consenting to researches' Shahroudi *et al.*, 2010), farm size (consent to researches' Shahroudi *et al.*, 2010; Fe'li *et al.*, 2010 and 2011; Mahboubi *et al.*, 2005; Onweremadu and Matthews-Njoku, 2007; Bandara d Thiruchelvam, 2008; Rezvanfar *et al.,* 2010; Demeke, 2003; Bayard *et al.*, 2006), land under wheat cultivation, amount of wheat produced per hectare (consenting to researches' Shahroudi *et al.*, 2010; Bandara and Thiruchelvam, 2008), distance between farm and agricultural service centers (consenting to researches' Lapar and Pandey, 1999; Chomba, 2004; Fe'li *et al.*, 2010), the amount of knowledge about FMSM (consenting to researches' Fe'li *et al.*, 2010; Rezvanfar *et al.*, 2010; Ghorbani and Kohansal, 2011; Mahboubi *et al.*, 2005), the amount of attitude toward FMSM (consenting to researches' Rezvanfar *et al.*, 2010; Demeke, 2003; Fe'li *et al.*, 2011), the amount of extension contacts about FMSM (consenting to researches' Shahroudi *et al.*, 2010; Pezeshki Rad *et al.*, 2010; Fe'li *et al.*, 2010 and 2011; Mahboubi *et al.,* 2005; Bayard *et al.*, 2006; Demeke, 2003; Matata *et al.*, 2008; Rezvanfar *et al.*, 2010; Chomba, 2004), the amount of social participation (consenting to researches' Shahroudi *et al.*, 2010; Cramb, 2004), the amount of social status (consenting to researches' Shahroudi *et al.*, 2010; Cramb, 2004), and the amount of farmers' adoption of FMSM. These results mean the more increase one variable, the more increase the other variable. Furthermore, there was a significant negative relationship between age and agricultural experiences in cultivating wheat and the amount of farmers' adoption of FMSM. These results mean the more increase one variable, the more decrease the other variable.

The results of multivariate regression analysis showed five variables namely 1) farm size; 2) knowledge about FMSM; 3) the amount of extension contacts about FMSM; 4) distance between farm and agricultural service centers; and 5) the amount of attitude toward FMSM explained 35.30 percent of the variance in the amount of farmers' adoption of FMSM. These results mean the more increase five mentioned variables, the more increase the amount of farmers' adoption of FMSM. Finally, according to standardized coefficients, it is clear that farm size in comparison with the other variables has a greater portion of the variance in the amount of farmers' adoption of FMSM.

## REFERENCES

1-Abedi, A. & Tabatabaee, H. (2007). Economic analysis of Watershed Management in Chaharmahal va Bakhtiyaree. The First Regional Conference Proceedings of Operation from Karoon and Zayanderood water resources. Iran: Shahr-e- Kord.

2-Alonge, J.A. & Martin, A.R. (1995). Assessment of the Adoption of Sustainable Agriculture Practices: Implications for Agricultural Education. Journal of Agricultural Education, 36 (3): 34-42.

3-Balali, M. & Afkhami, M. (2006). Soil Management and Farmer Field School Approach. Jihad Magazine, 25 (269): 98-107.

4- Bandara, D. & Thiruchelvam, S. (2008). Factors Affecting the Choice of Soil Conservation Practices Adopted by Potato Farmers in Nuwara Eliya District, Sri Lanka. Tropical Agricultural Research & Extension, 11: 49-54.

5- Bayard, B., Jolly, M.C. & Shannon, A.D. (2006). The Adoption and Management of Soil Conservation Practices in Haiti: The Case of Rock Walls. Agricultural Economics Review, 7 (2): 28-39.

6-Blanco, H. & Lal, R. (2008). Principles of Soil Conservation and Management. Springer Science + Business Media B.V.

7- Chizari, M., Karimi, S., Lindner, R. J. & Pezeshki-Rad, G. (2003). Perception of Soil Conservation Competencies among Farmers in Markazi Province, Iran. Journal of International Agricultural and Extension Education, 10 (3): 13-19.

8- Chomba, N. G. (2004). Factors Affecting Smallholder farmers' Adoption of Soil and Water Conservation Practices in Zambia. Unpublished Thesis in Agricultural Economics, Michigan State University.

9- Cramb, A.R. (2004). Social Capital and Soil Conservation: Evidence from the Philippines. Contributed Paper 48th Annual Conference Australian Agricultural & Resource Economics Society Melbourne, 10-13 February 2004.

10- Davenport, T. E. (2003). The Watershed Project Management Guide. London: Lewis publishers.

11- Debarry, P. A. (2004). Watersheds: Processes, Assessment and Management. New Jersey: John Wiley & sons Inc.

12- Demeke, B. A. (2003). Factors Influencing the Adoption of Introduced Soil Conservation Practices in Northwestern Ethiopia. Institute of Rural Development, University of Gottingen.

13- Dregne, H.E. (1992). Erosion and Soil Productivity in Asia. Journal of Soil and Water Conservation. 47 (1): 8-13.

14- Ebrahimi, N.G., Godoisi, J., Gandom Kar, A. & Pour Matin, A. (2003). Assessing the Role of Land Management in Soil Erosion. Tehran: Research Center of conserving Soil and Watershed Management.

15-Fe'li, S., Bondarian, N., Baghaei, M. & Mirzaei, A. (2010). Effective Factors on Adoption of Soil Testing for Farm Fertilizer Nutrition in Shahreza Township of Esfihan Province, Iran. Research Journal of Soil and Water Management, 1 (2): 38-44.

16- Fe'li, S., Bondarian, N., Mirzaei, A. & Baghaei, M. (2011). Factors Affecting Farmers' Knowledge about Sampling Concepts from Farming Soil in Shaheza Township, Esfihan province. Journal of extension and Education researches, 3 (2): 95-108.

17-Ghorbani, M. & Kohansal, M. (2011). Effective Factors on the Wheat Farmers' Participation in Green Subside Program and the Use of Soil Conservation Methods. Journal of Agricultural Economics and Development, 24 (1): 59-71.

18-Hinkle, D.E., Wiersma, W. & Jurs, S.G. (1988). Applied Statistics for the Behavioral Sciences. Boston, MA: Houghton Mifflin Company.

19- Isife B., Emah G. & Akahome A. (2006). Evaluation of the Agricultural Extension Programs of Shell Petroleum Development Company in Rivers State, Nigeria. International Journal of Agriculture and Rural Development, 7: 12-120.

20- Krejcie, R.V. & Morgan, D.W. (1970). Determining Sample Size for Research Activities. Educational and Psychological Measurement, 30: 608-610.

21- Lal, R. (2003). Cropping Systems and Soil Quality. In, Shrestha, A. (ed.). Cropping systems: trend and advances. Food Products Press, NY, pp. 33-52.

22- Lapar, M.L.A. & Pandey, S. (1999). Adoption of Soil Conservation: the Case of the Philippine Uplands. Agricultural Economics, 21 (3): 241-256.

23-Mahboubi, M., Iravani, H., Rezvanfar, A., Kalantari, K. & Mohseni, M. (2005). Effective Factors on the Adoption of Soil Conservation Technologies in Watershed Zone of Zarin Gol, Golestan province. Journal of Iran Natural Resources, 57 (4): 595-605.

24- Maiangwa, G.M., Ogungbile, O.A., Olukosi, O.J. & Atala, K.T. (2007). Adoption of Chemical Fertilizer for Land Management in the North-West Zone of Nigeria. Tropical Agricultural Research & Extension, 10: 33-46.

25- Ma'roof, F. (2007). Soil erosion. Ghods Newspaper, No.1298.
26- Matata, Z.P., Ajayil, C.O., Oduol, A.P. & Agumya, A. (2008). Socio-economic Factors Influencing Adoption of Improved Fallow Practices among Smallholder Farmers in Western Tanzania. International NGO Journal, 3 (4): 68-73.
27- Onweremadu, U.E. & Matthews-Njoku, C.E. (2007). Adoption Levels and Sources of Soil Management Practices in Low – input Agriculture. Nature and Science, 5 (1): 39-45.
28- Parvizi, Y. (2005). Optimize Farming Management in Sustainable Exploit of Soil and Water Resources. Zeyton Magazine, 158: 2-11.
29- Pezeshki Rad, G., Fe'li, S. & Chizari, M. (2010). The Impact of Wheat Farm Advisors on the Adoption of Technologies about Conserving Soil in Wheat Farmers of Tehran province. Iranian Journal of Agricultural Economics and Development Research, 40-2, 2: 53-61.
30- Qassim, A. (2003). Sprinkler Irrigation: A Situation Analysis. Department of Natural Resources and Environmental. State Government Victoria. (Online), Available at WWW: URL :http:// www. wcainfonet. org/servlet/BinaryDownloaderServlet?filename = 1058150347631_SPRINKLER_IRRIGATION.pdf
31-Rezvanfar, A., Shiri, N. & Mohammadi Kanigolzar, M. (2010). Exploring Factors Affecting Application of Soil Conservation Practices by Iranian farmers. Annals of Biological Research, 3 (5):2383-2389.
32- Shahroudi, A., Chizari, M. & Pezeshki Rad, G. (2010). Effective Factors on the Beet Farmers' Behavior about Sustainable Soil Management Methods. Iranian Journal of Agricultural Economics and Development Research, 40-2, 3: 101-115.
33-Srivastava, Y.C. & Pandey, A.P. (1999). Knowledge and Attitude of Small and Marginal Farmers Towards Soil Testing. Agricultural Extension Review, 11(6): 3-6.
34- Yadav, S.P.V., Raman, S.R. & Kumar, R. (2006). Knowledge and Attitude of Farmers Towards Soil Testing Practices. Indian Research Journal of Extension Education, 6 (3):1-3.

# 3

# Social Factors Influencing Adoption of Integrated Pest Management (IPM) Technologies by Paddy Farmers

*Fatemeh Razzaghi Borkhani*, Ahmad Rezvanfar, Hossein Shabanali Fami and Mahtab Pouratashi*

Abstract

This study aimed to identify the factors influencing adoption of Integrated Pest Management (IPM) technologies by paddy farmers in Sari County of Iran. A sample of 260 farmers who selected randomly, participated in this study through a survey questionnaire. Findings of the study revealed a statistically significant positive correlation between adoption of IPM technologies and farmers' participation in extension activities, Farmers' Field School, local associations, and influence of opinion leaders. Regression analysis indicated that nearly 43 % of variation in the IPM technologies adoption could be explained by participation in extension activities, local associations, and influence of opinion leaders. Findings have implications for designing extension programs that can encourage adoption of IPM technologies by farmers.

*Keywords:*
*Integrated pest management, Paddy farmer, FFS program, Extension activities*

*Department of Agricultural Extension and Education, Faculty of Agricultural Economics and Development, University of Tehran, Karaj, Iran.*
** Corresponding author's email: razzaghi.fatemeh@gmail.com*

## INTRODUCTION

With population growth, agriculture will need to produce enough supply of food to feed an expected more than eight milliard people by 2030 (FAO, 1992). Hence, it is necessary to decrease crop losses (which pests are one of agents to cause crop losses) in order to increase food safety and access to sustainable agricultural development (Van Huis and Meerman, 1997; Oerke *et al.,* 1994).

Rice is the world's most important food crop and a primary source of food for more than half of the world's population (Khush, 2004). Rice as the most important strategic product of Mazandaran province, with the cultivation size of 201,793 hectares, has the highest amount of rice production in Iran (Ministry of Agricultural Jihad, 2006). Considering chemical pesticides usage in Iran indicates the highest consumption of chemicals in the rice fields. Since pesticides can pose serious threats to human health and the environment (WHO, 1990; citing in Pouratashi and Iravani, 2012), there is the question of how can effectively handle agricultural pests. Integrated Pest Management (IPM) is a pest management strategy that focuses on long-term prevention or suppression of pest problems with minimum impact on environment, human health, and non-target organisms (Louise Flint *et al.,* 2003). IPM applies a combination of practices including biological, chemical, and cultural (Tette *et al.*, 1987). So, it enables farmers to reduce their reliance on pesticides while increasing yields, crop quality, and profitability (Mauceria, 2004). Biological control of pests is one technology of integrated pest management which deals with these issues. National Academy of Sciences (1987) defined biological control as "the application of natural or modified organisms, genes, or gene products to decrease the effects of undesirable organisms (pests), and to favor desirable organisms such as crops, trees, animals, and beneficial insects and microorganisms." In Iran, since 1990s the use of integrated pest management technologies in rice fields of north Iran was started with applying Trychograma bee, and now, by establishing field farmer school (as one of extension-participatory methods), these technologies are introducing to paddy farmers. Farmers are important in the process of adoption and application of IPM technologies. Therefore, it is important to find social factors affecting the adoption of IPM technologies by farmers.

### Reviews of Literature

Ferguson (1995) found that participation of farmers in IPM activities affected adoption of not chemical methods and IPM technologies. Muthuraman and Sain (2002) found that lack of knowledge about pest management strategies and lack of community action were among the major barriers in the adoption of IPM technologies by farmers.

Palis *et al.* (2002) found that kin networks, neighborhoods, membership in a farmer's association cause adoption of IPM technologies. Bonabana-Wabbi (2002) indicated that membership in farmers' association had positive effect on level of adoption of IPM practices.

Luther *et al.* (2005) found significant difference between farmers who participated in extension activities and farmer's field schools (FFS) for IPM adoption compared to those who did not participated in these activities. Barrera *et al.* (2005) found that information sources had a positive impact on the adoption of IPM technologies. FFS program was the main determinant in IPM adoption. Other factors were field days, pamphlets, and exposure to FFS-participants.

Asghari and Hadi (2009) found that social participation of farmers, membership in rural associations, participation in extension activities, and communication with extension experts had significant correlation with the adoption of biological control by farmers. Erbaugh *et al.* (2010) found that participation of farmers in farmer field schools (FFS) programs influenced increasing IPM knowledge of farmers. Also, IPM knowledge was the major factor in the adoption of IPM technologies. Noorhosseini Niyaki *et al.* (2010) found that the main important factors of adoption of biological control were farmers education level, family size, experience in rice cul-

Table 1: Summary of social factors influencing adoption of IPM technologies

| Variable | Source |
|---|---|
| Membership in rural- agricultural cooperatives | Asghari and Hadi (2009), Muthuraman and Sain (2002), Noorhosseini Niyaki *et al.* (2010), Palis *et al.* (2002). |
| Participation in extension-education activities | Asadpour (2011), Bonabana-Wabbi (2002), Boughton and de Faran (1994), Luther *et al.* (2005), Mariyono (2007), Noorhosseini Niyaki *et al.* (2010 ), Truong Thi (2008), |
| Participation in FFS programs | Rejaul and Bakshi (2005), Ridgley and Brush (1992). |
| Participation in local associations | Asai and Tokunaga (2007), Barrera *et al.* (2005), Erbaugh *et al.*(2010), Luther *et al.* (2005), Mauceria (2004), Nabirye *et al.* (2003). |
| Influence of opinion leaders | Abd-Ella *et al.* (1981), Adesina *et al.* (2000), Bonabana-Wabbi (2002), Clearfield *et al.* (1986), Strauss et al. (1991), Truong Thi (2008). Asghari and Hadi ,(2009)Jacob (1982). |

ture, rate of participation in educational - extensional activities. Asadpour (2011) found that important factors in the adoption of IPM technologies were risk aversion of paddy farmers, the value of yield per hectare, number of pieces of land, the number of release Trichogramma, use the technology in the neighboring land, level of cultivation, experience in application of technology, age of farmers and participation in extension classes. Age and number of pieces of agricultural land technology had negative correlation with adoption of biological control. Table 1 displays the summary of social factors influencing adoption of IPM technologies.

## Purpose and objectives

The purpose of this study was to investigate social factors influencing adoption of integrated pest management (IPM) technologies by paddy farmers. The specific objectives of the study were to:

1- Identify characteristics of the respondents;

2- Identify levels of farmers' participation with local associations;

3- Identify the influence of opinion leaders on adoption level of IPM technologies;

4- Determine the correlation between social factors and the extent of IPM technology adoption by farmers

5- Regression analysis for the extent of IPM technologies adoption by independent variables.

# MATERIALS AND METHODS

## Population and sample

This study was a descriptive-correlation research, carried out in Sari County. The population of the study consisted of paddy farmers (N= 28170) in five districts of Sari County (Chardangeh, Dodangeh, Markazi, Kelijanrestagh, and Miandorod). By calculation Cochran's formula, a sample of 260 farmers was selected by using proportional random sampling method. This formula is:

$$d = t\frac{s}{\sqrt{n}} \qquad n = \frac{N(t.s)^2}{Nd^2 + (t.s)^2}$$

In this formula, (n) is the number of sample, (N) is the number of population, (s) is standard deviation, and (t) is equal to 2. Table 2 displays the statistical population and sample size of this study.

## Instrumentation

A questionnaire divided into five parts was used to collect data from the target group. Part one, asked farmers to specify their demographic information such as age, educational level, and agricultural experience. Part two, was asked farmers' membership in rural- agricultural cooperatives. Part three, assessed farmers' participation in extension-education activities and FFS programs (divided them in two groups: participants and non-participants). Part four, as-

Table 2: Statistical population and sample size of the study

| County | District | No. of paddy farmers per district | Sample size |
|---|---|---|---|
| | Chardangeh | 3928 | 36 |
| | Dodangeh | 2121 | 20 |
| | Markazi | 12178 | 112 |
| Sari | Kelijanrestagh | 3815 | 35 |
| | Miandorod | 6125 | 57 |
| | Total | 28170 | 260 |

Table 3: Adoption level of IPM technologies

| Level | Frequency | % |
|---|---|---|
| Low | 52 | 20.0 |
| Relatively low | 64 | 24.7 |
| Relatively high | 101 | 38.8 |
| High | 43 | 16.5 |
| Total | 260 | 100.0 |

Mean= 100.19 SD= 15.16

sessed participation of farmers in local associations, using likert-type scale (0 = none, 1 = very low, 2 = low, 3 = intermediate, 4 = high and 5 = very high). Part five, assessed level of opinion leaders influence on adoption of IPM technologies, using likert-type scale (0 = none, 1 =very low, 2 = low, 3 = intermediate, 4 = high, and 5 = very high).

The extent of IPM practices application by paddy farmers was measured in three parts including application of IPM practices for rice pest control (20 questions), application of IPM practices for rice disease control (10 questions) and application of IPM practices for rice weed control (8 questions) (accordance to studies done by Bonabana-Wabbi, 2002; Asai and Tokunaga, 2007; Truong Thi, 2008). All these parts were measured on a Likert-type scale ranged from 0 to 5 (0=none, 1=low, 2=very low, 3=intermediate, 4=high and 5=very high). To categorize the extent of IPM practices application by paddy farmers, the following formula was applied:

Min<A<Mean-SD: A= Negative attitude

Mean- SD <B<Mean: B = Relatively negative attitude

Mean <C<Mean+ SD: C= Relatively positive attitude

Mean+ SD <D<Max: D = Positive attitude

### Validity and reliability

Validity of the instrument was obtained by Agricultural Jihad experts of Sari County and some faculty members at University of Tehran, Department of Agricultural Extension and Education. Reliability of the instrument was measured by calculating Cronbach's Alpha coefficient, a measure of internal consistency. The reliability for various parts was more than 0.7, which showed the acceptable level. Data were collected through face to face interviews with farmers at their farms.

### Data analysis

Data were analyzed using Statistical Package for the Social Sciences (SPSS). Descriptive and inferential statistics were used to analyze the collected data. Descriptive statistics included frequency, percentage, mean, and standard deviation. Correlation coefficient and multiple regression analysis were used in the inferential analysis section.

## RESULT AND DISCUSSION

### Characteristics of the sample

According to the findings, respondents were on average 49 years old. About 26.5.0% of respondents were between the age of 41 and 50 years. While, 12.0%, 13.5%, 24.2%, 21.1% and 2.7% of respondents were <31, 31-40, 51-60, 61-70 and >70 respectively. 71.9% of the respondents were literate and 28.1% were illiterate. Respondents' experience in agricultural activities was 29 years on average. Findings showed that more than half of the respondents (61.9%) had lands less than 2 hectares in size for cultivation of rice. While, 25.8%, 10.4% and 1.9% of respondents had lands 2-4, 4-6 and > 6 hectares in size respectively. The average income of paddy farmers was 84.96 million Rials annually.

According to the findings, more than half of the respondents (58.4%) had high and very high access to rice pesticides. While, 37.3 percent and 4.3 percent of those reported that they had intermediate and low and very low access to rice pesticides respectively.

### Adoption level of IPM technologies by farmers

The extent of IPM practices application by paddy farmers was measured in three parts including application of IPM practices for rice pest control, application of IPM practices for rice disease control, and application of IPM practices for rice weed control. Table 3 shows the level of IPM technologies adoption by farmers. 38.8 percent of respondents adopted IPM technologies at high levels. (Table 3).

Membership in rural-agricultural cooperatives

The findings (Table 4) showed that a majority of farmers were members of rural- agricultural cooperatives (79.2 %) and only 20.8 percent of

Table 4: Farmers' membership in rural-agricultural cooperatives

| Membership in rural-agricultural cooperatives | Frequency | % |
|---|---|---|
| Yes | 206 | 79.2 |
| No | 54 | 20.8 |

farmers were not members of rural-agricultural cooperatives. The findings showed that respondents' membership in rural-agricultural cooperatives was on average 15 years.

## Participation in extension-education activities and FFS programs

Table 5 showed respondents' participation in extension-education activities and FFS programs. According to the findings, 42.3 percent and 88.1 percent of farmers had never participated in extension-education activities and FFS programs on IPM technologies, respectively.

## Participation in local associations

Table 6 shows farmers' level of participation in local associations. 51.2 percent of farmers had low and relatively low participation in local associations. In contrast, 48.8 percent of farmers had high and relatively high participation in local associations.

## Influence of opinion leaders on farmers' adoption of IPM technologies

Table 7 shows farmers' view about the level of opinion leaders' influence on adoption of IPM technologies. As it can be seen, opinion leaders had low and relatively low influence on 42.7 percent of farmers for adoption of IPM technologies. In contrast, opinion leaders had high and relatively high influence on 57.3 percent of farmers for adoption of IPM technologies. Also, the findings revealed that local experts and extension agents had the most influence on farmers' adoption of IPM technologies.

## Correlation analysis

Correlation for independent variables and the extent of IPM technologies adoption by respondents are presented in table 8. The results indicated that there was positive and significant correlation between IPM technologies adoption and variables including participation in exten-

Table 5: Farmers' participation in extension-education activities and FFS programs

| Variable | Groups | Frequency | % |
|---|---|---|---|
| Participation in extension-education activities | participant | 150 | 57.7 |
| | non participant | 110 | 42.3 |
| Participation in FFS | participant | 31 | 11.9 |
| | non participant | 229 | 88.1 |

Table 6: Farmers' level of participation in local associations

| Level | Frequency | % |
|---|---|---|
| Low | 41 | 15.8 |
| relatively low | 92 | 35.4 |
| High | 88 | 33.8 |
| Relatively high | 39 | 15.0 |
| Total | 260 | 100 |

Table 7: Level of opinion leaders' influence on adoption of IPM technologies

| Level | Frequency | % |
|---|---|---|
| Low | 44 | 16.9 |
| relatively low | 67 | 25.8 |
| High | 124 | 47.7 |
| Relatively high | 25 | 9.6 |
| Total | 260 | 100.0 |

Table 8: Results of correlation analysis

| Variable | Frequency |
|---|---|
| Membership in rural- agricultural cooperatives | -0.104 |
| Participation in extension-education activities | 0.602** |
| Participation in FFS programs | 0.236** |
| Participation in local associations | 0.496** |
| Opinion leaders | 0.174** |

** $p< 0.01$

Table 9: An overview of stepwise model

| Model | R | R Square | Adjusted R Square |
|---|---|---|---|
| 1 | 0.496 | .246 | 0.243 |
| 2 | 0.602 | 362 | 0.359 |
| 3 | 0.653 | 0.427 | 0.420 |

Table 10: Regression analysis

| Description | B | Beta | t | Sig. |
|---|---|---|---|---|
| Constant | 80.175 | | 32.021 | 0.000 |
| Participation in extension-education activities | 2.633 | 0.500 | 9.736 | 0.000 |
| Participation in local associations | 4.114 | 0.214 | 4.102 | 0.000 |
| Opinion leaders | 2.534 | 0.129 | 2.646 | 0.009 |

sion-education activities, participation in FFS programs, participation in local associations and opinion leaders. According to the findings, there was no significant correlation between IPM technologies adoption and membership in rural-agricultural cooperatives.

### Regression analysis

In order to explain variation in the extent of IPM technologies adoption by farmers, stepwise regression analysis was applied. The R Square value of 0.427 reveals that 42.7 percent of variation in the extent of IPM technologies adoption could be explained by three variables including participation in extension-education activities, participation in local associations, and opinion leaders.

Considering the results shown in the table 10, regression equation in standard situation will be as follow:

$$Y = \text{constant} + B1X1 + B2X2 + B3X3 \quad (1)$$

Y = Extent of IPM technologies adoption, $(X_i)$ is independent variable that included X1= participation in extension-education activities, X2= participation in local associations and X3= opinion leaders. Also, $(B_i)$ is the coefficient of independent variable.

The findings showed that participation in extension-education activities (Beta= 0.500) could explain the most variation in the extent of IPM technologies adoption by farmers.

## CONCLUSION

The findings revealed that 38.8 percent of paddy farmers had high level of IPM technologies adoption. The findings indicated that there was positive and significant correlation between farmers' IPM technologies adoption and participation in local associations. This result is accordant to the results of Clearfield *et al.* (1986), Abd-Ella *et al.* (1981), Strauss *et al.* (1991), Adesina *et al.* (2000), Bonabana-Wabbi (2002) and Truong Thi (2008).

There was positive and significant correlation between IPM technologies adoption by farmers and opinion leaders. This result is accordant to the results of Jacob (1982) and Asghari and Hadi (2009). Opinion leaders for their popularity are one of the best channels for delivering and application of technologies, including IPM. This result shows the importance of identification and training of opinion leaders by extension agents for effective IPM technology adoption.

According to the findings, participation in extension-education activities and FFS programs were important factors in adoption of IPM technologies by farmers. These results are accordant to the results of Ridgley and Brush (1992), Boughton and de Faran (1994), Bonabana-Wabbi (2002), Rejaul and Bakshi (2005), Luther *et al.* (2005), Mariyono (2007), Truong Thi (2008) and Noorhosseini Niyaki *et al.* (2010).

## RECOMMENDATION

Results suggest the effective role of strengthening communication with farmers to extension employees (extension contacts) for best adoption IPM technology. Therefore, it is recommended to establish extension workshops to increase farmers' knowledge toward IPM practices.

Regression analysis indicated that about half of variation in the extent of IPM technologies adoption could be explained by participation in extension-education activities, participation in local associations, and influence of opinion leaders. The findings revealed that participation in extension-education activities could explain the most variation in the extent of IPM technologies adoption by farmers. It shows the important role of agricultural extension agents which can affect paddy farmers' perceptions and behaviors to adopt and apply IPM technologies. In this regard, use of extension-participatory methods such as farmer field schools is proper strategy for creating positive attitude of farmers towards IPM technologies. Finally, according to the findings and with regard to the impact of opinion leaders on farmers, local leaders' adoption of new technology is important for other farmers to adopt and apply these technologies. So, it is recommended that identification and training of opinion leaders be considered by extension agents for effective IPM technology adoption.

## REFERENCES

1- Abd-Ella, M., Eric, O.H. & Warren, R.D. (1981). Adoption Behavior in Family Farm Systems: An Iowa Study. Journal of Rural Sociology, 46 (1): 42-61.

2- Adesina, A., Mbila, D., Nkamleu, G.B. & Endamana, D. (2000). Econometric Analysis of the Determinants of Adoption of Alley Farming by Farmers in the Forest Zone of Southwest Cameroon. Agriculture, Ecosystems, and Environment, 80: 255-265

3- Asadpur, H. (2011). Socio-economic Factors Affecting the Development of Biological Technologies Stem Borer Pest in Rice Fields of Mazandaran. Agricultural Economic and Development, 19 (76): 231-252.

4- Asai, M. & Tokunaga, S. (2007). A Study on Integrated Pest Management (IPM) Programs in Thailand: A Case Study of Saraburi Province. Studies in Regional Science, 37(3): 855-866.

5- Asghari, S. & Hadi, F. (2009). The Study of Extension Effects on Adoption of Biological Control in Soybean Farmer at Moghan (Ardebil province, Iran). 1st National New Technology in Agriculture and Natural Resources, Rasht, p. 982-996.

6- Barrera, V., Norton, G.W., Alwang, J.R. & Mauceri, M. (2005). Adoption of Integrated Pest Management Technologies: A Case Study of Potato Farmers in Carchi, Ecuador. American Agricultural Economics Association Annual meeting, July.

7- Bonabana-Wabbi, J. (2002). Assessing Factors Affecting Adoption of Agricultural Technologies: The Case of Integrated Pest Management (IPM) in Kumi District, Eastern Uganda. MS thesis, Virginia Polytechnic Institute and State University.

8- Boughton, D. & de Faran, B.H. (1994). Agricultural Research Impact Assessment: the Case of Maize Technology Adoption in Southern Mali. MSU International Development Working Paper No, 41. Department of Economics, Michigan State University.

9- Clearfield, F. & Osgood, B.T. (1986). Sociological Aspects of the Adoption of Conservation Practices. Soil Conservation Service. Washington, D.C.

10- Erbaugh, J. M., Donnermeyer, J., Amujal, M., & Kidoido, M. (2010). Assessing the Impact of Farmer, Field, School Participation on IPM Adoption in Uganda. Journal of International Agricultural and Extension Education, 17(3): 5-17.

11- FAO. (1992). A Commodity System Assessment Methodology. FAO study. Rome.

12- Ferguson, W. & Yee, Y. (1995). A Logic model of Cotton Producer Participation in Professional Scout Programs. Journal of Sustainable Agriculture, 5(3).

13- Jacob, P.V. (1982). Study of Factors Associated with the Adoption Recommended Farm Practices in a Nigerian village. Agricultural Administration 9: 17-27.

14- Khush, G.S. (2004). Harnessing Science and Technology for Sustainable Rice-based Production System. In: Proceedings of FAO Rice Conference. International Rice Commission Newsletter, 53: 17-21.

15- Louise Flint, M., Daar, Sh. & Molinar, R. (2003). Establishing Integrated Pest Management Policies and Programs: A Guide for Public Agencies. University of California, Division of Agriculture and Natural Resources. ANR Publication 8093. http:// www.anrcatalog.ucdavis.edu.

16- Luther, G.C., Harris, C., Sherwood, S., Gallagher, K., Mangan, J. & Gamby, K.T. (2005). Development and Innovations in Farmer Field School and Training of Trainers. Globalizing IPM. Black Well Publishing, P. 275.

17- Mariyono, J. (2007). Adoption and Diffusion of Integrated Pest Management Technology: A Case of Irrigated Rice Farm in Jogjakarta Province, Indonesia. Journal of Agricultural Technology, 3(1): 39-50.

18- Mauceri, M. (2004). Adoption of Integrated Pest Management Technologies: A Case Study of Potato Farmers in Carchi, Ecuador. Master of Science in Agricultural and Applied Economics, Virginia Polytechnic Institute and State University, Blacksburg, Virginia.

19- Ministry of Jihad-e Agriculture. (2006). Agriculture Database. Tehran, Ministry of Agricultural Jihad, Bureau of Statistics and Information Technology. (In Farsi).

20- Muthuraman, P. & Sain, M. (2002). Sociology of Integrated Pest Management in Rice. Resources Management in Plant Protection during Twenty First Century, Hyderabad, India, 2. 14-15.

21- Nabirye, J., Nampala, P., Ogenga-Latigo, M.W., Kyamanywa, S., Wilson, H., Odeke, V., Ceduna, C. & Adipala, E. (2003). Farmer Participatory Evaluation of Cowpea. Integrated Pest Management Technologies in Eastern Uganda. Crop Protection, 22. 31-38

22- National Academy of Sciences. (1987). Report of the Research Briefing Panel on Biological Control in Managed Ecosystems. Washington, DC: National Academy Press. 206 p.

23- Niyaki, A., Radjabi R. & Allahyari, M.S. (2010). Social Factors Critical for Adoption of Biological Control Agents Trichogramma Spp. Egg Parasitoid of Rice Stem Borer Chilo suppressalis in North of Iran. American-Eurasian Journal of Agricultural & Environmental Science, 9(2): 133-139.

24- Noorhosseini Niyaki, A., Radjabi, R. & Allahyari, M.S. (2010). Social Factors Critical for Adoption of biological control agents Trichogramma Spp. Egg Parasitoid of Rice Stem Borer. Chilo suppressalis in North of Iran. American-Eurasian Journal of Agriculture & Environment Science, 9 (2): 133-139.

25- Oerke, E.C., Dehne, H.W., Schonbeck, F. & Weber, A. (1994). Crop Production and Crop. Protection: Estimated Losses in Major Food and Cash Crops. Amsterdam, Netherlands: Elsevier.

26- Palis, F.G., Morin, S. & Hossain, M. (2002). Social Capital and Diffusion of Integrate d Pest Management Technology: A Case Study in Central Luzon, Philippines. Paper presented at the Social Research Conference, CIAT, Cali, Columbia, September 11- 14, 2002.

27- Pouratashi, M. & Iravani, H. (2012). Farmers' Knowledge of Integrated Pest Management and Learning Style Preferences: Implications for Information Delivery. International Journal of Pest Management, 58(4): 347-353.

28- Rejaul, H. & Bakshi, K. (2005). Pest Management, Productivity and Environment: A Comparative Study of IPM and Conventional Farmers of Northern Districts of Bangladesh. Pakistan Journal of Sciences, 3(8): 1007- 1014.

29- Ridgley, A.M. & Brush, S.B. (1992). Social Factors and Selective Technology Adoption: The Case of Integrated Pest Management. Journal of Human Organization, 51(4): 367-378.

30- Strauss J., Barbosa, M., Teixeira, S., Thomas, D. & Junior, R.G.(1991). Role of Education and Extension in the Adoption of Technology: A Study of Upland Rice and Soybean Farmers in Central-West Brazil. Agricultural Economics, 5: 341-359.

31- Tette, J. P., Kovach, J., Schwarz, M. & Bruno, D. (1987). IPM in New York Apple orchards-development, demonstration, and adoption. New York State Agricultural Experiment Station, Geneva, a Division of the New York State College of Agriculture and Life Sciences, Cornell University, Ithaca. New York Food and Life Sciences Bulletin. (119).

32- Truong Thi, N.C. (2008). Factors Affecting Technology Adoption among Farmers in the Mekong Delta through the Lens of the Local Authorial Managers: An analysis of qualitative data. Omonrice, 16: 107-112.

33- Van Huis, A. & Meerman, F. (1997). Can We Make IPM Work for Resource-poor Farmers in Sub-Saharan Africa? International Journal of Pest Management, 43(4): 313–20

34- WHO, (1990). Public Health Impact of Pesticides Used in Agriculture. Geneva, Switzerland World Development Report 2008: Agriculture for Development. World Bank Publications.

4

# Drought Risk Vulnerability Parameters among Wheat Farmers in Mashhad County, Iran

*Mojtaba Sookhtanlo, Hesamedin Gholami * and Seyyed Reza Es'haghi*

Abstract

Identification and analysis of farmers' vulnerability associated with their risk aversion degree is one of the necessary requirements for planning and reducing impacts of drought in Iran. So, this study was investigated three risk vulnerability parameters (economic, social and technical) among wheat farmers categorized in accordance with their risk aversion degree in the Mashhad County (Iran) between drought years of 2009-2011. Vulnerability parameters were determined by Delphi technique. For measuring vulnerability and risk aversion degree, formula of Me-Bar and Valdes and method of Safety First Rule were applied respectively. Findings revealed that in social vulnerability indicators; education level, collaboratively farming activities and dependency on government and in technical vulnerability; irrigation method, cultivation method and type of cultivation; risk averse farmers have had the highest vulnerability level under drought conditions. While respecting economic vulnerability, risk neutral farmers (in insuring for crops, sale prices of crops and the type of land ownership), have had the highest vulnerability level.

***Keywords:***
***Drought, Wheat farmers, Vulnerability, Risk aversion degree***

*Department of Agricultural Extension and Education, University of Tehran, Karaj, Iran.*
** Corresponding author's email: hagholami@ut.ac.ir*

## INTRODUCTION

Drought is a slow-onset disaster that has economic, social, and environmental consequences. In Iran, drought is a re-current phenomenon and current drought management strategies in Iran are based on crisis management. For example, when drought occurs in different parts of the country, a state of emergency is declared and thus all resources are mobilized in that particular region.

However, this type of drought management strategy is proved to be ineffective. It is considerable that the year 2011 was the 13th continuous year that drought had been occurred in Mashhad County, northeast of Iran, and this has impacted most of the socioeconomic and technical dimensions of agricultural and rural sectors (The Agriculture Organization of Khorasan-e-Razavi Province, 2011). Therefore, this study is an abiding interest in how farmers cope with and overcome agricultural crises such as drought or natural disasters and provides a new and realistic vision for identifying of risk vulnerability indicators in drought. Drought risk is best defined as a combination of a location's exposure to drought and its vulnerability to drought (Ajijola *et al.,* 2011) and vulnerability is identified as the exposure to uninsured risk leading to an unacceptable level of well-being among farmers (Hoddinott and Quisumbing 2003, Hoogeveen *et al.,* 2005). Many studies (Gwimbi 2009, Deressa, 2010, Wilhelmi *et al.*, 2002, Kapoor and Ojha, 2006, Barbier *et al.*, 2008, Mongi *et al.,* 2010, Keshavarz *et al.*, 2011) highlight geographical situations and rainfall level as key factors on farmers' vulnerability. However, people within a locality and same area are not evenly vulnerable to drought (Slegers 2008). So, there is a growing appreciation that other factors such as farmers' characteristics including levels of their risk aversion (internal risk factors) have directly influenced drought vulnerability parameters but this still has been considered by rare studies (Hoogeveen *et al.*, 2005, Franke *et al.*, 2005, Brondizio and Moran, 2008, Ajijola *et al.*, 2011). Farmers' capacities to cope with drought, depending on ownership or access to a wide variety of resources such as land ownership, farmers' incomes, farming lands size, education level, access to governmental and bank credits (loans), crops insurance, technical assistance and information, social networking, and public support programs (Scoones 1998; Ellis 2000; St. Cyr 2006) are categorized in this study in three parameters of social, economic and technical. Eakin *et al.* (2006), Deressa (2010), Ajijola *et al.* (2011), Keshavarz *et al.* (2011) and Sharafi and Zarafshani (2010 and 2011) examined the impact of risk attitudes (level of risk aversion) on poverty and vulnerability level among rural farmers. The variety of information on household human resources and income sources, production and losses to climate hazards and pests, crop and livestock management practices, commercialization practices, input and machinery use, farmers' risk mitigation practices, landholding size, and farm profit, loans, selling of crop outputs, low income level, credits, irrigation method, household extension packages and farmers' access to resources and use of them considered importance of adaptation (technology, technical assistance, credit and insurance) on farmers' capacities to respond to stress and uncertainty conditions (drought). So, the main purpose of this study was to identify the most vulnerable farmers regarding their risk aversion degree in the Mashhad County (Iran). Particular interests are as follows:

- To identify wheat farmers based on economic, social and technical vulnerability indicators.
- To calculate wheat farmers' vulnerability.
- To calculate wheat farmers' risk aversion degree and categorize them according to it.
- To determine wheat farmers' vulnerability in each category of risk aversion degree.

## MATERIALS AND METHODS

This study was conducted in the Mashhad County (rural areas) located in Khorasan-e-Razavi Province, Iran. The capital of Mashhad County is the Mashhad City. This County that is located in North East of Iran and is the most populous county in *Khorasan-e-Razavi* Province. This county is 992–1184 meters above sea level. The area of this county is 1490 km$^2$. This county consists of 591 villages. The cultivated land of the county is 56615 km$^2$. This county with an arid- semi-arid and arid climate receives an average rainfall of 256 mm (The agricultural statistics and information office of *Khorasan-e-Razavi* Province, 2009). Wheat is

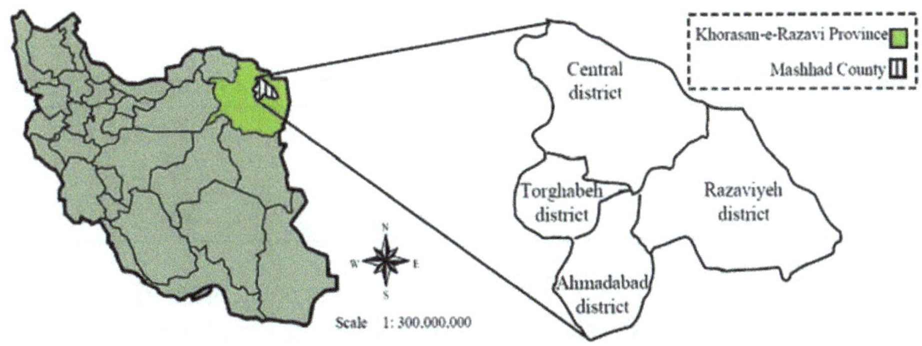


Figure 1: Area of study (Mashhad County, Khorasan-e-Razavi Province, Iran)
Reference: Statistical Centre of Iran (2012).

Table 1: Sample size in each district (Mashhad County)

| Districts (Bakhsh) | Statistical population | Sample size |
|---|---|---|
| Central | 2574 | 125 |
| Razaviyeh | 1320 | 65 |
| Ahmadabad | 1086 | 53 |
| Torghabeh district | 960 | 50 |
| **Total** | **5940** | **293** |

the dominant crop in the region, so the statistical sample of this study consisted of wheat farmers who live in *Mashhad* County. The selected region was severely affected by drought during the year 2009-2011. Mashhad County is divided into four districts (Bakhsh), with their capitals (Figure 1): Ahmadabad (capital: Malekabad), Central (capital: Mashhad), Razaviyeh (capital: Razaviyeh) and Torghabeh (capital: Torghabeh).

A proportional stratified random sampling was applied to access the respondents and using Cochran's test the size of sample was determined 293 wheat farmers (Table 1).

Two questionnaires and methods of interview were designed and used to gathering data. The first questionnaire included open questions to determine the most important socioeconomic and technical vulnerability indicators in the Mashhad County by Delphi technique. The second questionnaire was consisted of three parts. The first part was to collect data about farmers' personal and professional characteristics. The second part consisted of risk aversion indicators (according to formula of Safety First Rule). The third part consisted of vulnerability indicators was obtained through the first stage (Delphi technique) to calculate vulnerability level of farmers. This study is conducted in two main stages.

First stage (Delphi technique): This stage includes usage of the Delphi technique to identify and weigh major indicators of vulnerability in the study region as used in many previous studies (Kaly and Pratt 2000; Dercon 2004; Deressa 2010). Snowball method was used to determine experts related to the study objectives. In other words, we asked the experts who were known in the research process to introduce other experts to us. Finally 45 experts were chosen among which, 31 experts resend the questionnaires and their data was used. They were people who had field research about drought or extension experts who were directly engaged in programs or activities related to drought in the area of study. A primary questionnaire including open-ended questions (i.e. determine the most important socioeconomic and technical vulnerability indicators at Mashhad County) were distributed among experts. In the next step, first questionnaire data were used to determine and categorize common major social, economic and technical indicators with the most frequency. Acquired data were used to design another questionnaire including the primary indicators which were edited to send again to the experts to be confirmed by them. In the third step, the questionnaire was consisted of final confirmed economic, social and technical indicators and also a section for determining the weight (relative importance) ($W_i$=1… n) of each indicator in farmers' vulnerability by experts. They could weigh the indicators from 0 (the lowest importance) to 10 (the highest importance). It was emphasized in the questionnaire that, weighing must express the relative importance of indicators, so the indicators could not be weighted the same. These indicators used to design the next

Table 2: Status of the respondents, by the risk- aversion degree.

| Risk- aversion coefficient | Status of wheat farmers | Frequency | percent |
|---|---|---|---|
| $0.1 \leq R j \leq 1$ | Risk taker | 52 | 18 |
| $-0.1 \leq R j \leq 0.1$ | Risk neutral | 63 | 21 |
| $-1 \leq R j \leq -0.1$ | Risk averse | 178 | 61 |
| **Total** | - | **293** | **100** |

stage questionnaire.

Second stage (determining farmers' risk aversion degree and vulnerability level): In this stage another questionnaire was used. Formula of Safety First Rule was used to calculate the farmers' risk aversion degree and also categorizes them in three groups (namely risk adverse, risk neutral and risk taker). Furthermore, farmers' social, economic and technical vulnerability amount was determined by method of Me-Bar and Valdez (2005).

To calculate risk aversion degree Safety First Rule formula was used. Randhir (1991), Parikh and Bernard (1988), Sekar and Ramasamy (2001), Ajetomobi and Binuomote (2006) and Ajijola *et al.* (2011) used this method in their studies in order to determine the risk-aversion degree of farmers. Because of lack of access to accurate data needed for other common methods and lack of valid and categorized databases in the studied region, the mentioned formula was applied in this study. In this formula:

$R j = [E^{*} j - E j] / [S j], \quad j = 1, 2\ldots, n$

*(R j: Risk-aversion degree of farmer (j), E*j: Critical income level of farmer (j), E j: Expected income of farmer (j), S j: The standard deviation of the farmer (j)'s annual income (in the past three years of agricultural and non-agricultural activities))*

$$E^{*} = 7955936\,(FAM - CHI / 2) + DPT - (UAR + UAR')$$

$$E = VP\,(1 + DMG) - TC$$

The weighted crop damage variable was defined as:

$DMG = (\Sigma k_i DMG_i) / (\Sigma k_i)$

The parts of the above formulas are as the following:

-7955936: The per capita cost of supplying the least calorie supply in one year in Rial (the standard rate in Iran).

-FAM: The household's farm size (Hectare).

-CHI: Number of children (active members of the family in working of agriculture).

-DPT: Farmer's debt amount to formal and informal institutions (IRR).

-UAR: The farmers' annual income from activities other than wheat cultivation (IRR).

-UAR': The beneficiaries' annual income from non-agricultural activities (IRR).

- Total value of wheat production (IRR).

-DMG: The proportion of farmer's damage due to losses and abnormal incidents as a weighted average.

-TC: Total wheat production cost in the same year (IRR).

Among vulnerability assessment methods, a formula suggested by Me-Bar and Valdez (2005) was considered to be appropriate for assessment of socio-economic and technical vulnerability parameters in this study. Me-Bar and Valdez (2005) stated that vulnerability is a qualitative concept for which comparing societies should be measured quantitatively. Mentioned formula based on subjective assessment of factors is affecting drought vulnerability. Considering the lack of reliable resources of data and

Information which is a prerequisite for applying other common methods in the studied region and its successful application in other regions of Iran in previous studies (for example the studies of Sharafi and Zarafshani (2011) in Kermanshah Province and Keshavarz *et al.* (2011) in Fars Province) the applicability and efficiency of this method for the country condition was proved. So, this formula was applied for vulnerability assessment.

$V = 1 / C_0 \sum (P_i W_i)$

(V= each farmer vulnerability amount, $C_0$= sum of total vulnerability weight, Pi= each parameter amount, Wi= each parameter weight)

Also, in this formula:

$C_0 = \sum W_i, \sum W_i = (W_{max} \times n) / 2, \; C_0 = (W_{max} \times n) / 2, \; C_0 < W_{max} \times n$

($W_{max}$: The maximum weight that can belong to each parameter (10), n: The number of each factor parameters)

## RESULTS

### Personal and professional characteristics

Among farmers, 84.5 % were men and 15.5 % were women. The most frequency of the wheat farmers education level was secondary education level which constituted 33 % of the sample and also 21 % of wheat farmers were illiterate and only 9 % of statistical population had a degree higher than diploma. The most experiences of wheat cultivation among the respondents were between 21 to 30 years. Also, by looking to the extent of farmlands, the highest frequency was related to the farmers who had 4 to 7 hectares. The average area of each farmer farmlands was 1.14 hectares and the most experiences of farming among the respondents were 31 to 40 years.

### Risk-Aversion degree of respondents

In the table (2), risk-aversion coefficient (Rj) was calculated according to Safety First Rule formula. Based on the findings, 61 % of the respondents were risk averse, 23 % were risk neutral and 18 % were risk taker.

### Parameters of vulnerability

Findings related to economic, social and technical parameters of vulnerability are shown in the tables (3, 4 and 5). First, for measuring indicators of any parameter, total vulnerability weight ($\sum W_i$) was calculated.

$$\sum W_i = (W_{max} \times n) / 2 = (10 \times 9) / 2 = 45$$

### Indicators weight of parameters:

Findings showed that experts believed that economic parameter indicators (insuring crops ($W_i$=6.12), regional extension experts with economic advices ($W_i$=5.46), and access to governmental and bank credits (loans) ($W_i$=5.41)), social parameter indicators (farming collaborative activities ($W_i$=6.06), attending in extension

Table 3: The amount and weight of economic parameter indicators in three farmers' groups

| Economic parameter indicators | Indicators weight ($W_i$) | Indicators amount in farmers groups ($P_i$) | | |
|---|---|---|---|---|
| | | Risk taker farmers ($P_{i1}$) | Risk neutral farmers ($P_{i2}$) | Risk averse farmers ($P_{i3}$) |
| Insuring crops | 6.12 | 1.98 | 3.49 | 2.87 |
| Extension agents' economic advices | 5.46 | 2.08 | 1.78 | 1.59 |
| Farmers' incomes | 4.95 | 2.12 | 1.54 | 2.17 |
| Amount of liquidity | 4.65 | 2.65 | 2.03 | 2.47 |
| Pre-sale crops to middlemen | 3.75 | 2.67 | 1.71 | 1.63 |
| Sale price of crops | 5.24 | 1.77 | 2.89 | 1.80 |
| Land ownership type | 4.82 | 1.54 | 2.75 | 3.26 |
| Farming lands Size | 4.60 | 1.63 | 1.57 | 2.04 |
| Access to governmental and bank credits (loans) | 5.41 | 3.25 | 2.68 | 2.74 |
| **Total** | **45** | - | - | - |

Table 4: Amount and weight of social parameter indicators in farmers groups

| Social parameter indicators | Weight of indicators ($W_i$) | Indicators amount in farmers groups ($P_i$) | | |
|---|---|---|---|---|
| | | Risk taker farmers ($P_{i1}$) | Risk neutral farmers ($P_{i2}$) | Risk averse farmers ($P_{i3}$) |
| Social esteem | 4.93 | 2.63 | 2.63 | 2.56 |
| Membership in rural associations / organizations | 5.10 | 1.98 | 2.40 | 1.84 |
| Dependency to government | 4.77 | 3.27 | 2.65 | 2.62 |
| Attending in extension education programs | 5.52 | 1.65 | 3.15 | 1.60 |
| Education level | 5.35 | 2.29 | 1.94 | 3.43 |
| Farming collaborative activities | 6.06 | 2.21 | 1.67 | 2.76 |
| Family members collaboration | 4.81 | 2.50 | 2.15 | 1.97 |
| The level of related to farming religious believe | 3.82 | 1.85 | 2.13 | 2.11 |
| Participation in rural development programs | 4.64 | 1.90 | 2.25 | 2.51 |
| **Total** | **45** | - | - | - |

Table 5: The amount and weight of technical parameter indicators in groups of wheat farmers .

| Technical parameter indicators | Indicators weight ($W_i$) | Indicators amount in farmers groups ($P_i$) | | |
|---|---|---|---|---|
| | | Risk taker farmers ($P_{i1}$) | Risk neutral farmers ($P_{i2}$) | Risk averse farmers ($P_{i3}$) |
| Cultivation type (rain-fed/ watery) | 6.06 | 3.03 | 2.37 | 3.00 |
| Cultivation pattern (spring / autumn) | 4.84 | 1.90 | 2.13 | 2.46 |
| Cultivation method (traditional/ mechanized) | 5.12 | 2.96 | 2.98 | 3.02 |
| Use of drought resistant varieties | 4.94 | 2.51 | 2.59 | 2.73 |
| Irrigation method | 5.65 | 2.46 | 3.11 | 3.42 |
| Planting, saving and harvesting times | 4.71 | 3.05 | 2.90 | 2.87 |
| Use of chemical fertilizers | 4.55 | 2.19 | 2.97 | 2.88 |
| Weeds, pests and diseases control | 5.29 | 2.64 | 3.19 | 2.90 |
| Tillage implements | 3.84 | 2.36 | 3.22 | 2.63 |
| **Total** | **45** | - | - | - |

education programs ($W_i$=5.52) and education level ($W_i$=5.35)) and technical parameter indicators (cultivation type (rain-fed / watery) ($W_i$= 6.06), irrigation method ($W_{ii}$=5.65) and weeds, pests and diseases control ($W_i$=5.29)), respectively were the most important indicators in order to explain parameters of vulnerability in target regions.

**Indicators amount in parameters of vulnerability:**

Considering the findings among risk taker farmers, the economic parameter indicators (access to governmental and bank credits (loans) (3.25), pre-sale crops to middlemen (2.67) and amount of liquidity (2.65)), the social parameter indicators (dependency to government ($P_{i1}$=3.27), social esteem ($P_{i1}$=2.63) and collaboration of family members ($P_{i1}$=2.50)) and the technical parameter indicators (planting, saving and harvesting times ($P_{i1}$=3.05), cultivation type (rain-fed /watery) ($P_{i1}$=3.03) and cultivation method (traditional / mechanized) ($P_{i1}$=2.96)), respectively were three indicators which had highest scores.

Among risk neutral farmers, the economic parameter indicators (insuring crops ($P_{i2}$=3.49), sale price of crops ($P_{i2}$=2.89) and land ownership type ($P_{i2}$=2.75)), the social parameter indicators (participation in rural development programs ($P_{i2}$=3.15), dependency to government ($P_{i2}$=2.65) and social esteem ($P_{i2}$=2.63)) and the technical parameter indicators (irrigation method ($P_{i3}$=3.42), cultivation method (traditional/ mechanized) ($P_{i3}$=3.02) and cultivation type (rain-fed / watery) ($P_{i3}$=3.00)), respectively had the highest intensity during drought. Also among risk averse farmers, the economic parameter indicators (Land ownership type ($P_{i3}$=3.26), insuring crops ($P_{i3}$=2.87) and access to governmental and bank credits (loans) ($P_{i3}$=2.74)), the social parameter indicators (education level ($P_{i3}$=3.43), farming collaborative activities ($P_i3$=2.76) and dependency to government ($P_{i3}$=2.62)) and the technical parameter indicators (Irrigation method ($P_{i3}$=3.42), cultivation method (traditional / mechanized) ($P_{i3}$=3.02) and cultivation type (rain-fed / watery) ($P_{i3}$=3.00)), respectively had the highest rank and means that during drought, these farmers have had the highest vulnerability in these indicators.

Total 6: total vulnerability amounts in farmers groups.

| Vulnerability amount | farmers groups | | |
|---|---|---|---|
| | Risk taker farmers | Risk neutral farmers | Risk averse farmers |
| Economic vulnerability | 2.13 | 2.34 | 2.30 |
| Social vulnerability | 2.25 | 2.32 | 2.38 |
| Technical vulnerability | 2.59 | 2.83 | 2.90 |
| Total vulnerability | 6.97 | 7.49 | 7.58 |

Table 7: A summary of indicators priority in farmers groups.

| Farmers groups | farmers groups | | |
|---|---|---|---|
| | **Economic vulnerability** | **Social vulnerability** | **Technical vulnerability** |
| **Risk taker farmers** | Access to governmental and bank credits (loans)<br>Pre-sale crops to middlemen<br>Amount of liquidity | Dependency to government<br>Social esteem<br>Family members collaboration | Planting, saving and harvesting times<br>Cultivation type (rain-fed / watery)<br>Cultivation method (traditional/ mechanized) |
| **Risk neutral farmers** | Insuring crops<br>Sale price of crops<br>Land ownership type | Attending in extension education programs<br>Dependency to government<br>Social esteem | Tillage implements<br>Weeds, pests and disease control<br>Irrigation method |
| **Risk averse farmers** | Land ownership type<br>Insuring crops<br>Access to governmental and bank credits (loans) | Education level<br>Farming collaborative activities<br>Dependency to government | Irrigation method<br>Cultivation method (traditional /mechanized)<br>Cultivation type (rain-fed / watery) |

**Total vulnerability**

Formula of Me-Bar and Valdez (2005) was applied to calculate total vulnerability amount. For example, economic vulnerability in risk averse farmers is calculated:

$V = 1 / C_0 \sum (P_i W_i) = (6.12 \times 1.98) + (5.45 \times 2.08) + (4.95 \times 2.12) + \ldots + (5.42 \times 3.25) = 95.63/45 = 2.13$

According to table 6, the highest economic vulnerability was among risk neutral farmers and the lowest was among risk taker farmers. With respect to social vulnerability, the highest vulnerability was among risk averse farmers and the lowest vulnerability was among risk taker farmers. Also in technical vulnerability, risk averse farmers were the most vulnerable groups and risk taker farmers were the least vulnerable groups.

## DISCUSSION

This paper describes an investigation of drought vulnerability in Mashhad County (Iran). The length of drought in studied region implies that it is a harsh reality of Iran agriculture and mitigation to the severe continuous impacts of that is critical. As the results pointed out, most of the farmers are vulnerable. Therefore, farmers are being extremely stressed to find alternative appropriate mechanisms to reduce their vulnerability. Although most of the farmers are risk averse, they hardly adopt the new advises with potential risks. This means that policy makers should significantly act different from what they currently do. Findings imply that the kind and amount of vulnerability among farmers with various risk aversion degree is different, so when we categorized them in three groups named risk taker, risk neutral and risk averse, they would completely had different and general vulnerability parameters and thus unspecific supports from these groups would be inefficient.

The interesting conclusion which could be made is that there is a relationship between farmers' risk aversion degree and their vulnerability level. In other words, various effects of drought on different farmers' groups have not been considered by policy makers and managers in the studied region and thus managerial, educational and support programs have not been appropriate for these groups. In sum, findings revealed that risk taker farmers had the least vulnerability in all three vulnerability parameters named economic, social and technical parameters, while risk neutral farmers were the only most vulnerable group in economic parameter. Risk averse farmers were the most vulnerable group, because they were the most vulnerable group in two parameters named social and technical parameters. So, it can be said that they are the most vulnerable farmers. Some other parts of findings are shown in table 7.

Considering the common indicators in all three farmers groups, it can be said that as emphasized by Vásquez-León *et al.* (2003) and Nelson and Escalante (2004), in order to manage and reduce negative impacts of drought economic vulnerability, mechanisms such as granting gratuitous or low interest loans based on farmers livelihood level, establishing small rural banks, more governmental attention to crops insurance fund (Hazell, 2004), and developing and enriching local credit funds should be regarded as high priority actions.

With respect to social vulnerability, findings revealed that among risk taker farmers, dependency to government (consistent with Sharafi and Zarafshani (2011)), social esteem and family members' collaboration have had the most effect on social vulnerability. Among risk neutral farmers, attending in extension education programs, dependency to government and social esteem and education level among risk averse farmers which is consistent with Vásquez-León *et al.* (2003), Sengestam (2009) and Deressa (2010), farming collaborative activities (consistent with Iglesias *et al.* (2009)) and dependency to government have had the most effect on economic vulnerability.

With respect to technical vulnerability indicators for risk taker farmers, planting, saving and harvesting times, cultivation type (rain-fed/watery) and cultivation method (traditional/ mechanized) have had the most effect on their technical vulnerability. Among risk neutral farmers, weeds, pests and diseases control and irrigation method and among risk averse farmers, irrigation method, cultivation method (traditional/ mechanized) and cultivation type (rain-fed/watery) have had the most effects on their technical vulnerability. Hence, identification and promotion of varieties and species which are suitable for each group of farmers and also compatible with continental conditions as substitutions for crops with high water requirements, providing infrastructures for sustainable development of water resources such as draining, pressured irrigation systems and helping farmers to control pests and common diseases during drought is recommendable. The results of this study can imply that drought relief programs should be based on the rate of socio-economic and technical vulnerability among farmers' groups in term of their risk aversion. Furthermore, an up-to-date vulnerability assessment helps extension agents to plan more effective content for their educational programs.

## REFERENCES

1- Ajetomobi, J.O., & Binuomote, S.O. (2006). Risk aversion among poultry egg producers in southwestern Nigeria. International Journal of Poultry Science, 5 (6), 562-565.

2- Ajijola, S., Egbetokun, O.A., & Ogunbayo, I.E. (2011). Impact of risk attitudes on poverty level among rural farmers in Ogun State. Journal of Development and Agricultural Economics, 3(12), 581-587.

3- Barbier, B., Yacouba, H., Karambiri, H., Zorome, M., & Some, B. (2008). Human vulnerability to climate variability in the sahel: Farmers' adaptation strategies in northern Burkina Faso. Environmental Management, 43,790-803.

4- Brondizio, E. S., & Moran, E.F. (2008). Human dimensions of climate change: the vulnerability of small farmers in the Amazon. Philosophical Transactions of the Royal Society B: Biological Sciences, 363(1498), 1803-1809.

5- Dercon, S. (2004). Growth and Shocks: Evidence from rural ethiopia. Journal of Development Economics. 74(2), 309-329.

6- Deressa, T.T. (2010). Assessment of the vulnerability of Ethiopian agriculture to climate change and farmers' adaptation strategies. Ph.D thesis in environmental economics, Department of Agricultural Economics, Extension and Rural Development. University of Pretoria, Pretoria, available from: http: //upetd.up.ac.za/thesis/available/etd-10232010-165854/unrestricted/thesis.pdf.

7- Eakin, H., Webhe, M., Ávila, C., Torres, G.S., & Bojórquez-Tapia, L. (2006). A comparison of the social vulnerability of grain farmers in Mexico and Argentina. AIACC Working Paper 29, 47-55, Assessment of Impacts and Adaptation to Climate Change in Multiple Regions and Sectors Program, Washington, DC. USA.

8- Franke, G., Richard, C.S., & Subrahmanyam, M.G. (2005). Incremental risk vulnerability. University of Konstanz. Journal of Economic Literature, 05-08: 1-19, available from:http://cofe.uni-konstanz.de/ Papers/ dp05 _08. pdf.

9- Gwimbi, P. (2009). Cotton farmers' vulnerability

to climate change in Gokwe District (Zimbabwe): impact and influencing factors. Journal of Disaster Risk Studies, 2(2), 81-92.

10- Hazell, P. (2004). Climate change and management of catastrophic risk. Environment and Policy, 39, 385-395.

11- Hoddinott, J., & Quisumbing, A. (2003). Methods for micro econometric risk and vulnerability assessments. Social Protection Discussion Paper Series, 0324,1-59. Social Protection Unit, Human Development Network. The World Bank. Washington D.C. USA.

12- Hoogeveen, J., Tesliuc, E., Vakis, R., & Dercon, S. (2005). A guide to the analysis of risk, vulnerability and vulnerable groups. Social Protection Unit, Human Development Network. The World Bank, Washington, DC. USA.

13- Iglesias, A., Garrote, L., Cancelliere, A., Cubillo, F., & Wilhite, D. A. (2009). Coping with drought risk in agriculture and water supply systems: Drought management and policy development in the Mediterranean. Springer Netherlands. Advances in Natural and Technological Hazards Research, 26 (2), 153-159.

14- Kaly, U., & Pratt, C. (2000). Environmental vulnerability index: development and provisional indices and profiles for Fiji, Samoa, Tuvalu and Vanuatu. Phase II. SOPAC Technical Report 306, 88-89.

15- Kapoor, S., & Raj Kumar, O. (2006). Vulnerability in rural areas: potential demand for microinsurance. Sage Publications. International Journal of Rural Management, 2(1), 67-83.

16- Keshavarz, M., Karami, E., & Zamani, G. (2011). Drought vulnerability of farm households: a case study. Iranian Agricultural Extension and Education Journal.

17- Me-Bar, Y., & Valdez, F. (2005). On the vulnerability of the ancient Maya society to natural threats. Journal of Archaeological Science, 32, 813-825.

18- Mongi, H., Majule, A.E., & Lyimo, J.G. (2010). Vulnerability and adaptation of rain fed agriculture to climate change and variability in semi-arid Tanzania. African Journal of Environmental Science and Technology, 4(6), 371-381.

19- Nelson, C.H., & Escalante, C.L. (2004). Toward exploring the Location-Scale condition: A constant relative risk aversion Location-Scale objective function. European Review of Agricultural Economics, 31(3), 273-287.

20- Parikh, A., & Bernard, A. (1988). Impact of risk on HYV adoption on Bangladesh. Agricultural Economics, 2 (2), 167–78.

21- Randhir, T.O. (1991). Influence of risk on input decisions in Tankfed farms of South India. Indian Journal of Agricultural Economics, 46 (1), 57-63.

22- Scoones, I. (1998). Sustainable rural livelihoods: A framework for analysis, Institute of Development studies, IDS Working paper. University of Sussex. Brighton. England, 72, 1-18.

23- Sekar, I., Ramasamy, C. (2001). Risk and resource analysis of rainfed tanks in South India. Journal of Social and Economic Development, 3 (2), 208-215.

24- Sengestam, L. (2009). Division of capitals- What role does it play for gender-differentiated vulnerability to drought in Nicaragua?. Community Development, 40, 154-176.

25- Sharafi, L., & Zarafshani, K. (2011). Economic and social vulnerability assessment among farmers towards drought (Case of study: Wheat farmers in Kermanshah, Sahne, and Ravansar townships). Journal of rural research, 1(4), 129-154. In Farsi.

26- Sharafi, L., & Zarafshani, K. (2010). The assessment of vulnerability in respect of drought mitigation and management for farmers whom cultivating wheat (case study: sarpol-e-zahab, islamabad, and javanrood). Arid regions geographic studies, 1(1), 106-120. (In Farsi).

27- Slegers, M.F.W. (2008). If only it would rain: Farmers' perceptions of rainfall and drought in semi-arid central Tanzania. Journal of Arid Environments, 72, 2106- 2123.

28- Statistical Centre of Iran. (2012). Population and housing census, administrative units of razavi khorasan and their populations.

29- St. Cyr, J. F. (2006). At risk: natural hazards, people's vulnerability, and disasters. Journal of Homeland Security and Emergency Management. Published by The Berkeley Electronic Press, 2(2),1-5. Available from: http:// www.geo.mtu.edu/volcanoes/06upgrade/SocialKateG/Attachments%20Used /AtRiskReview.pdf

30- The Agricultural statistics and information office of Khorasan-e-Razavi Province. (2009). Agricultural statistic annals in 2009, Department of Planning and Economic Affairs. Pp, 1-164.

31- The Agriculture organization of Khorasan-e-Razavi Province. (2011). The eighth leaf. Magazine of news and Information Science, Public relations of Agricultural Jihad Organization of Khorasan-e-Razavi Province, 10, 1-12.

32- Vásquez-León, M., West, C.T., & Finan, T.J. (2003). A comparative assessment of climate vulnerability: agriculture and ranching on both sides of the US-Mexico border. Global Environmental Change, 13, 159-173.
33- Wilhelmi, O.V., & Wilhite, D.A. (2002). Assessing vulnerability to agricultural drought: a Nebraska case study. Natural Hazards, 25, 37-58.

# 5

# Livestock Farming Systems and Cattle Production Orientation in Eastern High Plains of Algeria, Cattle Farming System in Algerian Semi Arid Region

***Lounis Semara* [1], *Charefeddine Mouffok* [2*] *and Toufik Madani* [3]**

Abstract

This study was an attempt to devise productive orientations of cattle herds in eastern high plains of Algeria. In this regard, 165 farms randomly identified were investigated. The selection of breeders was based to existence of cattle on the farm, and the farmer proposed to investigation must have at least two cows. The approach taken was to identify all systems adopted by farmers in a region through the analysis of the relationship between the maintenance of different types of cattle and preferred marketing policies. The model has been emerged as a result of functional typology established using the procedure categorical principal components analysis (CATPCA) of optimal coding in SPSS [19. 2010]. Following this approach, five types of cattle productive orientation have been identified, the balanced mixed system (dairy-beef), beef mixed system, dairy mixed system, dairy system and beef system. These results showed that the breeders were oriented towards specialization (dairy or beef) in less than 20% of situations. Farmers in our context prefer mixed systems when beef mixed system was the model type frequently encountered in the region (over than 50% of farms).

*Keywords:*
*Livestock, typology, Farming system, Cattle, Management*

[1] *Ph.D Student of Animal Production, Department of Agriculture and Animal Science, Setif 1 University, Algeria.*
[2] *Assistant Professor, Department of Agriculture and Animal Science, Setif 1 University, Algeria.*
[3] *Professor, Department of Agriculture and Animal Science, Setif 1 University, Algeria.*
* *Corresponding author's email: mouffokcharefeddine@yahoo.fr*

## INTRODUCTION

Livestock and their products provide direct cash income and the animals are living the assets for many farmers (FAO/ILRI, 1995). Breeding cattle assume also the roles of job creation and income very important for social stability (Srairi *et al.*, 2009). In Algeria, animal production especially dairy cattle were always at the center of occupation of public authorities, well as several policies and actions have been applied. However, the dairy industry operates almost with imported powdered milk where 60% of milk requirements are imported. Several researchers have attempted to explain the poor performance of the cattle sector in the Algerian context by a constraint which opposes the development of a strong dairy activity, in particular, problems of adaptation of exotic breeds in different agro-ecological zones of countries (Madani and Mouffok, 2008) and the lack of fodder production required for intensive dairy farm.

Researches on livestock have always been guided by the search for efficiency improvement activity (Dedieu, 2009). Madani and Mouffok (2008), provide that the deficiency of milk production in Algerian's farms requires changes in technical choices and especially the type of animals and livestock systems implanted. Systemic vision on cattle farms prospection is therefore essential to understand better the factors influencing the elaboration of performances. Many authors suggest two conceptual approaches, one focused on the analysis of farmers' practices, it comes to technical, economic and social farmers practical (Chapman *et al.,* 2008; Dufumier, 1996) and the other attempts to understand how farmers make their decisions (Shalloo *et al.,* 2004).

This research can be considered as a contribution to characterization diversity of cattle farming systems. Its aims through the adjustment of some technical and economic practices in of Algerian Eastern high plains farms to analyze the organization of cattle production systems and identify pathways to explain management and planning strategies adopted.

## MATERIALS AND METHODS

### Methodological approach

Investigation in a single passage was conducted among farmers and herders of cattle. Selection of farmers' was based on existence of cattle breeding activities and farm proposed for investigation must have at least two cows. In study region, these categories of farms correspond to plus than 90% of all breeders. In this regard, a sample of 165 farms were randomly chosen and visited. Livestock farms that are the subject of our investigation were located in two provinces of the eastern high plains of Algeria, Setif and Bordj Bouarraridj departments (Figure 1).

We also selected scale of aridity gradient which increases from north to south. The investigations have been developed in a questionnaire consisting of three components (socio economics of farmers family, structure and resources of farms and functionning practices of cattle herd) with more than 150 questions. The objective of this survey was to collect among those surveyed as much information about the livestock, but

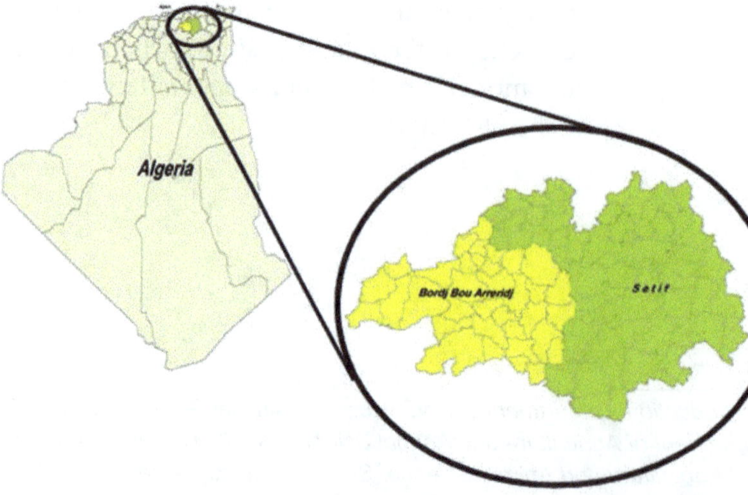


Figure 1: Localisation of studied area

Table 1: Data of cattle categories number in all farms'

| | Cattle (LU) | Cows | Heifer | Beef | Male Calf | Femelle Calf |
|---|---|---|---|---|---|---|
| **Mean** | 12.63 | 7.69 | 2.53 | 1.53 | 2.02 | 2.13 |
| **Standard error of mean** | 0.78 | 0.46 | 0.22 | 0.32 | 0.20 | 0.17 |
| **Standard deviation** | 10.00 | 5.49 | 2.80 | 4.01 | 2.55 | 2.10 |
| **Minimum** | 02.00 | 2.00 | 0.00 | 0.00 | 0.00 | 0.00 |
| **Maximum** | 71.45 | 45.00 | 14.00 | 40.00 | 18.00 | 10.00 |

LU : Livestock Unit

also variables related to the production environment and the diversity of functionning practices of cattle and strategies developed.

### Diagnostics tools

A graphics typology was established using Categorical Principal Components Analysis (CATPCA) optimal coding procedure of SPSS (19.2010) software. This procedure was most appropriate to research aims to analysis the relationship between quantitative variables describing the structure of cattle herd (dairy cows effective, beef cattle effective, heifers effective, effective male and female calves) and the different modalities of qualitative variables describing practice-policies adopted by farmers such as type de breeding choice and cattle product commercialization (amount of sold milk and calves sale age). This categorization was reinforced by two-step cluster automatic classification procedure. All variables was presented by means and standard deviation.

## RESULTS

### Overall characteristics of farms

### Effective and structure of cattle herd

The descriptive analysis of cattle herd size by farm was summarized in table 1. Results show that all farms exploit average herd of 12.6±10.0 LU. The number of cows was 7.6 ± 5.4 by farm and this category represents more than 60% of the total cattle population. Those farms mark the permanent presence of 2.5±2.80 heifers' and 1.5±4.0 young beef. A large standard deviation recorded reflects a high divergence in the compositional structure of cattle herd between farms, which was the first indicator of the diversity of cattle production policies.

### Description of economic practices

Theses farms were mostly cattle farms alone (46.7%) or cattle-sheep (41.8%). The association of cattle-sheep-goat was observed in less than 5% of cases and about 6.7% producers have developed a new trend to the association

Table 2: Farming system and economic practices of farms

| Variable | Modality | Percentage (%) |
|---|---|---|
| **Breeding species** | C.S.G | 4.8 |
| | C.S | 41.8 |
| | C.P | 6.7 |
| | C | 46.7 |
| **Milk soled** | Total | 45.4 |
| | Part of | 49.7 |
| | Never | 4.9 |
| **Age of calf sale** | Pre weaning | 12.5 |
| | After weaning | 13.5 |
| | Old age | 64.4 |
| | As needed | 10.0 |

C : Cattle ; S : Sheep ; G : Gaot ; P : Poultry.

Table 3: Model parameters' of CATPCA

| Dimension | Alpha of Cronbach | Proper Value | Explain variance |
|---|---|---|---|
| **1** | 0.691 | 2.52 | 31.55 |
| **2** | 0.409 | 1.55 | 19.42 |
| **Total** | 0.809 [a] | 3.41[b] | 42.67 [b] |

a. The total value of Alpha of Cronbach is based on the total proper value
b. Due to the presence of multiple nominal variables. the proper value and the total percentage of explained variance does not correspond to the sum on the dimensions

of cattle with intensive poultry production.

The analysis of economic practice (Table 2) shows that 45.4% and 49.7% of farmers commercialize respectively all or part of milk produced in the local and regional market and only 4.9% of producers refuse the sale of milk. Therefore, a large part of farmers surveyed (64.4%) announce that calves were sold at later age (more than one year of age). Therefore, about 12.5% of farms visited declare that the sale of male calves was programmed early before weaning and 13.5% of cases a marketing of male calves were done shortly after their weaning. In addition, 10% of producers using calves as saving money to mobilize when their economic needs (selling as needed).

## Multivariate analysis

### Statistical model presentation

Categorical Principal Components Analysis (CATPCA) was defined two axes with 43% of total variance. First axis represented about 32% of total variation. It was interpreted as an axis of dairy orientation. It was highly correlated to variables related to dairy activities such as number of cows' and heifers. The second axis explains 19% of total variation and was positively correlated to beef number and negatively correlated to dairy parameters' (Table 3).

### Cattle farming systems identified

The approach adopted enhanced with two step classification has demonstrated five types of cattle system according to productive orientation of cattle herd (Figure 2 and table 4). The term "dairy farms" includes different levels of orientation, development and integration of milk production.

### Type 1. Dairy system

Cattle breeders in this system (about 15% of all farms visited) prefer to exploit the potential of animals in dairy production. The principal concern of these farmers was the commercialization of all milk produced on their farms in order to ensure highest possible income. Male calves born on the farms were for this category of breeders a co-product that gets rid rapidly before their weaning. Livestock is mostly specialized (cattle alone) or associated with intensified production of poultry. Animal material is formed by a lower size of herd (10±8.5 LU/farm) characterized by a large dominance of cattle (over 90%). The milk production was ensured by the presence of 7±6 dairy cows.

### Type 2. Dairy mixed system (Dairy-beef cattle oriented milk)

This type covers about 20% of cattle farms in the region. In this livestock system, farmers adopt strategies of mixed cattle production but producing and sales milk was their essential income. The fattening of theirs calves was an unplanned act used to cope with economic uncertainties (sale of calves according to the economic needs). The livestock exploited was in order of 17±12 LU. Cattle alone breeding mark this type of farms in 65 % of situation and cattle herd represents more than 80% of all exploited ruminants. This system promotes the highest number of dairy cows (10±8) and a low number of beef (1.2 ±1.3) per farms' due to sold of calves at an early age.

### Type 3. Balanced mixed system (Dairy and beef cattle)

It was recorded in only 4% of total farms surveyed. In this first model, the cattle complete two different functions, complementary and reasonably balanced, milk and beef production. This system represents a small sample which are generally a large farms distinguished by prac-

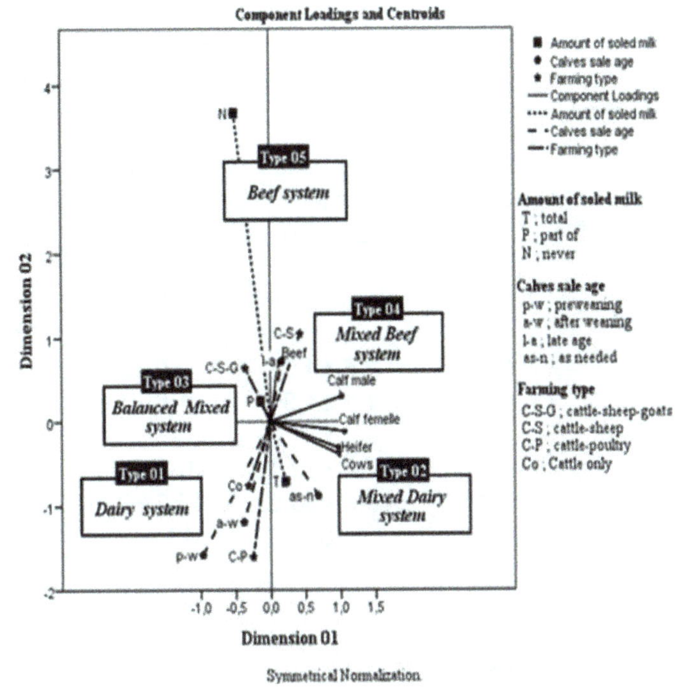


Figure 2: Graphical presentation of obtained model

tices of the partial marketing of milk and early sale of male calves. These farms do not give special attention to the type of production compared to the other due to large diversification of crops and livestock offered a multiple returns. In this regard, livestock herd was important (18.5 LU/farm in average) marks the breeding of fifty ewes and about ten goats near the cattle herd. Cattle represents about 50 % of all animals exploited characterized by the presence of 6.14±3.02 cows and 1.57±0.98 beef per farm.

### Type 4. Beef Mixed System (Dairy-beef cattle oriented beef)

This system dominates the study region and farms shown here make up the model frequently found in the context of the eastern high plains of Algerian (more than 56% of farms). These farmers adopt policies of mixed cattle system but more directed to the beef production. Suckling calves was a priority in farming practices as far as farmers reasoning was based on the earnings of beef compared to milk. A ruminant livestock contains 18±13 LU per farm. Over 60% of breeders who belonging to this group combine the cattle with sheep have subsequently more than thirty sheep, the rest of farmers exploits mostly cattle alone. Approximately 70 % of total livestock exploited per farm was formed by the cattle herd defined by the presence of 7.42±5.21 cows and 2.44 ±2.78 beef.

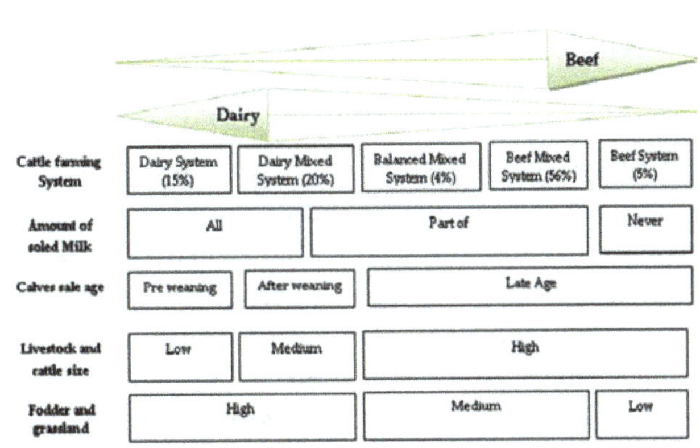


Figure 3: Characteristics of cattle farming systems'

Table 4: Characteristics of cattle farming system

| Variable | Modality | Cattle Farming System | | | | |
|---|---|---|---|---|---|---|
| | | Type 1 Dairy system (14.5%) | Type 2 Dairy Mixed System (20.0%) | Type 3 Balanced Mixed System (4.2%) | Type 4 Beef Mixed System (56.4%) | Type 5 Beef System (4.8%) |
| **Land** | **Arable Land** | 24.7 ±23.9 | 23.6 ±29.9 | 41.5 ±71.9 | 23.9 ±33.2 | 11.9 ±9.9 |
| **Fodder production** | **Fodder land** | 2.2 ±4.7 | 2.5 ±4.6 | 2.0 ±2.6 | 2.5 ±4.3 | 1.0 ±1.4 |
| | **Grass land** | 1.5 ±2.5 | 1.1 ±1.9 | 1.6 ±1.8 | 0.9 ±1.6 | 0.4 ±0.7 |
| **Livestock** | **Livestock** | 10.4 ±8.6 | 17.2 ±12.1 | 18.3 ±6.0 | 18.0 ±13.2 | 19.4 ±23.7 |
| | **Unit** | 7.9 ±22.6 | 18.9 ±36.5 | 45.3 ±22.8 | 35.4 ±47.4 | 12.7 ±15.6 |
| | **Ewes (head)** | 0.0 ±0.0 | 0.0±0.0 | 13.6±7.8 | 0.3 ±1.9 | 0.0 ±0.0 |
| | **Goats (head)** | | | | | |
| **Cattle** | **LU Cattle** | 9.2 ±7.5 | 14.3 ±10.2 | 10.2 ±3.7 | 12.7 ±8.9 | 17.5 ±23.2 |
| | **Cows (head)** | 6.9 ±5.6 | 9.9 ±7.9 | 6.1 ±3.0 | 7.4 ±5.2 | 6.4 ±6.4 |
| | **Beefs (head)** | 0.8 ±1.2 | 1.2 ±1.3 | 1.9 ±1.4 | 3.0 ±2.6 | 5.7 ±13.9 |
| | **Heifers (head)** | 1.9 ±1.6 | 3.2 ±2.9 | 1.9 ±1.5 | 1.7 ±3.0 | 0.0 ±0.0 |
| | **Calf Male** | 0.8 ±0.9 | 1.6 ±1.7 | 1.6 ±1.0 | 2.4 ±2.8 | 2.9 ±4.9 |
| | **Calf Femele** | 1.4 ±1.7 | 2.1 ±1.9 | 2.0 ±1.0 | 2.5 ±2.3 | 1.0 ±2.1 |

LU: Livestock unit

**Type 5. Beef system**

This category of farms encountered with a low frequency (less than 5% of farmers) is similar to the model of suckle cattle system in temperate regions. If reproduction and fattening calves are the center of interest of policy makers on these farms, the milk can not be sold. It was valued in suckling of future beef. In these situations, ¾ breeders exploit in parallel sheep if were not specialized farmers (cattle only). It has a biggest animal herd (more than 19 LU). The cattle herd separately form more than 90% of the overall livestock distinguished operated by breeding 6.8 ± 6.4 cows and 5.7 ±13.9 beefs.

## DISCUSSION

The study showed that more than 80% of farmers in the context of Algerian semi arid area adopt mixed cattle farming systems (Dairy-Beef). These producers operate in an unfavorable agricultural environment characterized by several economic and technical problems (instability of farm product price in the internal market and lack of technical backstopping). In this particular environment where milk production is low and unfavorable to these constraints, the profitability of cattle livestock specializes in the production and marketing of milk is unsecured. Only the profits generated by the fattening of calves born on the farm can encourage these farmers to continue their activities. On these farms, lack areas of grassland and feeble forage production for various reason preventing farmers to achieve a satisfactory of intra-farm level of forage autonomy (Figure 3). In several situations, forest grazing or cereals residues are sources of food for herds which housed under these livestock systems. Such obstacles are pushing farmers logically to avoid the attachment of the productivity of theirs farms to a single product (milk or beef). In this condition maximizing production is a secondary goal after the survival of the farm (Abbas, 2004). However, in temperate countries with high predisposition to the specialization activities Chatellier and Jacquerie (2004) reported that 25% and 20% of farms respectively in Belgium and Austria are mixed (Dairy-Beef) due to various reasons. In Tunisia, according to Jaoad (2004), a mixed dairy-beef systems can be observed in medium-sized farms. The reason for this is that beef production in combination with milk can be carried out with fewer animals than in beef production systems. Milk and beef production systems are closely connected and changes in milk production systems will cause alterations in beef production systems (Christel and Magnus, 2003).

The mixed system (Dairy-Beef) but more oriented towards beef production is the system model dominant cattle breeding in the study area (over 56%). Logical approach which these farmers have melted their policies are still economic profitability of beef production against milk and the efficiency of work organization in farms. Partial sale of milk guaranteed the coverage of life family expenses and daily cost of livestock activities, while the selling of beef and calves promotes the creation of funds using for new investment and modernization of farm. A difficulty of integration on milk collection networks and lack of milk conservation instruments in farms have also contributed in these policies. In France for example and according to a report established by Livestock Institute, meat produced comes in 35% from dairy cattle, and mixed and dairy farms supply 50% of young cattle for fattening.

The mixed system (Dairy-Beef) but more oriented towards the production of milk corresponds perfectly to farms ''cattle alone''. These farms sold all milk produced to the public or private dairies to benefit a subvention for milk production and other advantages. An important number of these farms were in mixed system (Dairy-Beef) oriented beef evolved gradually to Diary-Beef system oriented milk in search of stability and consistency of income provided by the sale of milk at a price substantially improved over the last years. Jaouad (2004) report that in Tunisia mixed systems can be found in small-scale irrigated farming which is predominantly oriented to dairy production.

The dairy system can be encountered only in farms with cattle-sheep breeding dominated by cattle herds or in cattle-poultry farms. These specialized farms are rarely only cattle farms. The early sale of young males born on the farm offers more facility in the sale of all milk produced by dairy cows. Farms structure factor is not to call into question but rather the search for stable sources of income that have guided this policy. In Morocco, the specialized diary system was observed in large farms (Srairi and Kessab, 1998) or in irrigated perimeter smallholders' (Srairi *et al.*, 2003) that 100% of arable land was used in fodder production. In this region only farms' directly committed to the way of specialization arrive at high economic performance.

In beef system, it was absolutely normal to accept that the sale of milk is never done on these farms for technical reasons relating to the valorization of milk producing in suckling of calves following the example suckling systems in temperate regions. However, cows' of local or cross breed in this environment are conducted in extensive on limited areas or without forage resources. It is reasonable also to assume that these practices are largely inflicted by traditional and socio cultural reasons. In Maghreb, the breeding of calves or beef fattening, are based on very limited areas (less than 5 ha) that much of the feed is purchased (Jemai and Saadani, 2000; Srairi *et al.*, 2003).

## CONCLUSION

Clearly, breeders in conditions of Algerian semi arid area prefer mixed systems. In cattle production strategies, it was the interaction of several factors that oriented breeders to favorite such system compared to the other. So it is logical to assimilate that the maximization of profits by reducing costs and optimizing production potential of herd, were the objectives of the breeder whatever manner with which it is organized. However, maximization of production per speculation was a secondary goal after the survival of the farm.

## REFERENCES

1- Abbas, K. (2004). La Jachère Pâturée dans les zones céréalières Semi-arides: Pour une approche de développement durable. Cahiers options méditeranniennes, 62, 169-173.

2- Chapman, D.F., Kenny, S.N., Beca, D., & Johnson, I.R. (2008). Pasture and forage crop systems for non-irrigated dairy farms in southern Australia. 1. Physical production and economic performance.

Agricultural Systems, 97(3), 108-125.
3- Chatellier, V., & Jacquerie, V. (2004). La diversité des exploitations laitières européennes et les effets différenciés de la réforme de la PAC de juin 2003. INRA Prod. Anim. 17 (4), 315-333.
4- Dedieu, B. (2009). Adaptation des systèmes d'élevage et incertitudes sur l'avenir. CRA-W & FUSAGx - Carrefour Productions animales 2009.
5- Dufumier, M. (2006). Diversité des exploitations agricoles et pluriactivité des agriculteurs dans le tiers monde. Cahiers d'Agriculture, 15 (6), 584-588.
6- FAO/ILRI. (1995). Livestock development strategy for low income countries. Proceedings of a Roundtable held at ILRI, Addes Ababa, Ethiopia, 27 Feb. to 02 March 1995.
7- Frensh Livestock Institute. (2005). La production de viande bovine en france. Le dossier economie de l'élevage.
8- Jaoad, M. (2004). Dynamique des cheptels bovins en Tunisie et contraintes alimentaires et fourragères. Cahiers Options Méditeranniennes, 62, 421-424.
9- Jemai, A., & Saadani, Y. (2000). Evolution des systèmes d'élevage dans les zones montagneuses du nord ouest de la Tunisie. Options méditerranéennes. 39, 39-56.
10- Madani, T., & Mouffok, C. (2008). Production laitière et perfomances de reproduction des vaches Mmontbéliardes en région semi aride algérienne. Revue Elev. Méd. Vét. Pays trop, 61 (2), 97-107.
11- Shalloo, L., Dillon, P., Rath, M., & Wallace, M. (2004). Description and validation of the moorepark dairy system model. Journal of Dairy Science. 87, 1945-1959.
12- SPSS, Inc. (2010). SPSS GLM 18.0. Chicago: M. J. Norusis.
13- Sraïri, M.T., Kiade, N., Lyoubi, R., Messad, S., & Faye, B. (2009). A comparison of dairy cattle systems in an irrigated perimeter and in a suburban region: Case study from morocco. Trop Anim Health Prod, 41 (2), 835–843.
14- Sraïri, M. T., & Kessab, B. (1998). Performances et modalités de production laitière dans six étables spécialisées au Maroc. Productions Animales-Paris-Institut National De La Recherche Agronomique-, 11, 321-326.
13- Sraïri, M.T., Leblond, J.M., & Bourbouze, A. (2003). Production de Lait et/ou de Viande: Diversité des Stratégies des eleveurs de bovins dans le périmètre irrigué du gharb au maroc. Revue d'élevage et de médecine vétérinaire des pays tropicaux, 56 (3-4), 177-186.

# 6

# Assessment of Rural Farming Households WTP for Fertilizers and Agrochemicals in Kwawara State, Nigeria

***Ademoye Emmanuel Akinboboye, Fadipe*, A.A., Adigun and Jubril Olayinka, Animashaun***

Abstract

Given the specific geographic and spatial location of rural areas in developing countries, to bring agrochemical to the rural farming households, it is argued, may have to come at a cost over and above the normal price it is sold in market. To this end, this work focuses on the willingness of rural farming households to pay more than the mean average regional retail price for agrochemicals in Kwara state, North Central Nigeria. Questionnaire was administered to 100 randomly sampled in the two agricultural development zones (ADP) in the study area. Descriptive statistics and the Logistic Binary Regression model were fitted to examine factors that influence respondents' decision to pay more than the prevailing average prices for access to agrochemicals in the study area. Findings indicated a high level of awareness of agrochemical use and modal responses to quantities used include: fertilizers (41-50Kg/ha), herbicides (<10liters/ha), and pesticides (<10liters/ha) on the cultivation of yam, cassava and maize, which were the dominant arable crops in the study area. About 90% of the respondents purchase agrochemicals from their personal savings and less than 10% of the respondents got their agrochemicals from other sources (governmental, developmental agencies, ADP and farmers' cooperative). Furthermore, 88% ($p<0.01$) of the respondents indicate willingness to pay more than the current average price to have access to agrochemicals and the logistic regression reveals that level of education ($P<0.01$), had a positive relationship with respondents' to willingness to pay more for agrochemicals in the study area. Recommendations were made towards encouraging expanded use of agrochemicals throughenhanced marketing strategies that will facilitate contact of marketing agents to prospective customers located in the rural communities.

***Keywords:***
***WTP, Agrochemicals, Fertilizers contingent valuation***

*Department of Agricultural Economics and Farm Management, University of Ilorin, PMB 1515, Ilorin, Nigeria .*
** Corresponding author's email: emmanfad@yahoo.com*

## INTRODUCTION

Agrochemicals refer to substance use to help manage an agricultural ecosystem or the community of organism in a farming area. They are important agricultural input useful for sustaining and increasing yields of agricultural products. Agrochemicals include fertilizer, pesticide, herbicide and rodenticide to eliminate the presence of living things that causes injury or disease to crops and also to increase soil nutrient (fertilizer addition) for improve crop production (Ayoola, 1990; Morris *et al.,* 2007).

However, low fertilizer use is a serious constraint to agricultural productivity growth in Nigeria. In Nigeria, consumption of fertilizer was only seven kilograms per hectare (kg/ha) of arable land in 2005, significantly lower than India's rate of about 121 kg/ha or Pakistan's 184 kg/ha (World Resources Institute, 2010). In addition, rural-based farmer demand for and utilization of fertilizer and agro-chemicals is still lower than what is experienced in advanced countries. This low use could be as a result of rural small scale farmers not being aware, unable to afforddue to higher cost attributed to longer distances or lack of access resulting from geographic and locale specific imposed constraints. Therefore, to effectively get agrochemicals to rural farmers it is argued, will have to come with an additional cost that will add up to its composite cost.

As agrochemical suppliers and marketers constantly seek for market outlets for the sales of their commodities, however, the geographic and spatial location of small-scale rural based farmers, who constitutes a significant portion of the agrochemical market, suggest that getting these products to them may come at a cost over and higher than the prevailing regional retail price. To estimate potential producers' demands, suppliers and marketers often rely on stated preference methods such as Contingent Valuation.

Initially, contingent valuation, a survey-based methodology, was developed to elicit the WTP for non-market goods and services like environmental services (Boyle, 2003; Carson and Hanemann, 2005; Zapata *et al.*, 2012), health economics (Diener *et al.*, 1998; Krupnick *et al.*, 2002), real estate appraising (Banfi *et al.*, 2008; Lipscomb, 2011), art valuation (Thompson *et al.,* 2002), and in agribusiness (Lusk and Hudson, 2004).

While majority of studies in the study area have holistically focused more on the adoption of fertilizers and agrochemicals, none of such studies critically examined farmers willingness to pay an extra cost that may be associated with getting the agrochemical inputs to them (Owusu *et al.*, 2007; Lagat *et al.*, 2007; Irene, 2012; Moser and Barret, 2003). In view of this, the purpose of this study was to identify the factors that determine rural farming households' willingness to pay an additional cost to have access to agrochemical in the Kwara State, Nigeria.

Specifically, it:

1- examines the rural farmers level of awareness about knowledge of the benefits of agrochemicals;

2- examines the quantity and cost of fertilizers and agrochemicals used by the rural farmers;

3- investigates the level of the farmers willingness to pay more for the access to the use of agrochemicals; and

4- identifies determinants of rural farmers willingness to pay to have access to agrochemicals.

Findings emanating from this study will add to the body of knowledge and assist governmental and non-governmental agencies with sound policy formulation particularly as its relates to agricultural development in Nigeria.

## MATERIALS AND METHODS

The study was conducted in Kwara State of Nigeria. It is located in the guinea savannah region of Nigeria and falls within latitudes $7^0$ 45! N and $9^030$!N and longitudes $2^030$!E and $6^0$ 25!E. Farming is the predominant occupation of residents in Kwara while some engage in craft activities such as weaving, blacksmithing, bricklaying, carpentry, welding etc. Fishing is also prominent along the lower River Niger Basin. The target respondents were rural small-scale farming households. A three stage sampling procedure was used in selecting our respondents. In the first stage, the list of four ADP zones in Kwara state was collected from ADP (Agricultural Developmental Program) of which two out of the four zones were randomly selected.At the second stage, 50 respondents were randomly sampled from each of the two zones. At the end, a total of 100 respondents were randomly selected and interviewed with a structured questionnaire.

Descriptive statistical analysis and the Logistic regression was fitted for to assess determinants of farmers' willingness to pay for access to agrochemicals, using the function:

$Y= \alpha+\beta X_1+ \beta X_2+ \beta X_3+ \beta X_4+ \beta X_5+\beta X_6+ \beta X_7+ \beta X_8+ \beta X_9+ \beta X_{10}+e$

Where

Y is a dichotomous response variable (1 for farmers that are willing to higher than the current determined market price for agrochemicals and 0 for otherwise). The study determined the hypothesized independent variables to include in the model after a review of relevant literatures on factors that influence WTP. The independent variables include:

$X_1$= age (years)

$X_2$= Yield of yam (kg/ha)

$X_3$= farming experience (years)

$X_6$= educational status (formal=1 and informal= 0)

$X_7$= income (Naira)

$X_8$= mode of action of pest

$X_9$= cost of agrochemicals (Naira)

$X_{10}$= yield of cassava (tonnes/ha)

$\alpha$= constant; e=Error term, it is assumed to be normally distributed with zero and constant variance and $\beta$ is the parameter to be estimated.

**Socio-economic characteristics of the respondent**

As shown in the Table 1 86% of our respondents are male compared to the 14% female. Accordingly, from the Table it reveals that 14% of the respondents are single, 79% are married, and 5% are widows while 2% are widowers. This could indicate that married people are more engaged in farming probably because they have support from their family members, in terms of labour unlike the rest of the respondents. The modal household size of our respondent was 6-9 (51%). Households with smaller sizes were more engaged in farming activities than those with larger farm sizes. The Table reveals that greater percentage (90%) of the respondents has one form of formal education and that about 10% that are with informal education. This makes it easier to conclude that the farmers in this area could improve on their ability to embrace new innovations if all their problems of availability and affordability of agrochemicals are solved.

Furthermore, the survey revealed that 25% of the respondents were engaged in farming alone while 75% of the respondents combine farming with other businesses (Table 1). Rural farmers always find it difficult to get capital required to buy inputs such as fertilizer and agrochemicals, and as such, they prefer to combine farming with other businesses so that they can source for funds. Capital earned by farmers in current production season would determine the extent to which the farmers would invest in the next session. The modal income class of the respondents ranges between 51,000-60,000 and the average income of the respondents is N 73,530 annually. This will enable the farmers to invest more in the next farming season (Table 1).

The survey showed that those that had experience less than 10 years were more in the study area which shows that they will be willing to accept or embrace new ideas ,since they will be looking for ways to improve their productivity (Table 1).

**Respondents Perception of the Benefits of Fertilizers and Agrochemicals**

The result of the respondents' awareness of the perceived benefits of agrochemicals was presented as their percentage response on the strongly agree, agree, disagree and strongly disagree.

The respondents' awareness of the benefits of agrochemicals would to an extent determine their willingness to pay for agrochemicals and their utilization of these agrochemicals. As shown in the survey (Table 2), about 69% of the respondents strongly agree that agrochemicals help to increase their yield while 27% agree, 2% disagree and 2% strongly disagree to this.

Also according to the Table, 62% of the respondents strongly agreed that control of pest and diseases is a benefit that can be gotten from the use of agrochemicals, while 35% agreed and 3% disagreed to this fact (Table 2).Majority of the rural farmers don't believe that use of agrochemicals provides for recreational areas, which is clearly shown from the response of the respondents with just having 1% of the respondents strongly agreeing and 33% agreed and another 33% disagree, with

Table 1: Socio economic distribution of respondents.

| Socio Economic variables | | Frequency | Percentage |
|---|---|---|---|
| Educational status | No formal | 10 | 10 |
| | Quranic | 9 | 9 |
| | Adult | 15 | 15 |
| | Primary | 30 | 30 |
| | Secondary | 30 | 30 |
| | Tertiary | 6 | 6 |
| Occupation | Farming alone | 25 | 25 |
| | Farming &others | 75 | 75 |
| INCOME(N) | ≤ 30,000 | 6 | 6 |
| | 31,000-40,000 | 16 | 16 |
| | 41,000-50,000 | 18 | 18 |
| | 51,000-60,000 | 22 | 22 |
| | 61,000-70,000 | 10 | 10 |
| | 71,000-80,000 | 12 | 12 |
| | 81,000 > | 16 | 16 |
| Experience (Years) | 1-10 | 33 | 33 |
| | 11-20 | 19 | 19 |
| | 21-30 | 16 | 16 |
| | 31-40 | 10 | 10 |
| | 41-50 | 8 | 8 |
| | 51 > | 14 | 14 |
| Sex | MALE | 86 | 86 |
| | FEMALE | 14 | 14 |
| Marital status | Single | 14 | 14 |
| | Married | 79 | 79 |
| | Widow | 5 | 5 |
| | Widower | 2 | 2 |
| Household size | ≤ 5 | 25 | 25 |
| | 6-9 | 51 | 51 |
| | 10-13 | 21 | 21 |
| | 14-17 | 2 | 2 |
| | ≥ 18 | 1 | 1 |
| Age (Years) | 21-30 | 11 | 11 |
| | 31-40 | 16 | 16 |
| | 41-50 | 18 | 18 |
| | 51-60 | 28 | 28 |
| | 61-70 | 13 | 13 |
| | 71-80 | 14 | 14 |

Source: Field Survey

24% strongly disagreeing (Table 2). The survey showed that 45% of the respondents were of the opinion that the use of agrochemicals eliminates storage pest, 31% agreed, 11% disagreed and 13% strongly disagreed probably due to the fact that some of these farmers are not accessible to the use of these agrochemicals which could be used for the elimination of pest that destroys their stored crops (Table 2).

**Estimation of Quantities and Cost of Fertilizers and Agrochemicals used by Respondents /ha**

As revealed in the Table, fertilizers remains the most widely used as shown by its spread among the respondents. Quantity of fertilizer (Table 3) purchased and used could be dependent on the farm size of the respondents as well as affordability and availability .The table shows that a larger percentage of the respondents use

Table 2: Respondents' awareness of perceived benefits of agrochemicals.

| Perceived benefits | Strongly agree (%) | Agree (%) | Disagree (%) | Strongly disagree (%) |
|---|---|---|---|---|
| Yield increase | 69 | 27 | 2 | 2 |
| Control of pest and diseases | 62 | 35 | 3 | 0 |
| Decrease in cost | 49 | 32 | 17 | 2 |
| Saves farmers money | 46 | 32 | 21 | 1 |
| Reduces drudgery | 28 | 35 | 25 | 12 |
| Helps farmers to grow more per unit area | 21 | 44 | 28 | 7 |
| Safe guard public health | 13 | 4 | 32 | 15 |
| Provides for recreational areas | 1 | 33 | 33 | 24 |
| Eliminates storage pest | 45 | 31 | 11 | 13 |

source: Field survey

Table 3: Estimation of total cost of production equation for greenhouse cucumber in Khahsh.

| Agro chemicals | Qty (kg, litre/ha) | | | Cost range (N/ha) | | |
|---|---|---|---|---|---|---|
| | | Freq | % | | Freq | % |
| **Fertilizer** | ≤10 | 33 | 33 | ≤1000 | 32 | 32 |
| | 20-Nov | 8 | 8 | 1100-2000 | 1 | 1 |
| | 21-30 | 7 | 7 | 2100-3000 | 5 | 5 |
| | 31-40 | 9 | 9 | 3100-4000 | 12 | 12 |
| | 41-50 | 22 | 22 | 4100-5000 | 11 | 11 |
| | 51-60 | 3 | 3 | 5100-6000 | 10 | 10 |
| | ≥61 | 18 | 18 | 6100-7000 | 3 | 3 |
| | | | | 7100-8000 | 1 | 1 |
| | | | | 8100-9000 | 3 | 3 |
| | | | | 9100-10,000 | 5 | 5 |
| | | | | 10,000 > | 17 | 17 |
| Total | 100 | 100 | | | 100 | 100 |
| **Herbicide** | ≤10 | 77 | 77 | ≤1000 | 57 | 57 |
| | 11-20 | 16 | 16 | 1100-2000 | 16 | 16 |
| | 21-30 | 6 | 6 | 2100-3000 | 12 | 12 |
| | 31-40 | 1 | 1 | 3100-4000 | 9 | 9 |
| | - | - | - | 4100-5000 | 6 | 6 |
| Total | | 100 | 100 | | 100 | 100 |
| **Insecticide** | ≤10 | 93 | 93 | ≤1000 | 93 | 93 |
| | 11-20 | 7 | 7 | 1100-2000 | 2 | 2 |
| | | | | 2100-3000 | 2 | 2 |
| | | | | 3100-4000 | 3 | 3 |
| Total | | 100 | 100 | | 100 | 100 |

Source: Field Survey

less than 10kg of fertilizer per hectare and a larger percentage of the respondents spend less than N1000 on the use of fertilizer per hectare which could largely be due to the fact that the fertilizer are either expensive or are not accessible to the respondents. The mean or average usage of fertilizer per hectare is 36.65kg and the average cost per hectare is N5148.07.The average use of herbicide and insecticide per hectare by the respondents is 6.28 litres and 2.58 litres respectively. While the average cost of usage is N1261.64 and N316.49 respectively (Table 3).

**Sources and accessibility of agrochemicals by respondents**

Table 4 shows the sources and accessibility of agrochemicals to the respondents'. From the table it shows that 58% of the respondents' re-

Table 4: Perception of accessibility of sources of agrochemicals by farmers.

| Accessibility | Very accessible (%) | Accessible (%) | Poorly accessible (%) | Not accessible (%) |
|---|---|---|---|---|
| Savings accessibility | 58 | 33 | 7 | 2 |
| Government accessibility | 1 | 6 | 1 | 92 |
| Developmental agencies accessibility | 0 | 0 | 12 | 88 |
| ADP accessibility | 1 | 3 | 1 | 95 |
| Farmers' cooperative | 4 | 11 | 1 | 84 |

Table 5: Statistical test to determine significance of respondents willingness to pay more.

| Willingness to pay | Frequency | Percentage | Standard error |
|---|---|---|---|
| Yes | 88 | 88 | 0.32 |
| No | 12 | 12 | 0.32 |
| Total | 100 | 100 | |

vealed that agrochemicals gotten from their personal savings were very accessible, 33% said it was accessible, 7% said it is poorly accessible and 2% said their personal savings were not accessible at all to get the agrochemicals. The table also reveals that accessibility of agrochemicals from other sources are poor, majority of the respondents were of the opinion that they are not accessible at all.

**Assessment of farmers willingness to pay for more for access to fertilizers and agrochemical**

Respondents were asked to state the price they bought their agrochemicals and fertilizers outside the rural areas. They were then asked to indicate if they would be willing to pay more to buy the inputs at their bases. About 88% indicated the affirmative and 12% of the respondents were negative. A test was carried out to examine the significance of there response. The result is presented in Table 5.

The Logistic regression was used to determine the respondents' willingness to pay for agrochemicals. The model parameters as shown by the Pseudo $R^2$ and the significance of the model chi $^2$ (<0.01) affirms the relative strength of the model as a good fit. As shown in the Table, possession of education is positively related to willingness to pay and it is significant at 1 % which indicates that the more educated

Table 6: Logistic regression analysis for the determinants of farmers willingness to pay.

| Predictor variable | Average Marginal Effect | Standard error | p>/z/ |
|---|---|---|---|
| Formal Education | 14.76 | 5.40 | 0.01*** |
| Income | 0.00 | 0.00 | 0.07 |
| Farming experience | -0.34 | 0.19 | 0.07* |
| Control of pest and disease | 14.65 | 8.55 | 0.09* |
| Saves farmers money | 2.36 | 1.29 | 0.07* |
| Drudgery | 2.42 | 1.62 | 0.14 |
| Cost of herbicide | 0.00 | 0.00 | 0.13 |
| Cassava yield | 0.00 | 0.00 | 0.06* |
| Yam yield | 0.00 | 0.00 | 0.10 |

Note: single asterisk refers to 10% significance and three asterisks refer to 1% significance.
Prob>chi=0.0023
Pseudo $R^2$=0.6326
Source: Field Survey

the farmers, the more they are willing to pay for agrochemicals. Income and knowledge of the benefit of control of pest and disease function of agrochemical are positively related to willingness to pay and significant at 10% which indicates that the higher the income and knowledge of benefits of agrochemicals, the more willing they will be to pay for it. Equally, respondents who record a higher yield from cassava are more likely to pay more for access to agrochemicals.

Conversely, farming experience shows a negative relationship with willingness to pay ($p<0.1$). This indicates that the higher the farming experiences of the farmer, the lower their willingness to pay.

## CONCLUSION AND RECOMMENDATION

The study was carried out to assess rural farming households' willingness to pay more than the central market price for agrochemicals. It was conducted in two ADP zones (Zone C and Zone D) in Kwara state. Findings show that majority of the respondents are aware of the benefits of fertilizers and agro chemicals and source it primarily from their savings.The mean or average usage of fertilizer per hectare is 36.65 kg and the average cost per hectare is N5148.07. The average use of herbicide and insecticide per hectare by the respondents is 6.28 litres and 2.58 litres respectively. While the average cost of usage is N1261.64 and N316.49 respectively.

Findings imply that a higher proportion of respondents (88%) are willing to pay higher prices if fertilizers and agrochemicals can be made available to them at their vicinity. This decision is however premised on certain socio-economic variables of which education, income from farming, knowledge of mode of action of agrochemical, farming experience and cassava yield were all significant in explaining this willingness decision.

Based on the findings of the study, the following recommendations were made:

1- The problem of bureaucracy should be eliminated by the government in distributing agrochemicals, so that the rural farmers can have equal access to it both at farm centres and Agricultural Development agencies.

2- Private business owners can take advantage of this willingness and situate marketing outlets at strategic locations across rural areas.

## REFERENCES

1- Ayoola, G.B. (1990). The marketing of agricultural pesticides in Nigeria workshop on the pesticides Industry in Nigeria, University of Ibadan, 6-8 February 1990.

2- Banfi, S., Farsi, M., Filippini, M., & Jakob, M. (2008). Willingness to pay for energy-saving measures in residential buildings, Energy Economics 30: 503-506.

3- Boyle, K.J. (2003). Contingent valuation in practice. In A primer on nonmarket valuation. Edited by Champ, P.A., Boyle, K.J., Brown, T.C. Kluwer Academic Publishers, Dordrecht.

4- Carson, R.T., & Hanemann, W.M. (2005). Contingent valuation. In Handbook on Environmental Economics, Vol 2. Edited by Mäler, K.G., Vincent, J.R. North-Holland, Amsterdam.

5- Diener, A., O'Brien, B., & Gafni, A. (1998). Health care contingent valuation studies: a review and classification of the literature. Health Economics, 7, 313–326.

6- Irene, S. E. (2012). Assessing the factor of adoption of agrochemicals by plantain farmers in Ghana using the ASTI Analytical framework. Kpmg: The growing importance of agrochemicals.

7- Krupnick, A., Alberini, A., Cropper, M., Simon, N., O'Brien, B., Goeree, R., & Heintzelman, M. (2002). Age, health and the willingness to pay for mortality risk reductions: a contingent valuation study of Ontario residents, Journal of Risk and Uncertainty, 24(2),161–186.

8- Lagat, J.K., Wangia, S.M., Njehia, B.K., & Ithinji, G.K. (2007). Environmental hazards African Agriculture: Factor influencing Application of agrochemicals in Nakuru district, Kenya. In Advances in Integrated Soil Fertility Management in Sub-SaharanAfrica: Challenges and Opportunities. Ed. A. Bationo. The Netherlands: Springer, 795- 804.

9- Lipscomb, C. (2011). Using contingent valuation to measure property value impacts. Journal of Property Investment and Finance, 29, 448-459.

10- Lusk, J.L., & Hudson, D. (2004). Willingness-to-Pay estimates and their relevance to agribusiness decision making. Review of Agricultural Economics, 26(2), 152–169.

11- Morris, M., Kelly, V.A., Kopicki, R.J., & Byerlee, D. (2007). Fertilizer use in Nigerian context.

12- Moser, C.M., & Barrett, C.B. (2003). The disappointing adoption dynamics of a yield- increasing low external input technology: The Case of SRI in Madagascar. Journal of Agricultural Systems, 76(3), 1085-1100.

13- Owusu-Bennoah, E., Anno-Nyako, F.O., Egyir, I.S., & Banful, B. (2007): Methodological framework: Analysing the agricultural science technology and innovation (ASTI) Systems in ACP Countries. The Ghana Case Study on Plantains. Wageningen, the Netherlands: Technical Centre for Agricultural and Rural Cooperation (CTA).

14- Thompson, E., Berger, M., Blomquist, G., & Allen, S. (2002). Valuing the arts: a contingent valuation approach. Journal of Cultural Economics, 26, 87–113.

15- World Resources Institute. (2010). Agriculture and food searchable database. <http:// earthtrends. wri.org/searchable_db/results.php?theme=8&years= & variable_ID=872&cID=&ccID=&years_rev=1>. Accessed May 10, 2010.

16- Zapata, S.D., Benavides, H.M., Carpio, C.E., & Willis, D.B. (2012). The economic value of basin protection to improve the quality and reliability of potable water supply: The Case of Loja, Ecuador. Water Policy, 14, 1-13.

7

# Analyze of Predictive Model of Innovation Management in Processing and Complementary Industries of Livestock Products

***Ahmad Reza Ommani***

**Abstract**

The purpose of this study was designing predictive model for innovation management in processing and complementary industries of livestock products. The method of research was correlative descriptive. The population of this research was managers in processing and complementary industries of livestock products of Khouzestan Province (N=486). By stratified random sampling, a random sample (n=125) was selected for participation in the study. A questionnaire was developed to gather data regarding Innovation Management in processing and complementary industries of livestock products. Data collected were analyzed using the Statistical Package for the Social Sciences (SPSS). According to the results, the innovation management level in processing and complementary industries of Khuzestan province is not desirable. Therefore, practitioners should be required to provide increasing levels of innovation management. Results showed that variables such as income, competitiveness, accountability, risk taking, tend to be creative, level of education and attitude to innovation management played a critical role in improving innovation management

***Keywords:***
***Innovation Management, Processing and Complementary Industries, Livestock Products***

*Associate professor, Department of Agricultural Management, Shoushtar Branch, Islamic Azad University, Shoushtar, Iran.*
** Corresponding author's email: a.omani@iau-shoushtar.ac.ir*

## INTRODUCTION

In the knowledge-driven economy, innovation has become central to achievement in the business world. With this growth in importance, organizations large and small have begun to re-evaluate their products and services to maintain their competitiveness in the global markets of today (Office for Official Publications of the European Commission, 2004).

Innovation is also a vital element in the success of small firms (e.g. those with less than 250 employees) (OECD, 2004). Such firms comprise the majority of businesses in most economies and have been recognized as a critical element in the national innovation system (OECD, 2004). Economic analysis of the relationship between research and development (R&D) activity, labor productivity and economic growth shows a significant correlation between these three elements (Crosby, 2000). Innovation allowing companies and economies to stay competitive in ever changing world markets. For all of the talk about the importance of innovation, innovation management and creativity in business, the topics are hardly generally well understood (Riederer *et al.*, 2005). Innovation in business has been studied by multiple researchers (Kleefl and Roome 2007; Kotelinkov, 2008; Kwamena, 2008).

This is recognized across the world and research suggests that a nation's innovation system is positively influenced by the level of investment in R&D, support for higher education, proportion of the workforce engaged in R&D, and the level of government support for innovation and commercialization (Porter and Stern, 1999). In order to be economically useful, innovation must translate into commercialization, which is the process of converting ideas and knowledge into applied outcomes typically as new products or business processes (Jolly, 1997). Also extension agents should also be encouraged to intensify innovation dissemination strategies so as to increase the level of adoption of available improved technologies of agricultural production (Nwaiwu *et al.,* 2012). Innovation can be classified using several different methods. Some apply in certain specific contexts, such as those frequently mentioned in our agricultural milieu (OECD, 2011).

Institutional innovation: These innovations, for our purposes, entail a change of policies, standards, regulations, processes, agreements, models, ways of organizing, institutional practices or relationships with other organizations, so as to create a more dynamic environment that encourages improvements in the performance of an institution or system to make it more interactive and competitive.

Technological innovation: This is the application of new ideas, scientific know how or technological practices to develop, produce and market new or improved goods or services, reorganize or improve production processes or substantially improve a service. Technological innovations are generally associated with changes in goods or productive processes; but technological innovations may also be applied to marketing processes or forms of organization by either producers or institutions.

Social innovation: This is the development or substantial improvement of strategies, concepts, ideas, organizations, goods or services, to bring positive changes in the way of meeting or responding to social needs or serving social purposes. Social innovations are constructed jointly by several different stakeholders for the well-being of individuals and communities; they may generate employment, consumption, participation or introduce some other change to improve the quality of life for individuals and that can be duplicated in other settings (OECD, 2011).

Investment in agricultural science and technology, generally in the form of research and extension services, has proved to be highly valuable for improving crop yields and lessening poverty in developing countries. Nevertheless, such investments should reflect all the parties' diverse needs for knowledge (Nwaiwu *et al.*, 2012). There is broad consensus that innovation is critically important for meeting the challenges that confront the human race, including the need to improve competitiveness, sustainability and equality in agriculture. Agriculture also needs to produce more food for a growing population, using a limited amount of farmland, while at the same

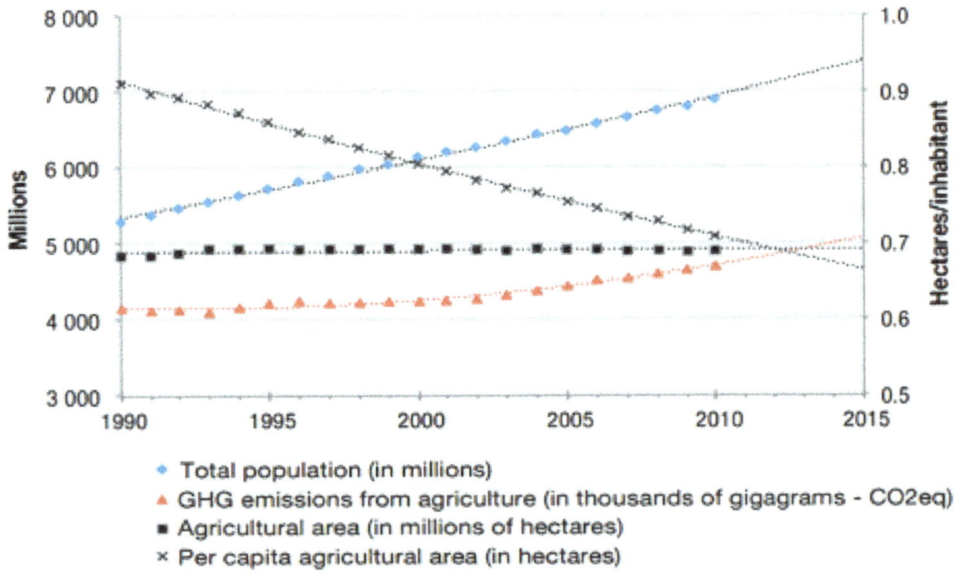


Figure 1: Evolution of selected variables (1990 - 2010) (IICA, 2014)

time reducing its greenhouse gas emissions to avoid worsening climate change (Figure 1). This suggests that agricultural production needs to use knowledge more intensively, which means it must innovate (IICA, 2014). Ryan and Oestreich (1991) observed incidences where a risk-averse attitude became the organizational norm due to an atmosphere of perceived fear, resulting in decreased innovation.

The purpose of this study was designing predicting model for innovation management in processing and complementary industries of livestock products.

## MATERIALS AND METHODS

The method of research was correlative descriptive. The population of this research was managers in processing and complementary industries of livestock products of Khuzestan province (N=486). By stratified random sampling, a random sample (n=125) was selected for participation in the study. The researchers used survey research methods in which data was gathered by the questionnaire. A questionnaire was developed to gather data regarding innovation management in processing and complementary industries of livestock products. Questions were generated from the literature review. The instrument consisted of two separate sections according to the purpose and objectives of the study. The first section was designed to gather data on personal characteristics of managers. The second section was designed to gather data regarding the innovation management with 26 items. For assessment level of innovation management in processing and complementary industries of livestock products were used seven subsystems of innovation management and items of each subsystem in Likert domain analyzed. Table 1 explained status of each subsystem. Managers were

Table 1: Status of all subsystems of innovation management in processing and complementary industries of livestock products

| Subsystems | Number of items | Mean | SD | Rank |
|---|---|---|---|---|
| **Designing and organizing of innovation** | 3 | 2.38 | 0.94 | 4 |
| **Infrastructure and financial supports** | 5 | 2.40 | 0.99 | 6 |
| **Research and development** | 4 | 2.42 | 0.89 | 3 |
| **Human development** | 3 | 2.67 | 1.02 | 7 |
| **Innovation Diffusion** | 5 | 2.48 | 0.88 | 2 |
| **Commercialization of researches and entrepreneurship development** | 3 | 2.05 | 0.98 | 5 |
| **Exploit of innovation** | 3 | 2.25 | 0.87 | 1 |

Table 2: Discriminate Analysis regarding predicting model of innovation management in processing and complementary industries.

| Variables | Structure Matrix | | Test | | Pooled within-group correlation matrix | | | | | | | | |
|---|---|---|---|---|---|---|---|---|---|---|---|---|---|
| | D1 | D2 | F | p-value | I | SJ | C | Ac | RT | TB | A | E | At |
| I | 0.392 | 3.255 | 1.46 | 0.178 | 1 | | | | | | | | |
| SJ | 0.364 | -0.151 | 10.47 | 0.000 | 0.46 | 1 | | | | | | | |
| C | 0.496 | 0.533 | 31.77 | 0.000 | 0.45 | 0.33 | 1 | | | | | | |
| Ac | 0.549 | 0.425 | 15.75 | 0.000 | 0.42 | 0.54 | 0.50 | 1 | | | | | |
| RT | -0.546 | -0.627 | 13.52 | 0.000 | 0.45 | 0.85 | 0.36 | 0.46 | 1 | | | | |
| TB | 0.644 | 1.838 | 13.26 | 0.000 | 0.34 | 0.56 | 0.45 | 0.64 | 0.73 | 1 | | | |
| A | 0.346 | -0.234 | 1.34 | 0.207 | -0.04 | -0.15 | 0.08 | 0.76 | -0.10 | -0.08 | 1 | | |
| E | 0.441 | 0.283 | 12 | 0.000 | 0.32 | 0.44 | 0.37 | 0.55 | 0.39 | 0.28 | -0.40 | 1 | |
| At | 0.165 | 0.173 | 88.97 | 0.000 | 0.13 | 0.18 | 0.35 | 0.05 | 0.24 | 0.14 | 0.09 | 0.35 | 1 |

I=Income, SJ=Second Job, C=Competitiveness, Ac=Accountability, RT=Risk Taking, TB=Tend to be Creative, A=Age, E=Education, At=Attitude

asked to rate their skills concerning items on a five point Likert-type scale: 1=very low, 2=low, 3=medium, 4=much and 5=very much. To determine the validity of the questionnaire, agricultural experts' comments were used. The questionnaire was pilot tested and reliability was estimated by calculating Cronbach's alpha. Reliability was calculated by Cronbach's alpha=0.89. Data collected were analyzed using the Statistical Package for the Social Sciences (SPSS).

## RESULTS

The ages of the respondents ranged from 24 to 67. The mean age was 42 (SD=7.87, n=125). The majority (38.4%, n=48) of respondent were 41-50 years old. Most of the respondents in the study were male (88%) and only 15 persons (12%) were female.

Three person of managers had a doctoral degree and 64% (n =80) of respondents were a bachelor's degree holders.

Table 1 present the descriptive statistics (Mean and standard deviation) for the subsystems of innovation management. As shown Table 1, "exploit of innovation" (M=2.25, SD=0.78) was believed to have the first priority to accomplish innovation management, followed by "innovation diffusion" (M= 2.48, SD= 0.88) and "research and development" (M= 2.42, SD= 0.89). Based on the results status of all subsystems of innovation management were moderate.

### Designing model of innovation management in processing and complementary industries of livestock products

Based on discriminant model (Table 2), different variables such as income, having second job, competitiveness, accountability, risk taking, tend to be creative, age, education and attitude were analyzed. For predicting adoption behavior of managers regarding innovation management the discriminate analysis was used. Based on results a discriminant function is:

$$D1 = 0.392X1 + 0.364X2 + 0.496X3 + 0.594X4 - 0.546X5 + 0.644X6 + 0.346X7 + 0.441X8 + 0.165X9$$

Wilks' lambda = 0.297 Chi square = 382.733 Sig = 0.000

Table 3: Grouping managers based on innovation management processing and complementary industries.

| Group | Number of Cases | Predicted Group Membership | | |
|---|---|---|---|---|
| | | G1 | G2 | G3 |
| G1 | 30 | 18<br>60% | 9<br>30% | 3<br>10% |
| G2 | 75 | 5<br>6.7% | 60<br>80% | 10<br>13.3% |
| G3 | 25 | 12<br>48% | 5<br>20% | 13<br>52% |

Not: 75% of the original cases was correctly classified

D2 = 3.253X1- 0.151X2+ 0.533X3- 0.425X4- 0.627X5+ 1.183X6- 0.234X7+ 0.283X8+ 0.130X9

Wilks' lambda = 0.334 Chi square = 43.665 Sig = 0.000

Wilks' lambda is used to test the significance of the discriminant function as a whole and the eigenvalue reflects the ratio of importance of the dimensions which classify cases of the dependent variable. The proportion of variance unexplained was 29.7% (Wilks' Lambda=0.297). The eigenvalue of 0.912 indicates that the discriminant function can explain 0.912 times as much as is not being explained. Also, the degree of association between the groups and the discriminant scores was expressed as a canonical correlation of 0.814. The Table 3 shows that the managers are the more accurately classified with 69.6% of the cases correct.

## CONCLUSION

According to the results, the innovation management level in processing and complementary industries of Khuzestan province is not desirable. Therefore, practitioners should be required to provide increasing levels of innovation management. Based on the results of the study variables such as income, competitiveness, accountability, risk taking, tend to be creative, level of education and attitude to innovation management played a critical role in improving innovation management. Ryan and Oestreich (1991) observed incidences where a risk-averse attitude became the organizational norm due to an atmosphere of perceived fear, resulting in decreased innovation. Also, innovation management correlation to a company's competitiveness and marketplace success are proven by several empirical studies (Mousavi, 2011).

Improve psychological characteristics such as risk taking, competitiveness, accountability for innovation through workshops, specialized training and scientific visits, will play an important role in the management of innovation.

## ACKNOWLEDGEMENT

This research was supported by the Shoushtar Branch Islamic Azad University of the Islamic Republic of Iran through a grant to Ahmad Reza Ommani. The author wants to express their gratitude for this support.

## REFERENCES

1- Crosby, M. (2000). Patents, innovation and growth. *Economic Record*, 76 (234), 255-262.

2- IICA (Inter-American Institute for Cooperation on Agriculture). (2014). Innovation in agriculture: a key process for sustainable development. Retrieved from http:// argus. iica. ac.cr/Esp/Programas/Innovacion/Documentos%20de%20Tecnologa%20e%20Innovacin/Innovation_PP_En.pdf

3- Jolly, V.K. (1997). *Commercializing new technologies: Getting from mind to market*, Harvard Business School Press, Cambridge MA.

4- Kleefl, J., & Roome, V. (2007). Developing capabilities and competences for sustainable business management as innovation: a research agenda, *Journal of Cleaner Production*, 15, 38-51.

5- Kotelinkov, V. (2008). Systematic innovation, the new holistic approach, Retrieved from http: // www.innovarsity.com/1000ventures/business_guide/innovation_systemic.html

6- Kwamena, A. (2008, May). Innovation and experience economy. Regional Studies Association Regions: The Dilemmas of Integration and Competition. Retrieved from http:// vbn. aau.dk/files/16527264/conference_paper.pdf

7- Mousavi, S. A. (2011). The factors hindering innovation at Iranian SMEs. *World Applied Sciences Journal,* 14 (11), 1635-1641.

8- Nwaiwu, I. U., Obasi, P. C., Orebiyi, J. S., Ibekwe, U. C., Korie, O. C., Henri-Ukoha, A., Amaechi, E. C. C., & Osuagwu, C. O. (2012). Comparative analysis of productivity of Cassava based crop mixture under modern and indigenous technologies in Anambra State, Nigeria. *International Journal of Agricultural Science, Research and Technology,* 2(1), 17-21.

9- OECD (2004). Promoting Entrepreneurship and Innovative SMEs in a Global Economy: Towards a more responsible and inclusive globalization; ICT, E-Business and SMEs. 2nd OECD Conference of Ministers Responsible for Small and Medium-Sized Enterprises (SMEs), Istanbul, Turkey 3-5 June 2004. Retrieved from http:// www.oecd.org/cfe/ smes/ 31919286.pdf

10- OECD. (2011). Fostering innovation to address social challenges. Workshop proceedings. Innovation Strategy. Retrieved from http://www.oecd.org /sti/inno/47861327.pdf

11- European Commission (2004). Innovation management and the knowledge-driven economy. European commission. Directorate-general for enterprise. Retrieved from http:// ftp:// ftp. cordis.europa.eu /pub/innovation-policy/ studies/ studies _innovation _management _final_report.pdf

12- Porter, M. E., & Stern, S. (1999). The new challenge to America's prosperity: findings from the innovation index. Washington D.C., Council on Competitiveness. Retrieved from http://www.econbiz.de /Record/the-new-challenge-to-america-s-prosperity-findings-from-the-innovation-index-porter-michael/ 10001460454

13- Riederer, J. P., Baier, M., & Graefe, G. (2005). Innovation management–An overview and some best practices. C-LAB Report, 4(3), 1-58.

14- Ryan, K. D., & Oestreich, D. K. (1991). *Driving fear out of the workplace.* San Francisco: Jossey-Bass.

# 8

# Analysis of Communicative Factors Affecting in Application of Sustainable Rice Farming Practices among Paddy Farmers of Mazandaran Province

*Laleh Salehi [1*], Ahmad Rezvanfar [2], Seyed Hamid Movahed Mohammadi [3] and Seyed Mahmood Hosseini [2]*

*Keywords:*
*Communicative factors, Communication, Paddy farmers, Sustainable Rice Farming Practices*

Abstract

The main purpose of this study was to analyze of communicative factors affecting in application of Sustainable Rice Farming Practices among paddy farmers of Mazandaran province. The statistical population of consisted of paddy farmers of Mazandaran province (N= 208839). According to Cochran formula, a sample of 240 farmers was selected through cluster sampling method. To collect data, a questionnaire was designed. Reliability of the questionnaire was measured by computing Cronbach's alpha ($\alpha \geq 0/7$) and its validity was approved by a panel of expert. Factor analysis revealed that communicative factors affecting in application of Sustainable Rice Farming Practices were four factors including: organizational communication, modern communication, local interpersonal communication and external interpersonal communication, respectively according to their importance. These factors explained 67.92 percent of the total variance.

[1] PhD of Agricultural Extension, Department of Agricultural Extension and Education, Faculty of Agricultural Economics and Development, College of Agriculture & Natural Resources, University of Tehran.

[2] Professor of Department of Agricultural Extension and Education, Faculty of Agricultural Economics and Development, College of Agriculture & Natural Resources, University of Tehran.

[3] Associate Professor of Department of Agricultural Extension and Education, Faculty of Agricultural Economics and Development, College of Agriculture & Natural Resources, University of Tehran.

* Corresponding author's email: lsalehi@ut.ac.ir

## INTRODUCTION

Despite long-standing introduction of sustainable agriculture in the globe, fast decline of critical agricultural resources due to erosion, increase the salinity of soils, desertification, extinction of species and environmental pollution are a matter of great concern in global reports (Shahvali and Moshfegh, 2005). This is due to different factors such as lack of farmers' awareness and knowledge about sustainable agriculture; their ignorance about consequences of natural resources destruction and most important, dominant perspective of applying conventional agricultural system which is often based on using external inputs and increasing commercial and export crops production are emphasized (Koucahki, 1997; Wauters *et al.,* 2010).

In this regards, paddy agriculture system, has been based on chemical external inputs since many years ago. This led to diseases and pests' resistance and higher production costs and environmental problems in rice production. Paddy farmers, have ignored, sustainable agricultural technologies and faces many challenges when try to adopt them.

Theoretical models have categorized adoption in three paradigm i.e. economic barriers of adoption paradigm (factors affecting adopter's behavior including natural resources, access to capital, learning costs, capital and attitude toward risk taking), innovation diffusion and adoption (factors affecting adopters' behavior such as access to information) and adopters' perception paradigm (factors affecting adopters' behavior including access to information, individual factors, human values, experience, training and level of understanding about sustainable agriculture) (Posthumus *et al.*, 2010). Innovation adoption and adaptors' perception paradigms that emphasis on access to information and affective factors on behavior have a wider application (Reimer and Prnokopy, 2014). In this regard, many researches (Ajzen, 2001; Ajzen and Fishbein, 2000; Briñol and Petty, 2009) showed that behavior is influenced by attitude and attitude is influenced by believes which is changed by information received from environmental communication channels and is modified by previous experiences (Kim *et al*., 2008) Then communication can change behavior through affecting believes and attitude. Therefore for changing behavior, content, source, contextual characters of message and communicational networks are very important (Petty and Cacioppo, 1996). Communicative networks' users, often faces a vast volume of contradicting information (Sparks *et al.,* 2013), and their perception, intention and behaviors about an issue is dependent of these information and their perception about attraction, information symbol and credit of source (Briñol and Petty, 2009; Hansen, 2005; Papathanassis and Knolle, 2011; Sigala, 2011; Sparks and Browning, 2011). Then, it can be said that individual behavior in adoption and application of sustainable Rice farming practices among paddy farmers affected by his/her communication and amount and type of information resources and communication channels used are very important. As Rogers (1995), believed that, communication, is the most important factor of agriculture modernization (Hejazi *et al.*, 2011). Then ignoring the communication program in any plan or project and supposing that communication will be naturally formed is a great mistake and an important opportunity for guaranteeing efficient communication will be missed (Ramirez and Quarry, 2004). So, the basic question is that what are communicative factors affecting Application of Sustainable Rice Farming Practices? Few studies have addressed this issue which will be reviewed in the next part.

Makokha *et al.* (1999) showed that participation of farmers in agricultural workshops and seminars and contacting with extension organization was effective to decision making on applying conservational operations in agriculture.

Mahboubi *et al.* (2005) showed that intending to consult with local leaders, awareness of soil conservation consequences, and number of training course participated and amount of information about conservation attained from radio and pamphlets, have a positive relationship with soil conservation operation.

Rezvanfar and Zare (2006) concluded that television, Colleagues and fishermen, friends and Relatives, and radio were respectively the most effective on adoption of GPS. In contrast, extension agents, newspaper, pamphlets and publications were respectively the least effective on adoption of GPS. They also categorized the communication channels whit respect to efficiency into five category i.e. cosmopolitan province channels, profit local channels, nonprofit local channels and television, through factor analysis. Bagheri *et al.* (2008) showed that education, contact with agricultural experts and presence in extension programs were the most important predicators of paddy farmers' aptitude toward applying sustainable agriculture technologies in Haraz region in Mazandaran province.

Shahvali *et al.* (2008) emphasized that in terms of communication and informing, radio and television were the most important information resources in the region and most of the farmers tried to meet their need for technical, political, economic and social information through these mass media. Furthermore wheat growers tried to upgrade their practices through interact with extension center and rural cooperatives and acquire information about weather and agricultural inputs distribution condition. Moreover, it is worth to take into consideration that computer and satellite receivers had special status among people of region so that these two technologies can be named as new gateway of informing in the region. Kutter *et al.* (2011), in a research entitled "the role of communication in adoption of precision agriculture is very important and communication factors such as phone, Email, internet, face to face meeting, webpages, fax and post, field day, fair, seminar, workshop, announces, demonstration farms and scientific journals had a great role in it.

Given to the importance of the issue and reviewed related literature, the main purpose of this study was to investigate the communicative effective factors in Application of Sustainable Rice Farming Practices among paddy farmers of Mazandaran Province, Iran. For meeting this aim following objectives were addressed:

1- Prioritization of information resources and communicative channels for Application of Sustainable Rice Farming Practices among paddy farmers of Mazandaran Province

2- Examining the level of using information resources and communicative among paddy farmers with respect to applying sustainable farming practices of rice.

3- Factor analysis of communicative factors affecting applying sustainable farming practices of rice.

## MATERIALS AND METHODS

Data for this survey study were collected using a structured questionnaire during 2013-2014, which addressed Paddy Farmers of Mazandaran Province. Face validity of the instrument was approved by a panel of experts. A pilot study was conducted with 30 farmers. The aim was to test and improve the questionnaire. Revisions were made based on the pilot study. Responses from the pilot test were not included in the final analysis. Reliability of the questionnaire was measured by computing Cronbach's alpha coefficient. Cronbach's alpha coefficient was calculated at 0.7 or higher for different part of the questionnaire, indicating an acceptable level of reliability.

The statistical population of the study consisted of 208,839 paddy farmers, working in Mazandaran province, Iran. The sample size was determined based on Cochran's formula, which consisted of 240 participants selected via a cluster random sampling approach.

The survey instrument contained a series of questions to which participants responded by indicating their level of application of communicative channels and recourses on a four-point scale ranging from never, to always (Not at all= 0, Rarely= 1, Often= 2, Always= 3).

Sustainable rice farming practices is sowing seeds in appropriate time and applying appropriate amount of main inputs i.e. Tillage practices, supplying nutrition elements through fertilizers, plant protection against pest, diseases and weeds and appropriate crop rotation (Hotfield and Karlen, 2007). Thus in this study sustainable rice farming practices was defined as producing rice process from seed to seed, considering appropriate time, amount and method of input consumption and agricultural practices.

For examining the level of paddy farmers ap-

Table1: Rankings of information sources use by paddy farmers.

| Information resources | Mean | SD | CV | Rank |
|---|---|---|---|---|
| Agricultural experts | 2.82 | 0.46 | 0.163 | 1 |
| Elite farmers | 2.66 | 0.54 | 0.203 | 2 |
| Neighbors | 2.53 | 0.65 | 0.256 | 3 |
| Spouse | 2.49 | 0.65 | 0.261 | 4 |
| Relatives | 2.40 | 0.72 | 0.300 | 5 |
| Children | 2.38 | 0.72 | 0.302 | 6 |
| Farmers from other villages | 2.11 | 0.71 | 0.336 | 7 |
| Orchardist (Kiwi and citrus | 2.32 | 0.85 | 0.366 | 8 |
| Agricultural inputs seller | 2.38 | 0.89 | 0.373 | 9 |
| Islamic council | 2.11 | 0.82 | 0.388 | 10 |
| Extension contacts | 1.88 | 0.73 | 0.388 | 11 |
| Village manger | 2.07 | 0.83 | 0.400 | 12 |
| Rural cooperative | 2.14 | 0.92 | 0.429 | 13 |
| Faculty members of agricultural colleges | 1.33 | 0.74 | 0.556 | 14 |
| Agricultural researchers | 1.36 | 0.76 | 0.558 | 15 |

(Scale: Not at all= 0, Rarely= 1, Often= 2, Always= 3)

plying different type of communication, sum of all items scores (the level applying each communicative channels and resources) was computed. For further investigation of this summative score, ISDM[1] was applied to categorize respondent into three class in terms of their use of communicative channels and resources. This classification conducted according to following formula (Gangadharappa *et al.*, 2007; Movahed Mohamadi *et al.*, 2012).

Weak: $A < (\overline{X} - ½ SD)$

Medium: $B (\overline{X} - ½ SD) - (\overline{X} + ½ SD)$

Strong: $C > (\overline{X} + ½ SD)$

## Findings

The average of respondents' age was 52.9 (SD=13.67). The average of agricultural experience was 34.69 (SD=15.95). 93.33 percent of the respondents were man and 6.66 percent of them were woman.

Coefficient of Variation (CV) was used to rank the communication channels and sources used by paddy farmers and results were summarized in Tables (1) and (2). Prioritization showed that agricultural experts, elite farmers, neighbors, spouse and relatives were the most used information sources for attaining information about sustainable farming practices of rice. Faculties of agricultural colleges and agricultural researchers were least used sources.

Furthermore, ranking of communication channels (Table 2) showed that face to face meetings, province TV network, question and answers sessions, training classes and announces were the most used channels for attaining information about application of sustainable farming of rice. Modern communication channels least used were internet and computer.

Classification of paddy farmers with respect to their total score of using different type of communication by Interval of Standard Deviation from Mean (ISDM) is reported in Table (3). Most of the respondents settled in medium category, that is, most of them use a medium level of communication.

Given to diversity of communication channels, factor analysis was applied to reduce the variables and categorize them into communicative affective factors of sustainable farming practices of rice.

To categorize communicative channels and information resources factors affecting application of rice sustainable farming practice and to determine the variance explained by each factor, an exploratory factor analysis approach was conducted. Using principal components approach, communicative channels and information resources were reduced into four factors. In order

[1] *Interval of Standard Deviation from Mean*

Table 2: rankings of communicative channels use by paddy farmers.

| Communicative channels | SD | CV | Rank | SD |
|---|---|---|---|---|
| Face to face meetings | 2.60 | 0.61 | 0.234 | 1 |
| Province TV network | 2.62 | 0.69 | 0.263 | 2 |
| Questions and answers sessions | 2.45 | 0.74 | 0.302 | 3 |
| Training classes | 1.53 | 0.50 | 0.326 | 4 |
| Announces | 1.53 | 0.50 | 0.326 | 4 |
| Cell phone | 1.25 | 0.43 | 0.344 | 5 |
| Phone | 2.20 | 0.77 | 0.350 | 6 |
| Group discussion | 2.26 | 0.81 | 0.358 | 7 |
| Playing instructional films by agricultural experts | 2.06 | 0.77 | 0.373 | 8 |
| National TV | 1.74 | 0.68 | 0.390 | 9 |
| Newspaper and magazines (The press) | 2.11 | 0.84 | 0.398 | 10 |
| Visiting research farms | 1.47 | 0.59 | 0.401 | 11 |
| Field day | 2.30 | 0.93 | 0.404 | 12 |
| Booklet, publication and brochure | 1.71 | 0.70 | 0.409 | 13 |
| Instructional agricultural CDs | 1.77 | 0.73 | 0.412 | 14 |
| Week of conveying findings | 1.60 | 0.66 | 0.412 | 15 |
| Video | 1.17 | 0.49 | 0.418 | 16 |
| Farmer field school | 1.74 | 0.73 | 0.419 | 17 |
| Workshops | 1.75 | 0.74 | 0.422 | 18 |
| Agricultural fairs | 1.44 | 0.62 | 0.430 | 19 |
| Field trip | 1.85 | 0.80 | 0.432 | 20 |
| Province radio | 1.90 | 0.85 | 0.447 | 21 |
| Poster | 1.85 | 0.83 | 0.448 | 22 |
| National radio | 1.60 | 0.96 | 0.600 | 23 |
| Research plan in continuous years | 1.27 | 0.78 | 0.614 | 24 |
| Research plan in 1 year | 1.42 | 0.90 | 0.633 | 25 |
| Book | 1.37 | 0.87 | 0.635 | 26 |
| Agricultural Message system (for cell phone) | 1.24 | 0.81 | 0.653 | 27 |
| Seminar and | 1.37 | 0.91 | 0.664 | 28 |
| Computer | 1.02 | 0.86 | 0.843 | 29 |
| Internet | 0.92 | 0.86 | 0.934 | 30 |

(Scale: Not at all= 0, Rarely= 1, Often= 2, Always= 3)

to test the appropriateness of data for factor analysis KMO (0.882) and Bartlett's test (4193.930, p<0.000) were applied. According to Kaiser Criteria there were four factors with Eigen values more than 1. These four factors explained 68.373 percent of variance .These results were summarized in Table (4).

Eigen values, variance percentage and the cu

Applying Varimax Rotation Method these factors were named: "Organizational communication", "Modern communication", "Local interpersonal communication" and "External inter-

Table 3: Frequency of paddy farmers in terms of their use of different types of communication for application of sustainable rice farming practices.

| Level of using communication | Frequency | Percent | Valid percent | Cumulative percent |
|---|---|---|---|---|
| Weak: < 82.875 | 57 | 23.75 | 25.22 | 25.22 |
| Medium: 89.705 – 82.75 | 95 | 39.58 | 42.04 | 67.26 |
| Strong: > 89.705 | 74 | 30.84 | 32.74 | 100.00 |
| Missing | 14 | 5.83 | - | |
| Sum | 240 | 100.00 | 100.00 | |

Mean=86.29 SD= 6.83 Min= 66 Max= 106

Table 4: The extracted factors along with the Eigen values, variance percentage and the cumulative variance percentage.

| Cumulative variance | Percent of explained variance | Eigenvalue | Factor |
|---|---|---|---|
| **22.948** | 22.948 | 5.278 | First |
| **44.689** | 21.741 | 5.000 | Second |
| **59.985** | 15.295 | 3.518 | Third |
| **68.373** | 8.388 | 1.929 | Fourth |

Bartlett's test =4193.930 ***p<0.000 KMO= 0.882

Table 5: Factors of communicative channels and information sources for application of sustainable rice farming practice explored and labeled.

| Factor name | Variables | Factor load |
|---|---|---|
| **Organizational communication** | Farmer Field School | 0.883 |
| | Workshop | 0.782 |
| | Visiting research farms | 0.842 |
| | Agricultural fairs | 0.804 |
| | Week of conveying findings | 0.789 |
| | Field trip | 0.782 |
| | Face to face meeting | O.770 |
| **Modern communication** | Playing instructional films by agricultural experts | 0.887 |
| | Instructional agricultural CDs | 0.825 |
| | Province TV network | 0.812 |
| | Announces | 0.797 |
| | Agricultural Message system (for cell phone) | 0.776 |
| | National TV | 0.773 |
| **Local interpersonal communication** | Local elite farmers | 0.778 |
| | Spouse | 0.748 |
| | Orchardist (Kiwi and citrus | 0.738 |
| | Children | 0.693 |
| | Relatives | 0.653 |
| | Neighbors | 0.590 |
| **External interpersonal communication** | Agricultural experts | 0.697 |
| | Agricultural inputs sellers | 0.652 |
| | Rural cooperatives | 0.510 |

Bartlett's test =4193.930 ***p<0.000 KMO= 0.882

personal communication" which were reported in Table (5).

## CONCLUSIONS

Ranking of amount of information resources use showed that paddy farmers mostly acquire information of sustainable farming of rice from agricultural experts, local elite farmers, neighbors, spouse and relatives. This can be due to convenient access to these information resources. Exploiting these resources is not usually expensive and time consuming. Furthermore, neighbors and relatives usually have common profits which make them reliable information resources of each other. Using agricultural experts as the first information resource is consistent with Bagheri *et al.* (2008); Makokha *et al.* (1999) and Shahvali *et al.* (2008). They pointed out the relationship of farmers with extension and agricultural experts. Importance of local elite farmers, neighbors, and relatives is consistent with Rezvanfar and Zare (2006). And the role of spouse is Consistent with Mahboubi *et al.*, (2005).

According to findings; face to face meeting, consistent with Kutter *et al.* (2011), province TV network, consistent with Rezvanfar and zare (2006) and Shahvali *et al.* (2008), questions and answers sessions, training classes and announces, consistent with Mahboubi *et al.* (2005), were respectively the most common communicative channels for acquiring information about sustainable rice farming practices. Furthermore, internet and computer were the least common modern channels. This is mostly due to age and education level of respondents which were mostly middle-aged and illiterate or under-literate; i.e. people who usually can not apply these channels.

Findings, also revealed that, more than half of the studied paddy farmers (70.42) have used different kind of communication at medium or high level, for applying rice sustainable farming practices. This shows that respondents have an appropriate access to communication for application sustainable forming practices of rice.

Factor analysis revealed that communicative resources and channels can be categorized into four factors i.e. organizational communication, modern communication, local interpersonal communication and external interpersonal communication. These factors explained 67.92 percent of total variance.

**Following suggestions are recommended:**

As finding showed, only 23 percent of studied the paddy farmers, used different kind of communication at low level and most of the respondent had appropriate use of communication related to sustainable farming practices of rice. Furthermore, communicative factors, explained 68 percent of total variance of applying rice sustainable farming practice. This is indicative of importance of communication for extension of sustainable farming practices of rice. Thus appropriate planning whit considering farmers need for enriching efficient communication with paddy farmers should be taken into account in order to increase their knowledge, information and awareness about rice sustainable farming.

Organizational factor explained the highest share of total variance. This factor consisted of items such as farmer field school, workshop, fair and week of conveying findings. These are experiential or visual methods of education. Then these kinds of methods are more efficient for training farmers about sustainable farming practices of rice.

Interpersonal communication consisted of two factors in factor analysis indicative of the importance of this type of communication. Moreover, should be attended and used modern communication channels such as educational films and CDs and other items along whit it. For example these items can be given to the audiences at the end of training courses. This is emphasis on transferring one message with several channels and will lead to more attention and application to message by audiences.

Given to role of local interpersonal communication and key role of spouse in paddy farming, paddy farmers should be trained with their spouses and families. Also, group training such as field trip, can enrich interpersonal communication which can be covered total person in paddy farmers' local communication network such as other farmers, orchardists, Islamic council members and others.

Providers of agricultural inputs are continuously contacted with farmers and have critical role in inputs consumption process. Thus they, as an important source of information about sustainable farming practices, should be trained. Furthermore, related organizations can use communicative print material such as poster, brochure, announce and leaflet which can be read easily and fast, in place of inputs sale to convey messages about sustainable farming practice such as method of inputs consumption and health and safety to farmers and sellers of inputs. Moreover, agricultural inputs sellers should be required to work under the supervision of agricultural experts at least in the season of inputs consumption.

## REFERENCES

1- Ajzen, I. (2001). Nature and operation of attitudes. *Annual Review of Psychology*, 52(1), 27-58.

2- Ajzen, I., & Fishbein, M. (2000). Attitudes and the attitude-behavior relation: Reasoned and automatic processes. *European Review of Social Psychology*, 11(1), 1-33.

3- Bagheri, A., Fami, H. S., Rezvanfar, A., Asadi, A., & Yazdani, S. (2008). Perceptions of paddy farmers towards sustainable agricultural technologies: case of Haraz Catchments area in Mazandaran Province of Iran. *American Journal of Applied Sciences*, 5(10), 1384.

4- Briñol, P., & Petty, R. E. (2009). Persuasion:

Insights from the self validation hypothesis. *Advances in Experimental Social Psychology*, 41, 69-118.
5- Gangadharappa, H., Pramod, K., & Shiva, K. H. (2007). Gastric floating drug delivery systems: a review. *Indian Journal of Pharmaceutical Education and Research*, 41(4), 295-305.
6- Hansen, T. (2005). Consumer adoption of online grocery buying: a discriminant analysis. *International Journal of Retail and Distribution Management*, 33(2), 101-121.
7- Hejazi, Y., Mirtorabi, M. S., & Hosseini, S. M. (2011). Communication factors influencing rural women's participation in post harvest activities (A case study of women in Asara, Karaj). *Journal of Woman and Society,* 1(4):19-32 (In Persian).
8- Hotfield, J. L., & Karlen, D. L. (2007). *Sustainable agriculture systems.* (Kouuchaki,E., Hossieni,M. & Khazaei,H. Trans.). Jahad Danshgahi Publication: Mashhad, Iran (In Persian).
9- Kim, D. J., Ferrin, D. L., & Rao, H. R. (2008). A trust-based consumer decision-making model in electronic commerce: The role of trust, perceived risk, and their antecedents. *Decision Support Systems,* 44(2), 544-564.
10- Kouchaki, A. (1997). Sustainable agriculture: insight or method? Journal of Agricultural Economics and Development, 5(20), 53-72 (In Persian).
11- Kutter, T., Tiemann, S., Siebert, R., & Fountas, S. (2011). The role of communication and cooperation in the adoption of precision farming. *Precision Agriculture,* 12(1), 2-17.
12- Mahboubi, M. R., Irvani, H., Rezvanfar, A., Kalantari, KH., & Mohseni Saravi, M. (2005). Factors affecting the adoption behavior regarding soil conservation technologies in the Zarrin- Gol watershed, Golestan province. *Iranian Journal of Natural Resources,* 57(4),595-606 ( In Persian).
13- Makokha, M., Odera, H., Maritim, H. K., Okalebo, J. R., & Iruria, D. M. (1999). Farmers' perceptions and adoption of soil management technologies in Western Kenya. *African Crop Science Journal,* 7(4), 549-558.
14- Movahed Mohammadi, H., Shiri, N., Arabi, R. and Karami, R. (2012). An analysis of the impact of social capital components on organizational entrepreneurship as regards agricultural staff (Darreh-Shahr 's Jahad Agriculture Organization). *Iranian Journal of Agricultural Economics and Development Research,* 43(1), 113-123 (In Persian).
15- Papathanassis, A., & Knolle, F. (2011). Exploring the adoption and processing of online holiday reviews: A grounded theory approach. *Tourism Management*, 32(2), 215-224.
16- Petty, R. E., & Cacioppo, J. T. (1996). *Attitudes and persuasion: Classic and contemporary approaches*. Westview Press: Colorado, United States.
17- Posthumus, H., Gardebroek, C., & Ruben, R. (2010). From participation to adoption: Comparing the effectiveness of soil conservation programs in the Peruvian Andes. *Land Economics*, 86(4), 645-667.
18- Ramirez, R., & Quarry, W. (2004). Communication for development: A medium for innovation in natural resource management. *International Development Research Center and Food and Agriculture Organization of the United Nations, printed in Canada*, 24p.
19- Reimer, A. P., & Prokopy, L. S. (2014). Farmer participation in US Farm Bill conservation programs. *Environmental Management*, 53(2), 318-332.
20- Rezvanfar, A., & Zare, A. (2006). Efficiency model of different communicative channels and information resources for adoption of GPS technology. *Iranian Journal of Agricultural Sciences*, 37(1), 49-60 (In Persian).
21- Rogers, E.M. (1995). *Diffusion of innovations*. Macmillan press, New York.
22- Shahvali, M., & Moshfegh, J. (2005). Evaluation of caricature function on Shiraz Jihad Agriculture Organization experts, attitude change concerning sustainable agriculture. *Journal of Science and Technology of Agricultural Natural Resources, Water and Soil Science, Isfahan University of Technology,* 9(1), 25-40 ( In Persian).
23- Shahvali. M., Keshavarz, M., Sharifzadeh, M., Forouzani, M., & Yazdanpanah, M. (2008). An identification of appropriate communication models for sustainable production of wheat in Marvdasht county of Fars. *Iranian Journal of Agricultural Economics and Development Research*, 39(1), 189-205 (In Persian).
24- Sigala, M. (2011). Special Issue on Web 2.0 in travel and tourism: Empowering and changing the role of travelers. *Computers in Human Behavior*, 27(2), 607-608.
25- Sparks, B.A., & Browning, V. (2011). The impact of online reviews on hotel booking intentions and perception of trust. *Tourism Management*, 32(6), 1310-1323.
26- Sparks, B. A., Perkins, H. E., & Buckley, R. (2013). Online travel reviews as persuasive communication: The effects of content type, source, and certification logos on consumer behavior. *Tourism Management,* 39, 1-9.
27- Wauters, E., Bielders, C., Poesen, J., Govers, G., & Mathijs, E. (2010). Adoption of soil conservation practices in Belgium: an examination of the theory of planned behavior in the agri-environmental domain. *Land Use Policy*, 27(1), 86-94.

# 9

# Extent of the Use of Drought Management Practices by Farmers in Tafresh County

***Maryam Bagheri* [1], *Mehrdad Nikanami* [2*] *and Hossein Shabanali Fami* [3]**

Abstract

The extent of utilization of drought management practices by the farmers of Tafresh County was studied in a descriptive-correlational research. Statistical population consisted of all farmers in Tafresh County which summed up to around 9061 people out of which 300 farmers were taken as the sample in accordance with Cochran's formula. The main tool of the study was a questionnaire. The validity of the questionnaire was confirmed by a panel of expert consisting of some faculty members of Islamic Azad University, Garmsar Branch and the University of Tehran, and the reliability of the questionnaire was estimated as to be 0.848 using Cronbach Alpha. The results of correlation test indicated a direct, significant relationship between the extent of utilization of drought management practices and the variables of agricultural activities experience and the level of farmers' social capabilities at 0.01 level. Moreover, there was a negative significant relationship among age, farming experience and the application of drought management practices at 0.01 level. Finally, the results of multiple regression analysis showed that three variables, i.e. extent of farmers' social capabilities, experience of agricultural activities and the age of the farmers, had the greatest influence on the extent of utilization of drought management practices.

***Keywords:***
***Drought, Drought management, Capability, Tafresh***

[1] *Graduate Student of Department of Agricultural Extension and Education, Garmsar Branch, Islamic Azad University, Garmsar, Iran.*

[2] *Department of Agricultural Extension and Education, Garmsar Branch, Islamic Azad University, Tehran, Iran.*

[3] *Department of Agricultural Development & Management, Faculty of Agricultural Economics & Development, College of Agriculture & Natural Resources, University of Tehran, Karaj, Iran*

* *Corresponding author's email: mehrdad.niknami@gmail.com*

## INTRODUCTION

Drought is a natural phenomenon that has happened in all eras. The vulnerability of the society to drought is increasing because of population growth, increased requirements and competition for limited water sources (Hejazizadeh and Joyzedeh, 2010). It is, also, projected that drought immigrant population will reach to one hundred million people in 2025, while the number of drought-related refugees was 23,5 million people in 2010 (David, 2011).

It is shown that drought has happened in almost all years in some parts of the country and even throughout the country (Fatehimarj, 2011) so that in total, out of the last 40 years, 15 years were wet, 5 years were normal and 20 years suffered from drought. In other words, at least 50% of the years were stricken by drought to varying degrees showing that this disaster has been a recurrent continental phenomenon in Iran (Seyfolahi and Shahabi, 2010).

This is while the recent droughts were so severe that their destructive results are still irreparable and has caused a lot of damage to rural societies. It has rung a warning alarm to farmer societies that experience maximum drought damages and become a vulnerable class (Sharifi and Zarafshani, 2010).

Since most people in these zones make living by farming, even a slight change in climate can immediately influence farming and cause more problems such as migration, dependence of country in supplying demanded foods and social, economic, cultural and even politic problems (Geravandi and Alibeigi, 2011; Panahi and Kheiri, 2009).

Accordingly, a new approach is proposed about disaster management especially drought and reducing vulnerability which is the study of farmers' characteristics and acclimation approaches in each region which needs the examination of farmers' capabilities from personal, social, … perspectives in applying appropriate management approaches to counteract the drought (Karpisheh, 2011).

Based on this, Rockstorm (2003) introduced management methods in small scales to decrease drought in homebred farming of semiarid zones and showed that water deficit of agricultural systems in semi-arid zones can be readily counteracted by managing systems that use water for supplementary irrigation helping the mitigation of drought losses.

In a study on approaches for the reduction of drought, Brown and Hansen (2008) concluded that crisis management should be replaced with pre-, during- and post-drought management integrated with farmers' local knowledge and modern knowledge in order for the destructive impacts of drought to be mitigated.

Rantakar and Govardahan's (2006) results about collaborative irrigation management in APWELL[1] project with the empowerment of human capacity theory showed that farmers' partnership in irrigation management during drought is very important and efficient.

According to what was said, we can summarize that drought management needs a special and comprehensive consideration to the effective variables in the management process although considering all the aspects is hard but not impossible. So the new approach for the management of natural disasters like drought and the reduction of their damages is oriented around local residents' capabilities for which farmers' capabilities including their personal and social abilities need to be assessed and appropriate management approaches for counteracting the drought need to be applied.

Accordingly, the present research was carried out to analyze personal and social capabilities of farmers in Tafresh County in order to facilitate the application of drought management practices and better management of drought.

## MATERIALS AND METHODS

The study was carried out as a survey with regression method. Population of this research contains all farmers of Tafresh county (N=9061) out of which 300 people were chosen by simple randomization using Cochran Formula (n=300). In this kind of sample choosing number of chosen samples are specified by each rural district (Bazerjan, Roodbar, Koohpanah and

[1] *The Andhra Pradesh Ground Water Irrigation Schemes*

Table 1: Farmers frequency distribution based on social institution membership

| Membership in Social institutions | Frequency | Valid percent |
|---|---|---|
| Islamic council | 21 | 7.0 |
| Village moot | 4 | 1.4 |
| Production cooperatives | 89 | 30.2 |
| None | 181 | 61.4 |
| Non-responses | 5 | |
| Total | 300 | 100 |

Kharazan) and urban district farmers.

A questionnaire consisting of three sections was used for data collection. The first section with three questions includes personal traits of Tafresh county farmers. The second section was related to farmers' social capabilities in using drought management operations (includes membership in social institution, interaction and social relation of farmers and cooperation morale and desire to work with others) and the third section included statements about using drought management practices and 10 statements that were filled by interview. The validity was confirmed by a panel of experts composed of faculty members of Gramsar Branch of Islamic Azad University and the University of Tehran and the reliability was estimated to be 0.848 based on the coefficient of Cronbach's alpha..

Data were statistically analyzed and all calculations were done by SPSS version 19 software package. In descriptive statistics part, percent, median, minimum, maximum and standard deviation and in inferential statistics part the correlation coefficient and multiple regression were used.

## RESULTS AND DISCUSSION

### Farmers' personal characteristics

According to the results, farmers' age range was between 24-80 years with the average of 51.2. Also, more than half of farmers (59%) were older than 45 and their average farming background was 25.6 years. Most of respondents (75%) were men and most of them (20.3%) were literate.

The fact that the majority of the farmers were elderly and they were in a low educational level implies the importance of adult's education to recognize drought management practices. These results are in agreement with Taghavinia (2009).

### Farmers' desire for membership in social institutions

According to Table 1, more than half of farmers (61.4%) did not have membership in any social institutions. It can be related to low awareness and weak relation of these social institutions with farmers and also the scant trust between farmers and institutions.

### Farmers' social Interaction and relation

Table 2 shows the extent of farmers' social interaction and relation based on 6-30-point domain categorized at five 5 levels with equal intervals. Results show that the level of most farmers' social interaction and relation was 31.7% (95 people) that was moderate.

Table 3 presents frequency distribution of the statements about farmer's social interaction and

Table 2: Farmers frequency distribution based on social interaction and relation rate

| Interaction and relation rate | Frequency | Valid percent | Cumulative percent |
|---|---|---|---|
| Very low(<10) | 24 | 8.0 | 8.0 |
| Low(10-15) | 83 | 27.7 | 35.7 |
| Middle(16-21) | 95 | 31.7 | 67.3 |
| High(22-27) | 59 | 19.6 | 87.0 |
| Very high(27>) | 39 | 13.0 | 100 |
| Total | 300 | 100.0 | |
| Median=20.20 | Standard deviation=6.14 | minimum=8 | maximum=3 |

Table 3: Frequency distribution and prioritizing farmers based on extent of social relation and interaction

| Interaction and relation | None | | Very low | | Low | | Middle | | High | | Very high | | Median | Priorities |
|---|---|---|---|---|---|---|---|---|---|---|---|---|---|---|
| | f | % | f | % | f | % | f | % | f | % | f | % | | |
| Farmers in neighborhood and local | 15 | 5.0 | 25 | 8.3 | 11 | 3.7 | 66 | 22.0 | 95 | 31.7 | 88 | 29.3 | 10.34 | 1 |
| Jihad –e- agriculture | 65 | 21.7 | 45 | 15.0 | 69 | 23.0 | 41 | 13.7 | 59 | 19.7 | 21 | 7.0 | 8.04 | 2 |
| Agricultural experts | 72 | 24.0 | 39 | 13.0 | 70 | 23.3 | 88 | 29.3 | 28 | 9.3 | 3 | 1.0 | 7.47 | 3 |
| Rural cooperatives | 117 | 39.0 | 39 | 13.0 | 49 | 16.3 | 46 | 15.3 | 35 | 11.7 | 14 | 4.7 | 7.16 | 4 |
| Islamic council | 173 | 57.7 | 39 | 13.0 | 35 | 11.7 | 28 | 9.3 | 15 | 5.0 | 10 | 3.3 | 6.91 | 5 |
| Village moot | 200 | 66.7 | 27 | 9.0 | 27 | 9.0 | 19 | 6.3 | 21 | 7.0 | 6 | 2.0 | 6.53 | 6 |

Scale: none=0, very low=1, low=2, middle=3, high=4, very high-5
(f = Frequency, % = Valid percent)

relation with mean of each statement. As is evident, a high percentage of respondents said that they had maximum interaction and relation with neighbor and local farmers.

## Farmer's desire for teamwork and cooperation with others

In this section, respondents expressed their opinions about cooperation and their desire for teamwork with others. According to the results presented in Table 4, 32% of farmers had high level of desire for cooperation and team-work, 30% had very high level of desire, 1.4% had no desire, 6.9% had very low desire, 2.8% had low desire and 26.9% had moderate desire for cooperation and team-work. The results mentioned are in agreement with TavakoliPoor and Ajili (2009) and Shokri *et al.* (2007).

Farmers' social capability in applying drought management practices was assessed in comparison with others given the scale range of the variables of farmers' social interaction and their desire for team-working. Therefore, the minimum point for the studied farmers' social capability was considered as 0 and the maximum as 7. Table (5) shows farmers' social capability in applying drought management practices in a range of 0-7 at three levels with equal intervals. It was revealed that most farmers' social capability (74.5% or 216 people) was at a low level. So it

Table 4: Farmers frequency distribution based on cooperation morale and desire for teamwork

| Cooperation morale and desire for teamwork | Frequency | Valid percent | Cumulative percent |
|---|---|---|---|
| None | 4 | 1.4 | 1.4 |
| Very low | 20 | 6.9 | 8.3 |
| Low | 8 | 2.8 | 11.0 |
| Middle | 78 | 23.3 | 37.9 |
| High | 93 | 32.0 | 67.9 |
| Very high | 87 | 30.0 | 100 |
| Non-responses | 10 | 3.6 | |
| Total | 300 | 100.0 | |

Table 5: Frequency distribution based on farmer's social capabilities in utilization of drought management practices

| Farmers social capabilities | Frequency | Valid percent | Cumulative percent |
|---|---|---|---|
| Low(<2) | 216 | 74.5 | 88.3 |
| Middle(2-6) | 40 | 13.8 | 13.8 |
| High(>6) | 34 | 11.7 | 100 |
| Non-responses | 10 | - | - |
| Total | 300 | 100 | |

Median: 5.51 standard deviation: 1.02 minimum: 1 maximum: 7

Table 6: Farmers frequency distribution based on using drought management levels

| Levels of using operations(score) | Frequency | Valid percent | Cumulative percent |
|---|---|---|---|
| Very low(<30) | 74 | 24.7 | 24.7 |
| Low(30-45) | 127 | 42.3 | 67 |
| Middle(46-61) | 66 | 22.0 | 89 |
| High(62-77) | 18 | 6 | 95 |
| Very high(>77) | 15 | 5 | 100 |
| Total | 300 | 100 | |

could be because of the low membership of farmers in social institutions located in county and low social relations that by itself could be the reason of weak social capabilities of farmers, weak partnership and cooperation between them and finally weakness of drought management by farmers of region. These results echo the findings obtained by Folkman (1984).

Utilization of drought management practices

Results of the range of the use of drought management presented in Table 6 show that most farmers (42.3%) are weak in the use of drought management practices seemingly an evidence of the shortage of technology and facilities.

On the other hand, computation of degree median of the rate of using drought management operation as presented in Table 7 shows that method of using modified and drought-resistant seeds that need less water, regular dredging of main and minor streams and the adjustment of irrigation time in accordance with crop water requirement had the maximum usages among farmers while farm insurance against drought, the use of training courses for the extension of drought management and under-plastic cultivation method to maintain soil moisture were prioritized in the last ranks. This situation shows that the use of drought-resistant seeds can be more available to farmers. Also according to the farmers' high tendency for group work, regular dredging of the main and subsidiary water rivulets and the joint use of one water channel will be more possible. Therefore, the items with lower priorities are costly. It is worth noting that regular dredging of main and minor streams is in agreement with Beik Mohammadi *et al.* (2005) and the use of modified and drought-resistant seed that need less water and covering them to avoid the loss of water is consistent with (Darijani *et al.*, 2011; Haddadi, 2002; Khabazzadeh, 2009; and Tavakolipoor and Ajili, 2009).

## Correlation analysis

According to Table (8), as the results of Spearman correlation coefficient show, there is a negative, significant relationship between the rate of utilization of drought management practices and the age at the 1% level and there is a direct,

Table 7: Farmer's priority based on rate of using drought management operation

| Operation | Mean | SD | Priority |
|---|---|---|---|
| Using modified and drought resistant seed that needs less water | 4.48 | 1.302 | 1 |
| Cleaning main-subsidiary sakes regularly between farms | 4.00 | 1.214 | 2 |
| Geminate usage of one transferor water channel | 3.45 | 1.457 | 3 |
| Controlling time of irrigation proportionate with water request of each harvest in different levels of improvement | 3.08 | 1.408 | 4 |
| Tubing and covering streams to avoid loss of water | 2.80 | 1.338 | 5 |
| Maintenance of chaff and stubble and herbal remains of the last farming year to keep moisture of the soil | 2.83 | 1.47 | 6 |
| Construct a water stockpile pool in the farm8-development of non-cultivated activities and shelter the crafts of county | 2.74 | 1.282 | 7 |
| Development of greenhouse cultivation | 2.68 | 1.791 | 8 |
| Changing the method of cultivation in the district proportionate | 2.61 | 1.243 | 9 |
| with drought strike | 2.51 | 1.430 | 10 |

Table 8: Correlation between rate of using drought management operation and random chosen variable

| Variables | r | p-value |
|---|---|---|
| Age | -0.222** | 0.001 |
| Agricultural experience | +0.217** | 0.000 |
| Level of farmers' capabilities of utilization of drought management practices | +0.248** | 0.000 |

**p<0.01

significant relationship between farmers' farming experience and their social capabilities so that the higher the farmers' social capabilities, the higher the level of the utilization of drought management practices. These results are in agreement with Shokri *et al.* (2007) and Taghavinia (2009).

## Regression analysis

Step-wise regression was used for measuring collective effects of independent variables on dependent variable whose results show that farmers' social capabilities only explains 36.5% of the variance of the rate of using drought management practices and it can be said that this variable is one of effective factors on the use of drought management practices by farmers. Then, social capabilities of farmers must be improved by partnership methods (Table 9), a point which was emphasized by Folkman (1998).

According to Table (9), linear equation from regression is as follows:

$$Y=27.78+0.121X_1+0.182X_2-0.082X_3$$

Where, Y= rate of using drought management operation, $X_1$= farmers' social capabilities, $X_2$= farming experience, $X_3$= farmers' age.

## CONCLUSION AND RECOMMENDATION

According to the results, given the dependency of agricultural activities on farmers' physical health and that majority of them are in the middle age, the managers of agricultural sector should find mechanisms for rural youths' engagement in farming and their takeover of the elder farmers' responsibilities. Given the high level of farmers' farming experience, appropriate decisions should be taken to get their beneficial experience and indigenous knowledge to successfully implement drought management practices in the area of study. These findings are in conformity with the results of research conducted by Taghavinia (2009).

Since majority of the farmers in the area of study had low level of education, appropriate different extension methods with emphasis on visual and demonstrative media should be used. Taghavinia (2009) found similar results. It was revealed that farmers with low social participation need to be empowered through mechanisms of social capital formation such as trust building. Since farmers were interested to cooperate with other farmers through team working, this capacity should be taken into account in participatory drought management. Hence, their social capital needs to be enhanced through appropriate measures. The findings of researches carried out by Folkman (1984), Shokri *et al.* (2007) and Tavakolipoor and Ajili (2009) confirm these results.

According to farmers' perception, solutions

Table 9: Model summery and coefficients

| Model | $R^2$ | B | β | Statistic (t) | p-value |
|---|---|---|---|---|---|
| Constant | - | 27.87 | - | - | - |
| Level of farmers social capabilities | 0.365 | 0.121 | +0.368 | +0.169** | 0.000 |
| Agricultural experience | 0.489 | 0.182 | +0.110 | -1.068* | 0.020 |
| Farmers age | 0.854 | -0.082 | -0.183 | +1.382** | 0.000 |

F=3.609 DF=13 R=0.432 $R^2$=0.854

**p<0.01, *p<0.05

such as the use of modified and drought-resistant seeds and the prevention of water wastes in canals are important for drought management. Darijani *et al.* (2011), Haddadi (2002), Khabazzadeh (2009) and Tavakolipoor and Ajili (2009) reported similar results. In addition, cleaning farm canals is also another drought management practice which was confirmed by Beik mohammadi *et al.* (2005). Given the results, it is necessary to provide suitable technologies, to establish infrastructures, to grant credits and to expand extension education for farmers in order to enable them to cope with the problems of drought management.

According to the regression analysis, social capability is a very important factor for drought management. Hence, this factors need to be enhanced. The finding is in conformity with Folkman (1998).

## REFERENCES

1- Beik mohammadi, H., Noori, H. A., & Bazrafshan, J. (2005). Effects of drought in 1998-2009 on the rural economy of Sistan and solutions to comfort it. *Journal of Geography and Improvement*, 3(5), 53-72 (In Persian).

2- Brown, C., & Hansen, J.W. (2008). Agricultural water management and climate risk. *International Research for Climate and Society,* April 2008, 1-14.

3- Darijani, A., Shah Hosein Dastjerdi, S., & Shahnooshi, N. (2011). Determination of priority of drought risk management in agricultural part of Gonbad e Kavoos county with using hierarchical techniques. *Journal of Agricultural Economic*, 2(3), 37-59(In Persian).

4- David, Rain. (2011). *Accra Ghana: A city vulnerable to and drought – induced migration.* Retrieved from: http://www.unhabitat.org/grhs/2011.

5- Fatehimaraj, A. (2011). *Monitoring drought indexes.* Paper Presented at the National Conference drought and change of climate in Iran, Esfehan: lack of water and drought in agriculture and natural sources research center, May 28, (In Persian).

6- Folkman, S. (1984). Personal control and coping processes: Theoretical analysis. *Journal of Personality and Social Psychology,* 46(4), 839-852.

7- Gardovani, SH., & Alibeigi, A.H. (2011). Determinant identification of agriculture risk management; case study farmers of Kermanshah County. *Journal of Economy and Development of Iran agriculture*, 2(42), 255-264(In Persian).

8- Haddadi, S. (2002). *Analysis of effective policy of agriculture/extension in reducing economic-social drought outcomes in Khodabandeh County from the perspective of managers and experts.* Unpublished thesis in Agricultural Extension and education, Faculty of agriculture, Islamic Azad University, Science and Research Branch (In Persian).

9- Hejazizadeh, Z., & Joyzedeh, S. (2010). *Introduction to drought and its indexes.* Publication: Tehran: study and codification of Liberal arts books Organization University (semat), Liberal arts Research and Development (In Persian).

10- Karpisheh, L. (2011). *Planning on drought management extension pattern based on combining local and new knowledge (case study Fars state)* Unpublished thesis in Agricultural Extension and education, Faculty of agriculture, Islamic Azad University, Science and Research Branch (In Persian).

11- Khabazzadeh, N. (2009). A*nalysis of drought management system of Khoozestan state.* Unpublished thesis in Agricultural Extension and education, Faculty of agriculture, Tarbiat Modares University (In Persian).

12- Rockström, J. (2003). Resilience building and water demand management for drought mitigation. *Physics and Chemistry of the Earth, Parts A/B/C,* 28(20), 869. Retrieved from www.sciencedirectcom

13- Panahi, F., & Kheiri, SH. (2009). Recovery of agricultural water resources management with total solution for confronting drought. Paper Presented at the Second National conferences about drought effects and its management solution, Agricultural Research Center and Isfahan Natural Sources, May 30-31 (In Persian).

14- Rantakar, R., & Govardahan Das, S. (2006). Participatory Anticipatory Irrigation Management in the APWELL Project: INDO-DUTCH TEAMWORK. *The Journal of Association for International Agricultural and Extension Education: 22nd Annual conference Proceedings Clearwater Beach*, Florida.

15- Seyfolahi, M., & Shahabi, S. (2008). Economic, social and environmental damages of drought, challenges and solutions to reduce that. Collection of journals of drought conferences, challenges and solutions September 2008, *Agricultural Research Center and Isfahan State Natural Sources, Total inspection and unexpected events office of Isfahan state* (In Persian).

16- Sharafi, L., & Zarafshani, K. (2010). Vulnerability measurement, solutions for drought management of wheat farmers case study: Sarpol zahab, west Islam Abad and Javanrood. *The Journal of Geographical Study of Dry Region,* 1(2), 106-120(In Persian).

17- Shokri, M., Hoseini, M., & Moghadas Farimani, SH. (2007). Analysis of environmental, economic and social –psychological of drought and relation of some farmers traits and outcomes above in agriculture part in Sistan Va Baloochestan, The Proceeding of First Conferences Compatibility With Lack of Water (In Persian).

18- Taghavinia, M. (2009). *Survey on damages of drought and farmers capabilities solution to confront with that in Dezful County*. Unpublished thesis in Agricultural Extension and education, Faculty of agriculture, Tehran university (In Persian).

19- Tavakolipoor, R., & Ajili, A.A. (2009). Arrangements of extension in drought. Second National Conferences of drought effects and its management methods. *Agricultural Research Center and Isfahan Natural Sources,* May 30-31 (In Persian).

10

# Factors Affecting Risk Management among Women Rice Farmers in Sari, Iran

***Mehdi Charmchian Langerodi***

Abstract

The present study investigated factors affecting risk management by women rice farmers in the township of Sari in Iran. The statistical population comprised 1677 women who cultivate rice in Sari; 248 women were selected for the study using stratified random sampling. The data was analyzed using descriptive and inferential statistics with SPSSWin16 software. The Chronbach's alpha was 0.84 and the ordinal theta was 0.89. It was found that the women rice farmers were more willing to consult with agricultural experts, were aware of appropriate planting and harvesting times and used collaborative rice farming, such as for planting, in their risk management. Risk management among women rice cultivators in Sari County was influenced by numerous factors. It was found that 29.3% of the variance in risk management by the women was determined by the extent of financial difficulty, their education levels and borrowing resources.

*Keywords:*
***Risk management, Paddy farmers, Borrowing resources, Sari***

[1] *Assistant professor of Agricultural Extension and Education Department, Sari Branch, Islamic Azad University, Sari, Iran.*
* *Corresponding author's email: Mcharmchian2004@yahoo.com*

## INTRODUCTION

Agriculture relies on nature and farmers face environmental, social, financial and legal issues that makes farming to be consistently and strongly associated with risk (Fraisse *et al.*, 2006). Agriculture can be considered to be the most risky economic venture. Some risks, such as the risk of natural occurrences, cannot be controlled by farmers, but other types of risk, such as the risk of weeds, can be controlled or eliminated (Ghorbani and Jafari, 2009). Of the 40 recognized types of natural disaster, 31 have been known to occur in Iran (Amini *et al.*, 2002). Given that the range of agricultural activities having critical risk are diverse and the climate can vary widely, more sophisticated services are needed to meet the demands, cope with hazards and decrease risk (Ertiaee and Chizari, 2006).

Risk is the assurance of future failure in relation to previous loss (Amini, 2011). Ahmadi (1999) listed several main sources of risk that farmers face. Economic risk includes price fluctuations for raw materials such as seed and fertilizer and for machinery at the pre-market delivery stage, difficulty obtaining bank credit and loans, price volatility for products in the post market delivery stage, inconsistent government policies for products and global fluctuations in the price of products. Social risk includes theft of crops, produce and agricultural machinery and war. Nature-related risk includes natural disasters, climate change, agricultural pests and disease. Market risk includes changes in the price of raw materials and products and increasing interest rates. The present study investigated factors affecting the risk management among women who cultivate rice in the township of Sari.

Women comprise more than half of the world population. Two-thirds of the global labor force is women, but only 10% of the global income and 1% of global assets are owned by women (Marofi and Hamidee, 2001). While bearing in mind the pivotal role that women play in care and maintenance of the family, they must be considered a key developmental element. The grim reality is that the position of women is undesirable on the international level (Sajadi, 2009). Women in rural Iran participate equally with men seeding, transplanting, and winnowing crops and maintaining livestock (Shahbazi, 2002). The current number of women in rice production in Sari and its services areas is 1677 (Jihad-e-Keshavarzi Management of Sari, 2012).

Understanding how women rice farmers deal with risk is essential for educators, agriculture-related industries such as insurance, and policymakers. If the attitudes of these women toward risk are accurately recognized, risk management strategies, risk-related educational programs and risk strategies can be designed to meet their needs. The goal of the present study was to model risk management by women rice farmers in Sari.

Tabatabaee *et al.,* (2013) studied the adoption of risk management among livestock breeders in Tehran. They found that 49% of breeders showed an average level of risk management. Correlation analysis indicated that there was a significant positive correlation between risk management and distance from ranch to home, amount of investment, degree of debt, annual income, and number of livestock and use of information resources. Linear multiple regression analysis revealed that the amount of investment, annual income and the awareness of the risk-related factors prompted 50.90% of breeders to change their approaches to risk management.

Alibigi *et al.* (2012) used a structural model to identify social capital factors affecting risk-taking by rapeseed farmers in Kangavar and concluded that the level of the components and farmer risk management was average. They found there was a significant positive relationship between social capital and the practice of risk management by farmers. Garavandi and Alibige (2012) identified factors affecting the behavioral reaction of corn farmers in Garmsar to risk and concluded that it was mainly devoted to risk aversion and was influenced by age, level of education, field ownership, and participation in extension classes. Age had the greatest effect on the behavioral response of farmers toward risk.

Alipour and Abdolaheefar (2011) maintain that at least 10 major types of risk threaten agricultural producers. Risk that can generally be controlled includes market risk, risk associated with production input, physical risk to the workforce, financial loss, risk associated with the information system and political risk. Risks such as natural disasters cannot be easily controlled and the human resources and capability needed to manage these risks is curtailed by the inadequacy and inefficiency of technological and scientific facilities.

Ghorbani and Jafari (2009) investigated factors influencing the degree of agricultural product risk for farmers in north Khorasan province and concluded that the age of the farmer, field ownership and working outside the farm had a significant negative effect on the degree of agricultural product risk. Participation in other agricultural activities, area under cultivation, extent of prior risk, and participation in a wheat supervision project had a significant positive effect on the adoption of risk management by farmers.

Zamani *et al.* (2009) considered agricultural insurance as an important mechanism for dealing with the inevitable risks to agriculture and the most appropriate strategy to decrease loss. Kohansal and Ghorbani (2008) argue that agriculture is inherently risky and insurance can be one way to decrease risk. Rostami (2007) asserts that factors such as high level of literacy, having a second non-agricultural job, increasing land ownership and full insurance coverage services increase the reliability and trust in farmers increases the farmer risk factor. Tiraee Yari (2002) showed that there was a significant positive relation between agricultural working record, area under cultivation, total agricultural land, communication with extension agents and experts and close contact with agricultural service centers and risk management.

Torkamani and Ezatabadi (2001) concluded that physical and financial assets, such as off-farm income, area under cultivation and education level had significant positive effects on farmer risk management. Ghorbani (2000) stud-

Table 1: Expert perspective on risk management methods

| Variables / Expert(s) | Consult experts | Participate in training courses | Use new technologies | Adopt extension recommendations | Use proper drainage | Change economic policies | Insure crops | Diversify investment | Promote maintenance and savings money for farmers | Membership in associations | Combatting weeds | Proper planting and harvest | Use new irrigation methods | Have adequate savings | Plant crops with lower risk | Use resistant, high yield and preterm varieties | Diversify production |
|---|---|---|---|---|---|---|---|---|---|---|---|---|---|---|---|---|---|
| Deshmukh and Khatri (2012 | | | | | | | * | | | | | | | | | | |
| Kahan (2008) | | | | | | | | | | | | | | | | | * |
| Hardaker (2006) | | | | | | | | | | | | | | | | * | |
| Anderson (2001) | | | | | | | * | | | | | | | | | | |
| Hueth and Forurtan (1994 | | | | | | * | * | * | | | | | | | | | |
| Shortreed *et al.* (2003) | | | | | | | | * | | | | | | | * | | |
| Khuu and Weber (2012) | | | | | | | * | | | | | | | | | | |
| Bader (2009) | | | | | | | | | | | | | * | | | | |
| Anderson and Mapp (1996 | | | | * | | | | | | | * | | | | | * | |
| Rosta (2009) | | | * | | | | | | | | | | | | | | |
| Tabatabaee *et al.* (2013) | * | * | | | | | * | | | * | | | | * | | | |
| Rostami (2007) | | | | | | | * | | | | | | | | | | |
| Yaghobi *et al.* (2007) | | | | * | | | * | | | | | | | | | | |
| Mirzad *et al.* (2013) | | * | | | | | * | | | | | | | | | | |
| Barimzadeh (2005) | | | | | * | | * | | | | | * | | | | * | |
| Bahrami and Agahi (2005 | | | | | | | | | * | | | | | | | | |
| Mazaheri (1998) | | | | | | | * | | | | * | | | | | | |

References: (Deshmukh and Khatri, 2012), (Khatri, 2012), (Hardaker, 2006), (Anderson, 2001), (Hueth and Forurtan, 1994), (Shortreed *et al.*, 2003), (Khuu and Weber, 2012), (Bader, 2009), (Anderson and Mapp, 1996), (Rosta, 2009), (Tabatabaee *et al.*, 2013), (Rostami, 2007), (Yaghobi *et al.*, 2007), (Mirzad *et al.*, 2013), (Barimzadeh, 2005), (Bahrami and Agahi, 2005) and (Mazaheri, 1998)

Table 2: Expert perspectives on factors affecting risk management

| Variables / Expert(s) | Take out loans | Connect with extension agents | Insurance | Total area under cultivation | Land under cultivation | Off-farm income | Amount of arable land | Communicate with service centers | Agricultural work experience | Social participation | Income | Off-farm income | Farm size | Participate in extension classes | Education level | Age | Use of information resources |
|---|---|---|---|---|---|---|---|---|---|---|---|---|---|---|---|---|---|
| Torkamani (1998) | | | | | | | | | | | | * | | * | | | |
| Monfared (1995) | | | | | | | | | | * | * | | * | | * | * | |
| Tiraee Yari (2002) | | * | | * | | | * | * | * | | | | | | | | |
| Torkamani and Ezatabadi (2001) | | | | | * | * | | | | | | | | | | | |
| Garavandi and Alibige (2012) | | | | | | | | | | | | | | | * | | |
| Rostami (2007) | | | | * | | * | | | | | | | | * | * | * | |
| Zamani *et al.* (2009) | | | * | | | | | | | | | | | | | | |
| Kohansal and Ghorbani (2008) | | | * | | | | | | | | | | | | | | |
| Tabatabaee *et al.* (2013) | * | | * | | | | | | | | * | | | | | | |
| Lu *et al.* (2008) | | | | | | | | | | | | | | | | | * |
| Stevens *et al.* (2003) | | | | | * | | | | | | * | | | | | | |
| Liu *et al.* (2010) | * | | | | | * | | | | | | | | | | | * |
| Velandia *et al.* (2009) | | | | | | * | | | | | | | | * | | * | |

References: (Torkamani, 1998), (Monfared, 1995), (Tiraee Yari, 2002), (Torkamani and Ezatabadi, 2001), (Garavandi and Alibige, 2012), (Rostami, 2007), (Zamani *et al.*, 2009), (Kohansal and Ghorbani, 2008), (Tabatabaee *et al.*, 2013), (Lu *et al.*, 2008), (Stevens *et al.*, 2003), (Liu *et al.*, 2010), (Velandia *et al.*, 2009)

ied the effect of insurance on wheat productivity of farmers in Mazandaran Province. He concluded that there was a positive effect for insurance on wheat production and productivity and that the level of insurance coverage by farmers participating in the wheat insurance project, optimal use of machinery and selection of high quality seed based on technical and agricultural advice were the major risk management practices.

Moghadasi (1997) studied risk-taking and concluded that most farmers practiced risk aversion. He recommended extension training about new technologies such as pesticides, fertilizers and improved seed to increase the level of production and as important factors in risk management.

Deshmukh and Khatri (2012) studied agricultural insurance in India and evaluated the development of agricultural insurance in India. They identified the types of risk that threaten crop yield in India and considered agricultural insurance to be a risk-reduction mechanism to decrease risk from natural disasters. Khuu and Weber (2012) concluded that western Australian farmers take risks for threats such as hail, fire and that paying insurance premiums increased when changes were made to insurance and risk management policies.

Liu *et al.* (2010) assessed the demand for weather index insurance among households in Anhou Province, China. They interviewed 660 households in 22 villages in the region about risks faced by households, mechanisms to deal with risk and the willingness to pay for weather index insurance. The results showed that the biggest weather-related risks were drought and floods. They found that farmers tend to pursue non-agricultural employment and borrow money from relatives and friends. Many farmers were willing to obtain weather index insurance. The farmers who were strongly affected by climatic conditions were more likely obtain insurance.

Lu *et al.*, (2008) reported that the farmers show different risk behaviors depending on their financial potential, type of farming, amount of arable land and income. The Organization for Economic Co-operation and Development (2009) researched agricultural risk management and found that providing a competitive business environment with guidelines and rules encouraged individual responsibility among farmers for risk management and facilitated the flow of information about risks for effective and efficient risk management.

Velandia *et al.* (2009) examined factors affecting use of an agricultural risk management mechanism and stated that off-farm income, education and age had meaningful relationships with adoption by farmers of risk management mechanisms. Olarinde *et al.* (2007) concluded that natural, social, economic and technical risks existed among the corn farmers of Nigeria.

Steven *et al.* (2003) argued that the best approach to risk management was to obtain information, especially the private sector. Meuwissen (2000) showed that farmers felt that price and production risks are the most important risks. They introduced insurance as a risk management approach and believed that geography, type of farm and organizational infrastructure increased farmer risk management.

Shrapnel and Davie (2000) showed that personal and social features are tools to identifying how people perceive risk. The present study identifies risk factors among women rice farmers in Sari. Tables 1 and 2 show risk management methods and factors influencing risk management and risk from the expert prospective, respectively.

## MATERIALS AND METHODS

The data were analyzed using descriptive and inferential statistics with SPSSWin16 software. The statistical population of the study was all women rice farmers in Sari. As reported by Jihad-e-Keshavarzi in Sari, there are currently 1677 women rice farmers in Sari and its surrounding areas (Jihad-e-Keshavarzi Management in Sari, 2012). Using the Cochran formula, the final sample size was estimated to be 248 subjects. Stratified random sampling and a proportional allocation formula was used to distinguish agricultural services.

To determine the validity of the questionnaire, copies were distributed to experts in the field and their comments were recorded and relevant corrections were made. Corrected content validity was determined by experts at the Agricultural Organization of Mazandaran Province, and then reliability of the questionnaire was assessed by random completion of 30 questionnaires by women rice farmers in Qaemshahr by interview and survey. Cronbach's alpha was 0.84 and ordinal theta was 0.89. SPSS 16 was used for data analysis.

The questionnaire contained questions about age, work experience, area of land owned, extent of financial difficulty related to rice farming, and educational level of the women rice farmers. The questionnaire contained questions about risk management approaches (17 questions), sources of information for women rice farmers in Sari (12 questions) and sources for borrowing funds (7 questions). The Likert scale was: none (0), very low (1), low (2), moderate (3), high (4), and very high (5).

## RESULTS AND CONCLUSIONS

Table 3 showed the frequency distribution for age and work experience of women rice farmers. The results indicated that the average age of women was 46.68 years and their average work experience was 26.17 years, which represents good work experience on paddy land. Table 3 reveals that 25.2% of women had high school diplomas and 23% had completed intermediate school.

The results showed that the average area of land owned by women was 2845.56 $m^2$. Because of inaccessibility of data, the mean was not a good indicator for making conclusions, thus, other central tendency measures were used. The mode and median for the area of rice land owned by women were 1,000 m2, which suggests that the area of land owned by women was low. Table 3 indicated that the average financial difficulty for the women was high.

Women rice farmers moderately adopted risk management methods. They were more willing to consult with agricultural experts, have awareness of appropriate planting and harvesting times and collaborate with other farmers for planting (Table 4). The most important sources of information were associations of women rice farmers, other woman rice farmers and participation in agricultural extension classes (Table 5).

The woman rice farmers showed an average level

Table 3: Frequency distribution for risk factors of women rice farmers

| Variable | | f | Valid percent (%) | Cum. (%) | Mean | SD | Min. | Max. |
|---|---|---|---|---|---|---|---|---|
| **Age (yr)** | ≤25 | 8 | 3.7 | 3.7 | | | | |
| | 26-40 | 55 | 25.1 | 28.8 | | | | |
| | 41-55 | 111 | 50.7 | 79.5 | 46.68 | 10.98 | 20 | 69 |
| | ≥56 | 45 | 20.5 | 100 | | | | |
| | No response | 29 | - | | | | | |
| | Total | 248 | 100 | | | | | |
| **Work experience (yr)** | ≤10 | 16 | 7.8 | 7.8 | | | | |
| | 11-20 | 61 | 29.8 | 37.6 | | | | |
| | 21-30 | 73 | 35.6 | 73.2 | | | | |
| | 31-40 | 47 | 22.9 | 96.1 | 26.17 | 9.60 | 2 | 45 |
| | ≥41 | 8 | 3.9 | 100 | | | | |
| | No response | 43 | - | | | | | |
| | Total | 248 | 100 | | | | | |
| **Education** | Illiterate | 32 | 14.4 | | | | | |
| | Can read and write | 42 | 18.9 | | | | | |
| | Primary school | 51 | 23 | | | | | |
| | Intermediate school | 38 | 17.1 | | | | | |
| | High school diploma | 56 | 25.2 | | - | - | - | - |
| | AA or higher | 3 | 1.4 | | | | | |
| | No response | 26 | - | | | | | |
| | Total | 248 | 100 | | | | | |
| **Area of land (m²)** | ≤500 | 47 | 26.1 | | | | | |
| | 501-1500 | 76 | 42.2 | 26.1 | | | | |
| | 1501-2500 | 26 | 14.4 | 68.3 | | | | |
| | 2501-3500 | 5 | 2.9 | 82.7 | | | | |
| | ≥3051 | 26 | 14.4 | 85.6 | 2845.56 | 5767.84 | 200 | 30000 |
| | No response | 68 | - | 100 | | | | |
| | Total | 248 | 100 | | | | | |
| **Financial difficulty** | None | 3 | 1.4 | | | | | |
| | Very low | 20 | 9 | 1.4 | | | | |
| | Low | 26 | 11.7 | 10.4 | | | | |
| | Moderate | 43 | 19.4 | 22.1 | | | | |
| | High | 56 | 25.2 | 41.5 | 3.58 | 1.36 | - | - |
| | Very high | 74 | 33.3 | 66.7 | | | | |
| | No response | 26 | - | 100 | | | | |
| | Total | 248 | 100 | | | | | |

Likert -type scale: none (0), very low (1), low (2), moderate (3), high (4), very high (5)

for obtaining loans. They were most likely to obtain loans from friends and neighbors and also to obtain financing from winning the lottery (Table 6).

Table 7 showed there was a statistically significant relationship between information resources, borrowing resources, degree of financial difficulty in rice cultivating, rice cultivation experience and education levels with risk management methods.

The use of risk management by women rice farmers in Sari was influenced by numerous factors; 29.3% of the variance in risk management was determined by degree of financial difficulty, education level and borrowing resources (Table 8). The regression line equation is:

$$Y = 0.843 + 0.395 \text{ (financial problems)} + 0.209 \text{ (education)} + 0.187 \text{ (borrowing resources)}$$

## DISCUSSION AND SUGGESTIONS

There was a significant positive relationship between information resources, borrowing resources, extent of financial difficulty, rice cultivation experience, and educational level with risk management methods. Tabatabaee *et al.* (2013) stated that the extent of borrowing affected risk management. Three main areas of government intervention for risk management and coping that have been studied broadly in developing countries are: insurance, saving/credit and safety nets. Credit provision may allow for better risk coping (Cervantes-Godoy *et al.*, 2013). Tiraee Yari (2002) pointed to agricultural experience affecting risk management. Garavandi and Alibige (2012), Monfared (1995) and Torkamani and Ezatabadi (2001) believe that educational

Table 4: Ranking risk management methods by women rice farmers in Sari

| Risk management method | M [1] | SD [2] | C.V [3] | R [4] |
|---|---|---|---|---|
| Consult with agricultural experts | 3.65 | 1.60 | 43.77 | 1 |
| Aware of proper planting and harvest times | 3.69 | 1.62 | 44.07 | 2 |
| Collaborate with other farmers | 3.55 | 1.58 | 44.48 | 3 |
| Product insurance | 3.52 | 1.65 | 46.98 | 4 |
| Collect information for decisions about crops | 3.38 | 1.62 | 47.9 | 5 |
| Use new technologies | 3.36 | 1.68 | 49.89 | 6 |
| Have an off-farm job | 3.38 | 1.70 | 50.4 | 7 |
| Having enough savings | 3.33 | 1.78 | 53.4 | 8 |
| Use of herbicides to combat weeds | 3.30 | 1.77 | 53.53 | 9 |
| Diversify crops besides rice | 3.41 | 1.88 | 55.08 | 10 |
| Implement advice of agricultural experts | 3.23 | 1.76 | 54.57 | 11 |
| Participate in educational programs | 3.14 | 1.81 | 57.72 | 12 |
| Use pesticides for diseases and pests | 2.96 | 1.83 | 61.73 | 13 |
| Use of resistant, high yield and preterm varieties | 2.93 | 1.88 | 64.24 | 14 |
| Practice crop rotation | 2.76 | 1.84 | 66.49 | 15 |
| Membership in Agricultural Association | 2.65 | 1.81 | 68.68 | 16 |
| Use government loans | 2.68 | 1.85 | 69.17 | 17 |

Likert -type scale: none (0); very low (1); low (2); moderate (3); high (4); very high (5)

Table 5: Ranking sources of information for women rice farmers in Sari

| Source of information | M [5] | SD [6] | C.V [7] | R [8] |
|---|---|---|---|---|
| Associations of women rice farmers | 3.33 | 1.80 | 54.05 | 1 |
| Other women rice farmers | 3.13 | 1.73 | 55.51 | 2 |
| Participate in agricultural extension classes | 3.04 | 1.75 | 57.49 | 3 |
| Read educational publications | 2.89 | 1.80 | 62.25 | 4 |
| Communicate with agricultural experts | 2.79 | 1.74 | 62.49 | 5 |
| TV | 2.92 | 1.85 | 63.21 | 6 |
| Private companies provide fertilizers and educational services | 2.67 | 1.76 | 65.92 | 7 |
| Farmers | 2.92 | 1.95 | 66.64 | 8 |
| Market | 2.53 | 1.94 | 76.62 | 9 |
| Radio | 2.56 | 1.99 | 77.68 | 10 |
| Jihad-e-Keshavarzi management and service center | 2.06 | 1.99 | 96.43 | 11 |
| Internet | 2.09 | 2.02 | 96.72 | 12 |

Likert -type scale: none (0); very low (1); low (2); moderate (3); high (4); very high (5)

level affects risk management, which is consistent with the results of this study. The farm management techniques that sustainably increase productivity and reduce production risk must be delivered to farmers, especially smallholders, by both the private and the public sector (Diouf *et al.*, 2012).

The following recommendations merit attention:

- Women should be encouraged to pursue off-farm income sources and occupations as risk management methods. Collaborative rice farming and consulting agricultural experts are other forms of risk management.
- There is a relationship between information resources and level of risk management. Recommended sources of information are women rice farmer associations, other women farmers, and participation in agricultural educational programs.
- Consulting with experts requires that these experts have technical and communicative skills and be accessible to the women farmers.
- There is a relationship between information resources and the application of risk management methods. Increasing the knowledge and awareness of women farmers about the sources of risk increases risk management. This should be regarded as a mission for agriculture extensions.

[1] *Mean,* [2] *Standard deviation,* [3] *Coefficient of variation,* [4] *Rank,* [5] *Mean,* [6] *Standard deviation,* [7] *Coefficient of variation,* [8] *Rank*

Table 6: Ranking of borrowing sources for women rice farmers in Sari

| Borrowing resource | M [1] | SD [2] | C.V [3] | R [4] |
|---|---|---|---|---|
| Friends and neighbors | 3.23 | 1.79 | 55.42 | 1 |
| Women rice farmers | 3.22 | 1.81 | 56.1 | 2 |
| Winning the local lottery | 2.99 | 1.85 | 61.98 | 3 |
| Relatives | 2.90 | 1.86 | 64.30 | 4 |
| Shopkeepers | 2.82 | 1.93 | 68.38 | 5 |
| Banks | 2.69 | 1.92 | 71.33 | 6 |
| Loan fund | 2.68 | 1.99 | 74.13 | 7 |

1: mean; 2: standard deviation; 3: coefficient of variation; 4: rank
Likert -type scale: none (0); very low (1); low (2); moderate (3); high (4); very high (5)

Table 7: Correlation between research variables and use of risk management methods

| Variable | $r_s$ | p-value |
|---|---|---|
| Information resources | 0.238** | 0.000 |
| Borrowing resources | 0.139* | 0.028 |
| Extent of financial difficulty | 0.447** | 0.000 |
| Rice cultivation experience | 0.231** | 0.000 |
| Education level | 0.165* | 0.014 |

**p < 0.01; *p < 0.05

Table 8: Risk management multiple regression for women rice farmers in Sari

| Model | Unstandardized coefficients | | Standardized coefficients | t | p-value |
|---|---|---|---|---|---|
| | B | Std Error | Beta | | |
| 1 (Constant) | 0.843 | 0.321 | - | 2.626 | 0.009 |
| Extent of financial difficulty | 0.395 | 0.056 | 0.448 | 7.004 | 0.000 |
| Educational level | 0.209 | 0.054 | 0.246 | 3.838 | 0.000 |
| Borrowing resources | 0.187 | 0.060 | 0.200 | 3.121 | 0.002 |

$R = 0.552$; $R^2 = 0.305$; Adjusted $R^2 = 0.293$; $F = 24.872$; sig = 0.000

• Borrowing sources, the extent of farmer financial difficulty, work experience and educational level should be considered when advocating risk management methods to women rice farmers.

## ACKNOELEDGMENTS

The paper was supported by the Sari Branch, Islamic Azad University.

## REFERENCES

1- Ahmadi, M. (1999). *Investigating factors affecting farmers' membership in rural production cooperatives based on an innovation extension model.* Unpublished thesis in Agricultural Extension and Education, Agriculture department, Tehran: Tarbiat Modarres University.

2- Alibigi, A. M., Aliabadi, V., & Garavandi, S. H. (2012). The structural model of social capital component affecting villagers' risk-taking, a case study on rapeseed farmers in Kangavar. *Rural Development and Space Economics Journal*, 1, 59-64.

3- Alipour, H., & Abdolaheefar, P. (2011). Risks in agriculture, Scientific. *Economic, Agriculture and Process Industry Journal,* 20(237), 34-38.

4-Amini, R. (2011). Insurance and Risk Management role in Agricultural Economics. *Agricultural Journal,* 76, 21-25.

[1] *Mean,* [2] *Standard deviation,* [3] *Coefficient of variation,* [4] *Rank*

5- Amini, A., Jamshidi, M., & Sadeghi, J. M. (2002). Risk related Factors and the herdsman' willingness to insure their cattle in East Azerbaijan Province. *Agricultural Economics and Development,* 10(39), 125-140.

6- Anderson, J. R. (2001). *Risk management in rural development: A review.* The World Bank. Washigton D.C.

7- Anderson, K. B., & Mapp, H. P. (1996). Risk management programs in Extension. *Journal of Resource Economics,* 21(1), 31-38.

8- Bader, T. M. (2009). Managing risk on the Gresham 80. Retrieved from https://www.ffa.org/documents/rme_essay_thomasbader.pdf

9- Bahrami, A., & Agahi, H. (2005). Risk management in agricultural fields. *Agricultural and Insurance Journal,* 2(8), 97-113.

10- Barimzadeh, V. (2005). *Risk management in agriculture.* Agricultural Education Publication, Tehran: Ministry of Agriculture.

11- Cervantes-Godoy, D., Kimura, SH., & Anton, J. (2013). Smallholder risk management in developing countries. *OECD Food, Agriculture and Fisheries Papers*, No. 61, OECD Publishing. Retrieved from http://dx.doi.org/10.1787/5k452k28wljl-en

12- Deshmukh, A. K., & Khatri, D. (2012). Agriculture insurance in India-A paradigm shift in Indian agriculture. *International Journal of Research in Economics and Social Sciences*, 2(2), 138-139.

13- Diouf, J., Nwanze, K. F., & Sheeran, J. (2012). *The state of food insecurity in the world: How does international price volatility affect domestic economies and food security*? Rome: FAO.

14- Ertiaee, F., & Chizari, M. (2006). Investigating extension experts' knowledge on agricultural product insurance. *Agricultural Insurance,* 3(10), 54-79.

15- Fraisse, C. W., Breuer, N. E., Zierden, D., Bellowc, J. G., Pazd, J., Cabrera, V. E., Garcia, A., Ingram, K. T., Hatch, U., Hoogenboomd, G., Jones, J. W. & Obrien, J. J. (2006). A climate forecast information system for agricultural risk management in the southeastern USA. *Computers and Elctronics in Agriculture,* 53, 13-27.

16- Garavandi, SH., & Alibige, A. M. (2012). Identifying factors influencing the corn farmers' behavior in Garmsar regarding the production risk. *Economics Research and Agricultural Development Journal,* 2-43(2), 293-300.

17- Ghorbani, M. (2000). The insurance impact on wheat productivity in Mazandaran province using analysis model. *Agricultural Economics and Development Journal*, 20, 73-91.

18- Ghorbani, M., & Jafari, F. (2009). Investigating influential factors on agricultural crops risk frequencies in North Khorasan Province. *Agricultural Economics and Development Journal,* 23(1), 41-48.

19- Hardaker, J. B. (2006). Farm risk management: past, present and prospect. *Journal of Farm Management,* 12(10), 593-612.

20- Hueth, D., & Forurtan, W. (1994). *The economics of agricultural insurance:* Theory and evidence. USA: Kluwer academic publishers.

21- Jihad-e-Keshavarzi Management in Sari. (2012). *Rice cultivating women in Sari classified by their relative agricultural service center.* Tehran: Ministry of Agriculture.

22- Kahan, D. (2008). Managing Risk in farming/ farm management extension guide. Rural Infrastructure and Agro- Industries Division Food and Agriculture organization of the united Nations Vialedelle Terme dicaracalla. Rome, Italy, (153), 38-75 .

23- Khuu, A., & Weber, E. J. (2012). How Ausralian farmers deal with risk? Australia: University of Western Australia.

24- Kohansal, M. R., & Ghorbani, V. (2008). Evaluating influential factors on cotton insurance adoption: A case study in North Khorasan Province. *Agriculture and Insurance Journal,* 5(15 and 16), 23-46.

25- Liu, B., L. I, M., Guo, Y., & Shan, K. (2010). Analysis of the demand for weather index agricultural insurance on household level in Anhui, China. *Agriculture and Agricultural Science Procedia*, 1, 179-186.

26- Lu, W., Xi, A., & Ye, J. (2008). *Disaster Risk reduction strategies and risk management practices: Critical Elements for adoption to climate change.* Report, 11 November 2008, Nairobi: UNFCCC.

27- Marofi, P. & Hamidee, A. (2001). *A Step towards Empowerment: A Guide to the use of micro-credit to start small income generating activities*, Tehran: Ronas Publication.

28- Mazaheri, D. (1998). *Mixed agriculture.* (2 ed.), Tehran:Tehran University publication.

29- Meuwissen, M. (2000). I*nsurance as a risk management tools for European Agriculture.* Unpublished thesis in Wangeningen University.

30- Mirzad, N., Abedi Sarvestani, A. & Abdolahzadeh, GH. H. (2013). Agricultural product insurance a new Approach to Risk Management, Eighth Symposium on Advances in Science and Technology, Mashhad: Khavaran Higher education institute. Retrieved from http://8thsastech.khi.ac.ir.

31- Moghadsi, R. (1997). Tendency toward risk-taking. *Markazi Insurance*, 1(1), 5-9.

32- Monfared, N. (1995). *Factors influencing the rice cultivating technologies adoption and their impact on rice producer women in Mazandaran province and Shiraz*. Unpublished thesis in Agricultural Extension and Education, Agriculture department, Shiraz University.

33- Movaahed, M. (2003). Women political participation and effective social factors on it (Case Study of Bandar Abbas). *Women Study Magazine*, 1 (3), 3-28.

34- Olarinde, L. O., Manyong, V. M., & Akintola, J. O. (2007). Attitudes towards risk among maize farmers in the dry savanna zone of Nigeria: some prospective policies for improving food production. *African Journal of Agricultural Research,* 2(8), 399-408.

35- Rosta, K. (2009). Evaluating the factors affecting farmers' ability to manage risk (a case study on wheat growers in Khorasan-e-Razavi). *Agricultural Economics and Extension Journal*, 1(1), 1-16.

36- Rostami, F. (2007). Wheat production risk management in farming systems (a case study in Harsin). *Iranian Agricultural Sciences Journal,* 37(2), 93-106.

37- Sajadi, S. (2009). Assessing the rural production cooperatives changes in post-regulation period in Ghaenat, Unpublished thesis, Birjand: Islamic Azad University of Birjand.

38- Shabazi, A. (2002). *Educated villagers*. Tehran: Cultural and Art Institute of Shaghaiegh Rusta Publishers, Inc.

39- Shortreed, J., Hicks, J., & Craig, L. (2003). Basic frame work for risk management. The Ontario Ministry of the Environment/Network for Environmental Risk Assessment and Management (NERM), 7, 74.

40- Shrapnel, M., & Davie, J. (2000). *The influence of personality in determining farmer responsiveness to risk.* Paper Presented to the International Workshop on Farm Management Decision With Climate Risk. Toowoomba: Department of Primary Industries.

41- Steven, W. M., Oscar, V., George, F. P., Keith, H. C., Thomas, O. K., & Alan, E. B. (2003). Extension educators′ perceptions of risk management training needs. *Journal of Extension*, 41(4), 1-9.

42- Tabatabaee, F., Pezeshgirad, G. H., & Chizari, M. (2013). Risk management adoption in industrial husbandry in Tehran province: Influential structures. *Iranian Journal of Agricultural Economics and Development*, 43(1), 135-142.

43- The Organization for Economic Co-operation and Development. (2009). Managing risk in agriculture: A holistic approach. Organization for Economic Cooperation and Development. Retrieved from www.oecd.org/agriculture/policies/risk.

44- Tiraee Yari. N. (2002). Factors affecting modern agricultural insurance models adoption applied by farmers in Khuzestan province. Unpubliahed thesis, Agriculture department, Tehran: Tarbiat Modarres University.

45- Torkamani, J. (1998). Establishing risk aversion degree, technical efficiency and its influencing factors: A Case Study in Fars Province. *Agricultural Economics and Development Journal,* 24, 49-68.

46- Torkamani, J., & Ezatabadi, M. (2001). The socio - economic factors impact on decision-making process associated with risk. *Agricultural Economics and Development periodicals*, 9 (33), 23-45.

47- Velandia, M., Rejetus, R. M. Knight, T. O., & Sherrick, B. J. (2009). Factors affecting farmer's utilization of agricultural risk management tools: The case of crop insurance, forward contracting, and spreading sales. *Journal of Agricultural and Applied Economics*, 41(1), 107-123.

48- Yaghobi, A., Chizari, M., & Fele, S. (2007). *Agricultural product Insurance: appropriate strategy in risk management.* Sixth Conference on Agricultural Economics, Mashad: Ferdowsi University of Mashad. Retrieved from http://confbank.um.ac.ir/index.php?module=conf_display&file=user&req=conf_display_main_view_abs&id=1613&conf_id=14

49- Zamani, G. H., Karami, A., & Yazdanpanah, M. (2009). Factors influencing farmers' satisfaction of agricultural crop insurance policy. *Agricultural Extension and Education Journal,* 4(2), 53-66.

# 11

# The Sustainability Practices among Dairy Farmers: The Case of Johor

*Zainab Mohd Karim [1], Nalini Arumugam [1*] and Boniface Bonaventure [2]*

**Abstract**

Sustainability concept in agriculture is becoming widespread throughout industries in many fields even in crop production, livestock production and etc. Sustainable in agriculture is believed can give present and future good benefit to the farmers and society. However, in dairy sector in Malaysia, the sustainability concepts and practices seems not fully aware by the farmers which result in low self- sufficiency level in milk production. The self-sufficiency level in Malaysia is merely 6% and this lead to the urgency of importing more milk from other countries such as Australia, Holland and others. This study attempts to describe the descriptive thoughts on sustainability among Malaysian dairy farmers as well as describing the socio economic characteristics of the dairy farmers which will be the indicator of adoption of sustainability practices among dairy farmers. The preliminary data was collected using questionnaires through conducting face to face interviews with 50 dairy farmers from Johor. The Likert-type scale was employed to determine the practices that adopted by the dairy farmers. The gathered data was analysed using SPSS. The findings indicated that the dairy farmers had a higher understanding about the farm sustainability aspect which gives the highest mean score of 4.41. The second highest mean score is the ecological aspect which is about 4.13. The mean score for economic aspect shows the mean score about 4.00. The social aspect of sustainability revealed the lowest mean score which was about 3.46. The results indicated that the farmers favour sustainable practice is farm sustainability system. Sustainability in dairy farming sector in Malaysia can be achieved if the three elements of sustainability which is environmental, economic and social aspect were implemented and adopted in this sector.

*Keywords:*
***Dairy farming, Farmers, Practices, Sustainability***

[1] School of Agriculture & Biotechnology, Faculty of Bioresource and Food Industry, Universiti Sultan Zainal Abidin (UniSZA)
[2] Centre for the Promotion of Knowledge and Language, Universiti Malaysia Sabah (UMS), Agricultural Campus
* Corresponding author's email: nalini@unisza.edu.my

## INTRODUCTION

Dairy farming sector in Malaysia is considered as small with the level of milk sufficiency about 6% but the government aim are to increase local milk production in the short term as well as the long term and to increase self-sufficiency level To achieve this, dairy farmers need to adopt sustainable practices in their farms. It is now accepted that economic development should be based on the concept of sustainability and also be environmental-friendly. Sustainable dairy farming involves protecting and improving the natural environment, animal welfare and conditions of the local community, while at the same time, being productive and efficient. A sustainable agriculture is one that, over the long term, enhances environmental quality and the resource base to provide for basic human food and fibre needs; is economically viable; and enhances the quality of life for farmers and society as a whole. American Society of Agronomy, (1989). The sustainability of the Malaysian agriculture sector has been on the national agenda for some time with the latest thrust being the National Agro-food Policy (2011 – 2020). The National Agro-food Policy programs were made as soon as the National Agriculture Policy ended by the year 2010. The agenda in the programs is maintained which is to encourage the development in Malaysian agriculture sector and secure the food security in Malaysia. National Agro-food Policy (2011 – 2020) functions to carry out previous agriculture programs such as Economic Transformation Program (ETP). Most of the dairy farmers in Malaysia consist of smallholders. In order to improve the dairy industry in Malaysia, it is important for dairy farmers adapting the sustainability practices in their management and farm operation. Implementation of sustainable practices in dairy should starts from the farm where as the farmers should understand the concept thoroughly as they can do the practices effectively in the long run. Sustainable practices also will lead to increase in profitability and enhance food security as well as sustaining the ecological and social aspects. This study attempts to discover the determinants in sustainability as well as describing the socio economic characteristics of the dairy farmers which will be the indicator of adoption of sustainability practices among dairy farmers.

### Literature review

Sustainability is known as the best technique to cultivate farming which ensure the continuity of the practices, guarantee the economics, social of the people and also preserved the good environment. Sustainable agriculture are refers to technique in producing food that is healthy for humans and animals, which did not threat the environment, humane for workers, considered on animal welfare, offers good earnings for the farmers and develops rural communities (Altieri, 1995). Another two popular and widely used definitions of sustainable development are identified by Van Calker *et al.* (2005) which are "development that meets the needs of the present without compromising the ability of future generations to meet their own needs" and "development that improves the quality of human life while living within caring capacity of supporting ecosystems".

Generally, sustainable agriculture is defined as an approach to ensure economic, social and ecological sustainability. Tatlidil *et al.* (2009) mentioned that to achieve sustainability, the development process should includes the three dimensions of environment, economic and social whereby the protection and effective management of natural resources provides environmental sustainability, long-term employment and income stability. This, in turn, will move towards economic sustainability, and strong participation from the farming community will thus lead to the attainment of social sustainability.

The concept of sustainability is becoming widespread throughout industries in many fields. Dairy farming is no exception and the farming systems will have to evolve in this direction. However, the issues of sustainability in dairy farming or in other agricultural field did not well touch in Malaysia.

Given facts that Malaysia`s dairy farmers are both small and large scale farmers, which they may also differ in terms of skills and experiences. It is supported by a study carried out by Boniface *et al.* (2010) which indicated that farmers are not homogenous but heterogeneous in nature. So that the farmers should not be treated equally with other farmers for they have their own husbandry practices, skills and expertise in operating their farm. Their awareness in sustainability practice also differ from each farmers which have different socio- economic characteristics and different way on seeking information of their decision making in their farm. Abolhasan *et al.* (2010) stated that the perceived importance of sustainability among farmers differs from farmer to farmer and is influenced by socio-economic characteristics as well as information-seeking behavior.

Studies on acceptance of sustainable agriculture are lacking for Malaysia. Hence, farmers in Malaysia did not really know what the sustainability terms itself and somehow they did not aware about the practices they done whether it can sustain their farm or not. This is supported by Azman *et al.* (2013) stated that there is mass of international studies which found several factors to impinge on the acceptance of sustainable among farmers, but the same scenario are lacking for Malaysia. Acceptance of sustainability practices in Malaysian dairy farm is believed can improved Malaysian dairy industry as well helping the farmers in dairy industry to understand and adopted the right practices in dairy sustainability which can enhance their livelihood in all aspects in sustainability.

D' Silva *et al.* (2011) mentioned that acceptance of sustainable agriculture practices will undeniably facilitate agricultural entrepreneurs to enhance their economic and social status, and thus, enable them to enjoy a better livelihood.

## MATERIALS AND METHODS

The data was collected using questionnaires through conducting face to face interviews with the 50 dairy farmers. In this research, all the structured questionnaires are prepared in Malays. The purpose was to guide the interview with respondents to avoid any misunderstandings that might be occurring in the process of obtaining data. The Likert-like scale was used to determine the practices that adopted by the dairy farmers. The gathered data was analysed using descriptive analysis by SPSS. This data were collected by utilizing both qualitative and quantitative methods. Focus Group Discussion (FGD) technique also was applied to gather the information on dairy farming practices from the farmers and other related stakeholders. It is an effective way to explore participants` perceptions and arguments. The participants themselves determine which topics they discuss in depth, showing where their interest lies. Quantitative methods ensure high levels of reliability of the gathered data. The quantitative method ensures high levels of reliability of the gathered data. In this research the population sample was dairy farmers from Johor to provide a representative of dairy farm operations in Malaysia as they represent various forms of marketing channels and scales of operation. The sample size was 50 since it were estimated that there were about 134 dairy farmers (DVS, 2014) in the Johor state. Random sampling method was used in this research as it chosen because of it is convenience and economical (Higgins, 2009). The research areas were selected based on their agricultural potential, location, population density, and the extent to which the represented the socio-economic characteristics. The Statistical Package for Social Science (SPSS 20.0) was used to analyze the data.

## RESULTS AND DISCUSSION

### Demographic profile

From the Table 1, the respondents were classified into six classes of aged grouped. The biggest percentage was formed by group between 31-40 years old with 40%. The second classes of 51-60 formed 22 % followed by group within age range of 41-50 with 16%. The group with age of more than 60 formed about 12 % while the second smallest percentage was given by group of respondent with the range of 21-30 and the smallest percentage was 4% which refers to respondent with less than 20

Table 1: Demographic profile of respondents (n= 50)

| Demographic profile | | Frequency | Percentage (%) |
|---|---|---|---|
| Age | <20 | 2 | 4 |
| | 21-30 | 3 | 6 |
| | 31-40 | 20 | 40 |
| | 41-50 | 8 | 16 |
| | 51-60 | 11 | 22 |
| | >60 | 6 | 12 |
| Gender | Male | 43 | 86 |
| | Female | 7 | 14 |
| Ethnic | Malay | 9 | 18 |
| | Chinese | 1 | 2 |
| | Indian | 40 | 80 |
| Marital Status | Married | 41 | 82 |
| | Single | 9 | 18 |
| Education | No Education | 2 | 4 |
| | Primary | 12 | 24 |
| | Secondary | 29 | 58 |
| | Tertiary | 7 | 14 |
| Involvement | Full Time | 42 | 84 |
| | Part Time | 8 | 16 |
| Experience | <10 Years | 15 | 30 |
| | 11-20 Years | 22 | 44 |
| | 21-30 Years | 13 | 26 |
| Income | RM 1,000- RM 2,000 | 11 | 22 |
| | RM 2,001-RM 3,000 | 15 | 30 |
| | > RM 3,001 | 24 | 48 |

years old. From the result obtained, 50% of the farmers who involved in dairy farming were between the ages of 15 to 40 years old. In Malaysia, people in this range of age are considered as youth (Bahaman *et al.*, 2010) Most of the respondents which are youth are willing to participate in dairy farming. It seems that they involved in this field are influenced by their family as they inherited the farm from their parents.

On gender category, 86% were male and 14% were female. In terms of ethnicity, 80% were Indian, 18% Malay and 2% were Chinese. Majority of respondent in this study were Indian as other races like Malays and Chinese are not really exposed in this farming. All the Indians normally involved in the dairy farming since childhood as it is their family heritage which done ancestrally. In marital status category, 82% were married while 18% were single.

The respondents were classified into four categories based on their educational background. The highest group was secondary level with 58%, the second highest group was primary level with 24% while tertiary level group was 4%. The smallest percentage was 4% with no formal education. Involvement of respondents was divided into two parts the full time respondents was higher at 84% as compared to part time at 16%.

On experience category, 44% were involved in dairy farming for between 10-20 years, 30% with less than 10 years of experience in dairy farming while the smallest percentage was 26% which refer to the respondent's experience between 21-30 years. The respondent with income higher than RM3001 constituted the biggest percentage at 48% while second highest were 30% by respondents with income between RM2001-RM3000. The smallest percentage was income group between RM1001-RM2000 at 22%.

**Dairy farming information seeking behavior**

Table 2 represent the information seeking behavior among the 50 dairy farmers surveyed.

Table 2: Dairy farming information seeking behavior (n= 50)

| Dairy farming information seeking behavior | | Frequency | Percentage (%) |
|---|---|---|---|
| Technology | Internet | 5 | 10 |
| | Agencies | 40 | 80 |
| | Members | 3 | 6 |
| | Others | 2 | 4 |
| Capital | Internet | 2 | 4 |
| | Agencies | 28 | 56 |
| | Members | 11 | 22 |
| | Others | 9 | 18 |
| Technical Problem | Agencies | 47 | 94 |
| | Others | 3 | 6 |
| Seminars/Workshop | Agencies | 45 | 90 |
| | Members | 2 | 4 |
| | Others | 3 | 6 |
| Labour | Agencies | 35 | 70 |
| | Members | 8 | 16 |
| | Others | 7 | 14 |

This study revealed the medium that the farmers use in information seeking in several aspect such as technology in dairy farming, how to get capital and labours, technical problem solving and also the seminars or workshops that they can attend, In technology aspect, about 80% of farmers tend to seek information from the agencies, 10% on internets, 6% from members and only 4% from others which was based on their own decision making or sometimes from their family and relatives. For capital aspect, 56% of the farmers seek information on capital source from agencies, 22% from members, 18% from others and 4% from the internet. While in technical problem aspect, most of the farmers seek information from agencies which is about 94% and only 6% find solution on others. In category of seminars/workshop, 90% of surveyed farmers seek information from agencies, 6% from others and last 4% from their members. For the last category of labour, about 70% of the farmers find the labour source information from the agencies while 16% from the members and last 14 % from others.

**Dairy Farming Information Seeking Behavior**

From the result obtained also, we figured that most of the farmers tend to seek information from government agencies which is by Department of Veterinary Services (DVS) and "Pusat Perkhidmatan Industri Tenusu (PPIT). It seems that the farmers trust the extension officers' ability and knowledge in every aspect in managing their farm. In aspects of technology, technical problems and event of seminars or workshops, most of them depend on the extension workers from the agencies. However, in labour seeking aspect, they more depend to the agency which provides foreign workers as the local workers are hard to be obtained. Some of them seek from members' help which were their friends and colleagues and also from their family. Another way on seeking information is from internets. However, only few farmers who seek information through internets as they mostly depend on agency. The farmers who seek information from internets mostly youth and quite educated. Some of the farmers still tend to do practices which told by the parent as their family involved in dairy farming for generations.

**Sustainable dairy farming**

From Table 3, the research findings showed that the dairy farmers understood more about the farm sustainability aspect which gives the highest mean score of 4.41. The second highest mean score was the ecological aspect with mean score of 4.13. The mean score for economy as-

Table 3: Mean score of sustainable dairy farming

| Sustainability of Dairy Industry | M | SD |
|---|---|---|
| Farming System | 4.41 | 0.38 |
| Environmental sustainability | 4.13 | 0.34 |
| Economic sustainability | 4.00 | 0.55 |
| Social sustainability | 3.46 | 1.00 |

pect was about 4.00. The social aspect of sustainability indicated the lowest mean score which is about 3.46. The results indicated that the most favourable sustainable practice is farm sustainability. The dairy farmers should have in depth knowledge in farm sustainability, economic sustainability, social sustainability and ecological sustainability to enhance the dairy farming industry.

### Sustainable dairy farming

The results indicated that the most favourable sustainable practice is sustainable farming systems. Almost, all the farmers agreed that sustainable farming system is the essential practice to be adopted to obtain more profit and reduce the cost of production. The section in sustainable farming systems comprise of basic knowledge on farming such as suitability of breeds, animal welfare and the management of dairy herd. Those farmers are more concern on the herd health, milking process and the management since they have been in the dairy farming for a long time. Result shows that most of the farmers are full time dairy farmers.

For the environmental sustainability, most of the farmers are aware about the consequences of the waste management to the environmental. They agreed that water quality and quantity is an essential part in order to produce good quality of milk. For other aspects such as odour and its management, the farmers seem not applying the right way in managing them. Most of the farmers just discard the waste product to drench. The farmers have favourable attitude toward sustainable practices. They tend to adopt the practices that really necessary and profitable to them. This is supported by Bagheri (2010) which found that farmers had favourable attitude towards sustainable practices such as resource conservation, negative effects of agrochemicals, pests' invasion arising from successive cultivation and have moderate attitude towards the negative environmental effects of modern agricultural technologies.

In economic sustainability aspects, the surveyed farmers mostly have their marketing channel. They sell their milk to the cooperatives (Koperasi Penternak Lembu Tenusu Negeri Johor Bhd) that will process and markets their products. Some of the farmers sell their own milk with higher price. They believe that economic sustainability can be achieved if the price of milk is increase. The higher price will help them to offset the cost of production since the feed stuff cost is increasing gradually every year.

Social sustainability gave the lowest mean score as the farmers having problems in finding workers with good dairy working experiences and willing to do the jobs. The management on dairy herds is more complicated and need more working hour. Most of the farmers start their work in the early morning. It is because the farmers need to milk their herds twice a day to ensure higher profit and prolonged lactation time of herds. Some of the farmers employed foreign workers to manage their farm as the local worker are not willing to be in this field. Most farmers use family labour to help in the farming as they face difficulty in hiring external skilled workers.

## CONCLUSION

In conclusion, on the information seeking behaviour aspect, dairy farmers tend to seek information and basically rely on government agencies which are DVS and PPIT. Dairy farmers in Malaysia also more favour and adopted the farming system aspect compared to other aspect as it believed able to reduce the cost of production and obtain more profits in this industry. Dairy farmers nowadays also are moving towards sustainability in some aspect and it is believed to be sustaining in becoming years. Government helps in creating awareness of the importance in adopting this sustainability practices in dairy farming should help the farmers to understand sustainability concept and encourage them to adopt the practices in dairy farming. However, to achieve the sustainability in our dairy industry in full aspect, the three elements of sustainability which is environmental, economic, and social aspect should be fully implemented and adopted. Studies in dairy farming sustainability should be deepened in every aspect such as economic, social and environmental in other to find the most beneficial way to help the farmers.

## ACKNOWLEDGEMENTS

The author wants to thanks Universiti Sultan Zainal Abidin (UniSZA) for providing the financial assistance and also to Department of Veterinary and Services, Milk Collection Centre for assisting throughout the study.

## REFERENCES

1- Abolhasan, S.S., Hosain, S.F., Ali Asadi & Abolghasem, S.S. (2010). Farmer's attitude on sustainable agriculture and its determinants: A case study in Behbahan Country of Iran. *Research Journal of Applied Sciences, Engineering and Technology.* 2 (5), 422-427.

2- Altieri, M.A. (1995). Agroecology: *The science of sustainable agriculture.* Westview Press, Boulder, CO.

3- American Society of Agronomy- (1989). "Decision reached on sustainable agriculture" Agronomy News, January, p. 15, Madison, Wisconsin.

4- Azman, A., D'Silva, J. L., Samah, B. A., Man, N., & Shaffril, H. A. M. (2013). Relationship between Attitude, Knowledge, and support towards the acceptance of sustainable agriculture among contract farmers in Malaysia. *Asian Social Science*, 9(2), 99.105

5- Bagheri, A. (2010). Potato farmers' perceptions of sustainable agriculture: the case of Ardabil province of Iran. Procedia. *Social and Behavioral Sciences*, 5, 1977-1981.

6- Bahaman, A.S., Jeffrey, L.S., Hayrol Azril, M.S. & Jegak, U. (2010). Acceptance, attitude and knowledge towards agriculture economic activity between Rural and Urban Youth: The Case of contract farming. *Journal of Applied Sciences,* 10(19), 2310-2315.

7- Boniface, B., Gyau, A., Stringer, R. &Umberger, W. (2010). Building producer loyalty in Malaysia's fresh milk supply chain. *Australasian Agribusiness Review,* 18(5), 66-84.

8- D'Silva, J.L., Man,N., Shaffril,H.A.M., & Samah, B.A. (2011). Acceptance of sustainable agricultural practices: the case of crop farmers. *American Journal of Agricultural and Biological Science,* 6: 227-230.

9- DVS. 2014. Livestock/livestock products statistics, Department of Veterinary Services, Malaysia.

10- Higgins, G.E. (2009). Quantitative versus qualitative methods: Understanding why quantitative methods are predominant in criminology and criminal justice. *Journal of Theoretical and Philosophical Criminology*, 1(1), 23-37

11- Tatlıdil, F. F., Boz, I., & Tatlidil, H. (2009). Farmers' perception of sustainable agriculture and its determinants: a case study in Kahramanmaras province of Turkey. *Environment, Development and Sustainability*, 11(6), 1091-1106.

12- Van Calker, K. J., Berentsen, P. B., Giesen, G. W., & Huirne, R. B. (2005). Identifying and ranking attributes that determine sustainability in Dutch dairy farming. *Agriculture and Human Values*, 22(1), 53-63.

12

# Factors Influencing the Gardening Management Skills among Citrus Farmers (Case of Iran, Sari)

***Mehdi Alikhani Dadoukolaei* [1], *Mehdi Charmchian Langerodi* [2] *and Amir Zameni* [3]**

Abstract

Production management is the best and foremost factor in exploiting the production factors more desirably and advantageously. Accordingly, the present study sought to evaluate the factors affecting the gardening management skills among citrus farmers in Sari, north of Iran. The statistical population for the study was 12,800 people. The stratified random sampling using the Cochran test was applied, and 250 subjects were selected for the study. The research instrument was a researcher-made questionnaire, the content validity of which was confirmed by the agricultural and extension experts and its reliability was estimated by Cronbach's alpha as to be 0.94. The data analysis results showed that the respondents' gardening management skills was at high (41.6%) and average levels (54.1%). Considering the research findings, there was a significant positive correlation at the one percent level between citrus farmers' gardening management skills and social, knowledge, attitude, educational factors, the total cultivated area and gardening experience. Also, there was a significant positive correlation at the five percent level between citrus farmers' gardening management skills and the garden area, citrus annual income and the distance of the garden to the agriculture center. Considering the stepwise regression results, 45.7% of the variation in respondents' gardening management skills was accounted for by three variables of knowledge, educational factor and the total area of cultivated land, serving as the best predictor of citrus farmers' gardening management skills in Sari.

***Keywords:***
***Skill, Gardening management, Citrus farmers, Sari***

[1] *Young Researchers and Elite Club, Sari Branch, Islamic Azad University, Sari, Iran*
[2] *Department of Agricultural Extension and Education, Sari Branch, Islamic Azad University, Sari, Iran*
[3] *MSc. Graduate in Agricultural Extension and Education, Sari Branch, Islamic Azad University, Sari, Iran*
* *Corresponding author's email: alikhany.mehdi@yahoo.com*

## INTRODUCTION

Agriculture is the most dynamic economic sector enjoying a distinctive niche in the socioeconomic development of the country, and like other sectors of the economy, requires people who are good managers and experts in productive activities. In order to augment the production besides providing the materials, machinery and other investments, it is required to consider the production management as the most important factor in more appropriate and more profitable use of the production factors (Soltani *et al.*, 1998). The management – along with three main factors of production including land, labor force and investment – can increase production (Nuthall, 2006). Management can be defined as the art or knowledge that combines the ideas, methods and resources to produce and sell the product more profitably and favorably (Borimnejad, 2006). The farmer's success as the production manager depends on the farmer's management ability (FAO, 2003) that cannot benefit from the economic activity associated with success without proper knowledge, skills and a rational management (Dehghanian, 2002). Rolls (2001) argues that farm management is the knowledge and ability to organize, control and maintain the desired specific resources in a way that benefits the farmer to maximize the profits and continuous return. The Britannica Encyclopedia defines farm management as the adoption and implementation of the decision in the organization and applying them in farms to maximize production and profit (Borimnejad, 2006). In general, farm management is a skill set allowing farmers to make sensible decisions and enable them to implement changes that will move operations toward predefined and expected objectives (Balasubramanian *et al.,* 1999).

The new farm management levels enable farmers and researchers to collectively apply the resources, implement the applicable innovations, and increase the efficiency and research effectiveness on the field (Dorward *et al.*, 2003). It should be noted that the extent to which farm management practices are implemented by farmers is referred to as garden management adoption (Shabanalifami, 2013). However, it seems that gardening management skills have not been adopted, and the burden is on the extension agents' and department shoulder to encourage farmers to adopt them (Asadi *et al.*, 2009). It is worth mentioning that various studies have been conducted on farm management skills, some of which will be referred to in the following paragraphs.

In a study on determining the factors affecting the adoption of farm management skills among rice farmers in Savadkouh, Hasani (2013) showed that 28.7% of the variation of respondents' adoption of farm management skills was accounted for by three factors of knowledge, planting pattern, and the seed types which are the best predictor of farm management skills adoption. Carli and Canavari (2013) conducted a study on introducing direct costing and activity based costing in a farm management system: a conceptual model. The results of the study indicated that this system offers precise information about crop costs with general costs allocation procedures based on the consumption of activities and enables sensitivity analyses.

Lantzosa *et al.* (2013) conducted a study on Farm Manager: an Android application for the management of small farms. Farmers working on small farms are now able with a low cost smart phone and the specialized software (in our case Farm Manager) to obtain facilities that could not have had on their hands before. The use of the Farm Manager software in a smart phone can overleap the grave difficulties of farm management requirements which had stood as obstacles for many years so far. Tasks such as field definition, task operations, listing and reporting and virtually, all farming data use can be submitted and carried on together in a smart phone at any farm working condition. In this paper, we present the Farm Manager which is an Android smart phone application and how it creates the management basis for recording and browsing the ground fields, field relations (occupied or rented land), cultivation and its tasks, equipment, employees and European cultivations reports and all of them to be performed by simply touching smart phone screen button. The software is currently available free of charge and there are more than one thousand farmers using it in Greece.

Oenemaa *et al.* (2011) conducted a study on participatory farm management adaptations to reduce environmental impact on commercial pilot dairy farms in the Netherlands. From 1998 to 2002, average nutrient surpluses on the pilot farms were decreased by 33% for nitrogen (N) and 53% for phosphorus (P). Important measures include reducing the use of inorganic fertilizer, optimizing the use of home-produced organic manure, reducing grazing time, reducing the number of replacement stock, and lowering crude protein content in the ration. Over the years, variation in N surpluses among farms (inter-farm variation) remained almost constant. Differences in farm management strategy could not unequivocally be related to farm typology (high/low N surplus; high/low production intensity). It was concluded that decisions by individual farmers on farm development are not always based on 'rational' arguments, but are co-determined by 'emotional' perceptions. In a study on analyzing farm management skills adoption in poultry production enterprises in Iran, Allahyari *et al.* (2011) concluded that among the studied management skills, marketing skill was placed as the least ranking means. In addition, poultry production operators had the best ability in technical skills. The average ability rate of poultry production operators in technical area was 4.08 indicating the high to very high ability rate of respondents in this area. According to the results, it is necessary to improve the marketing and farm management skills of farm operators through extension and participation in training activities.

Veisi *et al.* (2010) categorized factors affecting farmers' behavior in integrated management technology into four groups of farmers' personal and family characteristics, farm biological and physical properties, financial aspects of farm management, and external factors. In a study on the effectiveness of management in agricultural production in Kenya, Onyuma *et al.* (2006) indicated that establishing incentives, attending educational classes, and empowering small-scale farmers would improve their managerial skills. Using a two-stage cluster sampling, interviews and questionnaires, Mohammadi (2001) analyzed the production management factors in corn farmers' farms in Fars province, and concluded that socio-structural factors such as the type of ownership, number of owned land, level of experience, education and the area under cultivation were the most important factors affecting the differences in performance among corn farmers.

Considering the above-mentioned facts, it can be stated that the main and key challenges in the agriculture, especially in horticulture sector rests on the lack of proper gardening management, and if such an important factor is maintained and stressed, many developments in agriculture will be resulted. To increase citrus production and ensure maximum sustainability as one of the strategic products, it is required to adopt management skills. Therefore, the study was carried out on factors affecting the gardening management skills among citrus farmers in Sari, Iran.

## MATERIALS AND METHODS

### Study area

The research geographical location refers to Sari Township. Sari is located to the north of Iran and to the South of the Caspian Sea and occupies an area of ca. 3,923 km$^2$. It has a population of 255,396 making it the biggest city in Mazandaran. The climate of Sari is generally temperate (Figure1).

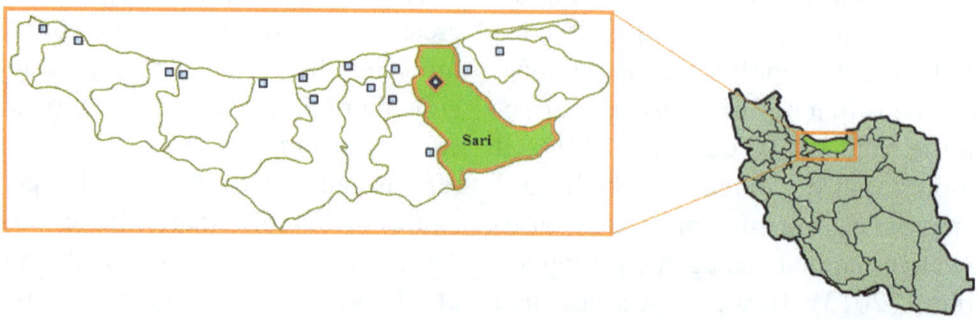


Figure 1: The geographical locations of the study area, Sari Township, Iran

## Methods

The research adopted an applied, descriptive-correlation approach. The statistical population consisted of all Sari citrus farmers (four areas: Holar, Panbezarkiti, Semeskandeh, and Aboksar) in 2015 (N = 12,800). Using statistical sampling in a stratified randomization method, 250 citrus farmers were selected and finally, 233 questionnaires were returned and analyzed.

$$n = \frac{N(ts)^2}{Nd^2 + (ts)^2}$$

where,
N= 12800
t=1.96
S=1.32

$$d = t . \frac{s}{\sqrt{n1}}$$

d=0.507 ÷3=1.69
n=230

A field study was used in order to obtain the required data by a researcher-made questionnaire. Data collection tool was a questionnaire with four sections. Section one was related to demographic information of the participants including age, work experience (gardening) and family member. Section two (six items) was designed to identify job-related factor (including total cultivated area, the distance of the garden to the agriculture center, yield performance, number of owned gardens, garden area, and citrus annual income). Section three was designed to determine the social (eight items), educational (11 items), knowledge (21 items), and attitudinal (12 items) factors. Section four (28 items) was designed to determine the gardening management skills. Three sections of the questionnaire were assessed on a five-point Likert-type scale ranging from very low (1) to very high (5). The content validity was confirmed by experts and agriculture extension experts and researchers. To estimate the reliability of questionnaire, a pilot test was carried out by distributing 30 questionnaires in a statistical population similar to the studied statistical society (Citrus farmers in Ghaemshahr), and then, Cronbach's alpha coefficients were estimated by SPSS Software Package as to be 0.94.

## Analysis of data

The results were assessed through inferential and descriptive statistics. The Spearman correlation coefficient was used to show the relationship between the variables, and the stepwise multiple regression coefficients was used for the correlation of the independent variable on the dependant variable.

# RESULTS

## Demographic features

Based on the results of the questionnaire and Table 1, the demographic characteristics of the respondents revealed that the majority of the subjects were male with the average age range of 45.97 years representing middle-aged farmers. The respondents' average annual income was about 330.49 million IRR, showing a good income level, and the gardening experience average was 22.84 years.

## Ranking of the research factors

As is evident in Table 2, the variation coefficient was used to prioritize the constituting questions on the variables from the respondents' viewpoints. Regarding items ranking the gardening management skills, the estimating citrus cost, soil drainage to improve soil quality, and mechanized spraying and fertilizing had the highest priorities. Wholesale purchase of gardening inputs for discounts, labor recruitment, and crop insurance had the lowest priority.

Table 1: Demographic features of the respondents

| Variable | Mean | Minimum | Maximum | SD |
|---|---|---|---|---|
| Age (year) | 45.97 | 24 | 70 | 10.86 |
| Gardening experience (year) | 22.84 | 1 | 40 | 9.73 |
| Yield performance (tons) | 37.51 | 2 | 100 | 15.39 |
| Garden area (ha) | 1.75 | 0.4 | 10 | 1.46 |
| Citrus annual income (million IRR) | 330.49 | 20 | 1800 | 320.53 |

Table 2: Ranking of the variables based on the respondents' viewpoints

| Variable | M[1] | SD[2] | CV[3] | R[4] |
|---|---|---|---|---|
| Estimating citrus cost | 3.78 | 0.65 | 17.18 | 1 |
| Soil drainage to improve soil quality | 4.05 | 0.71 | 17.58 | 2 |
| Mechanized spraying and fertilizing | 3.94 | 0.76 | 19.17 | 3 |
| Proper use of recommended fertilizers | 4.13 | 0.82 | 19.77 | 4 |
| Applying pruning for aeration | 4.08 | 0.92 | 22.44 | 5 |
| Using agricultural practices such as deep plowing (tiling) and weeding | 3.74 | 0.85 | 22.65 | 6 |
| Monitoring and controlling working conditions | 3.48 | 0.82 | 23.73 | 7 |
| Spotting a proper place for planting seedlings | 3.79 | 0.94 | 24.78 | 8 |
| Applying biological control approaches | 3.88 | 1.01 | 25.93 | 9 |
| Limited application of pesticides to control weeds in the garden | 2.91 | 0.81 | 27.86 | 10 |
| Recording the gardening affairs | 3.55 | 1.10 | 30.97 | 11 |
| The use of non-chemical fertilizer and livestock manure | 3.25 | 1.01 | 31.00 | 12 |
| Using machinery to reduce costs | 3.64 | 1.14 | 31.48 | 13 |
| Choosing the best time to sell a product | 3.79 | 1.21 | 31.99 | 14 |
| Estimating production costs and the income from production | 3.42 | 1.15 | 33.49 | 15 |
| Maintaining financial and garden records | 2.87 | 0.97 | 33.77 | 16 |
| Using spaced planting records | 3.40 | 1.22 | 35.90 | 17 |
| Applying garden soil and water testing | 2.98 | 1.10 | 36.72 | 18 |
| New data analysis for planning | 2.63 | 1.00 | 37.97 | 19 |
| Collecting information on seeds, fertilizers and pesticides price | 3.54 | 1.40 | 39.52 | 20 |
| Using standard, resistant and improved varieties | 2.92 | 1.20 | 41.27 | 21 |
| Repairing machineries and equipment | 3.22 | 1.43 | 44.31 | 22 |
| Analyzing the products demand, supply, pricing and marketing | 2.45 | 1.09 | 44.64 | 23 |
| Health behavior | 2.77 | 1.27 | 45.71 | 24 |
| Water Saving | 2.57 | 1.18 | 45.94 | 25 |
| Wholesale purchasing of gardening inputs for discounts | 2.73 | 1.32 | 48.34 | 26 |
| Labor recruitment | 2.48 | 1.25 | 50.41 | 27 |
| Crop insurance | 2.03 | 1.21 | 59.55 | 28 |

1- Mean 2- Standard Deviation 3- Coefficient of Variation 4- Rank
Likert scale: very Low (1), low (2), average (3), high (4), very high (5)

The dependent variable of gardening management skills is comprised of 28 questions. As is seen in Table 3, the results showed that 54.9 % of respondents considered their gardening management skills at an average level and that 41.6 % reported their gardening management skills at a high level.

## Normality test

Based on the results of Table 4, the results of Kolmogorov–Smirnov test (K–S test) test were significant for all variables. Therefore, the non-parametric test was used.

## Correlation coefficient

The Spearman correlation coefficient was used to determine the relationship between the selected research variables and the respondents' gardening management skills. As shown in Table 5, there was a significant positive correlation at 1% error level between gardening management skills and farmers' social, knowledge, attitude, educational factors, the total cultivated area and gardening experience. Also, there was a significant positive correlation at 5% error level between the respondents' gardening management skills and the

Table 3: The respondent's frequency distribution based on gardening management skills

| Gardening management skills | Frequency | Valid percent | Cumulative percent |
|---|---|---|---|
| Age (year) | 45.97 | 24 | 70 |
| Gardening experience (year) | 22.84 | 1 | 40 |
| Yield performance (tons) | 37.51 | 2 | 100 |
| Garden area (ha) | 1.75 | 0.4 | 10 |
| Citrus annual income (million IRR) | 330.49 | 20 | 1800 |

Table 4: Determining the normal distribution of the variables by the Kolmogorov–Smirnov test (K–S test)

| Variable | M | SD | K-S test | p-value |
|---|---|---|---|---|
| Gardening management skills | 91.9871 | 10.71676 | 1.221 | 0.011 |
| Social factor | 23.9736 | 6.40249 | 1.582 | 0.013 |
| Education factor | 24.1189 | 7.74718 | 3.709 | 0.000 |
| knowledge factor | 81.1057 | 13.09388 | 1.809 | 0.003 |
| Attitude factor | 60.7372 | 5.33024 | 1.859 | 0.002 |

garden area, citrus annual income and the distance of the garden to the agriculture center. Therefore, the null hypothesis was rejected.

## Regression

The stepwise linear regression was used to determine the of different factors in respondents' gardening management skills. Among all the selected variables included in regression, the knowledge, total cultivated area, and education were put in three steps in the regression equation. The results revealed that 45.7% of the variation of the respondents' gardening management skills was determined by the factors knowledge, total cultivated area, and education and all the other changes were accounted for by the other factors which were not mentioned in this study. Regarding the coefficient in Table 6, the multiple linear regression equation in the final step would be measured through the following formula:

$$Y=46.826+0.477x_1+1.217x_2+0.169x_3$$

## DISCUSSION

Data analysis results showed that the respondents' gardening management skills were at high and average level (96.5%). A significant positive correlation was found between respondents' gardening management skills and their social factor at 1% error level meaning the greater the social features, the more the gardening management skills and vice versa. This finding is consistent with the results reported

Table 5: Correlation between the selected research variables and the respondents' gardening management skills

| Variables | Correlation coefficient | p-value |
|---|---|---|
| Social factor | 0.172** | 0.008 |
| Education factor | 0.341** | 0.000 |
| Knowledge factor | 0.561** | 0.000 |
| Attitude factor | 0.171** | 0.009 |
| Total cultivated area | 0.224** | 0.002 |
| Gardening experience | 0.190** | 0.009 |
| Garden area | 0.158* | 0.032 |
| Citrus annual income | 0.178* | 0.017 |
| Distance from the garden to the agriculture center | 0.146* | 0.048 |
| Age | 0.056 | 0.446 |
| Family member | 0.037 | 0.614 |
| Number of owned garden | -0.044 | 0.554 |
| Performance | 0.104 | 0.157 |

*p<0.05, **p<0.01

Table 6: The variables coefficient on regression equation

| Variable | B | Beta | t | p-value |
|---|---|---|---|---|
| (Constant) | 46.826 | - | 12.374 | 0.000 |
| Knowledge factor | 0.477 | 0.572 | 9.432 | 0.000 |
| Total cultivated area | 1.217 | 0.188 | 3.283 | 0.001 |
| Education factor | 0.169 | 0.124 | 2.071 | 0.040 |

Sig= 0.000 F= 51.133** $R^2_{AD}$=0.457 $R^2$=0.466

by Hasani (2013) and Mohammadi (2001). There was a significant positive correlation between the educational factor and gardening management skills at 1% error; that is, the higher the educational level, the more the gardening management skills and vice versa. The result of the study conducted by Allahyari *et al.* (2011), Lantzosa *et al.* (2013) and Onyuma *et al.* (2006) is consistent with the results of this study. There was a significant positive correlation between the knowledge factor and gardening management skills at 1% error; that is, the higher the knowledge factor, the more the gardening management skills and vice versa. The result of the study conducted by Hasani (2013) and Veisi *et al.* (2010) is consistent with the results of this study. There was a significant positive correlation at 1% error between attitudinal factor and gardening management skills; i.e., the higher the attitudinal factor, the more the gardening management skills and vice versa. It is in agreement with Hasanpour (2010). A significant positive correlation was observed at 1% error between total cultivated area and gardening management skills; in other words, the higher the total cultivated area, the more the gardening management skills and vice versa.

Given the results of stepwise regression, 45.7% of the variation in respondents' gardening management skills was explained by three variables of knowledge, educational factor and the total area of cultivated land serving as the best predictor of citrus farmers' gardening management skills in Sari.

## RECOMMENDATIONS

Regarding the results of descriptive and inferential statistics, following recommendations are drawn:

Given the results of the study, the total area of cultivated land had a positive impact on the gardening management skills which seems necessary to be observed. It is worth noting that knowledge with highest regression had the highest influence on gardening management skills. So, to promote citrus growers' knowledge, it is recommended that agricultural organizations organize training courses and Farmer Field Schools (FFS) as the first priority on issues such as citrus irrigation in a timely manner, removal of weeds in citrus garden, and proper planting of the citrus seedlings. Also, due to the impact of educational features on gardening management skills, it is suggested that agriculture organizations in a mutual cooperation with the public television organization communicate experts' or extension agent' recommendation to citrus growers through radio broadcasting media to enable farmers to use them and encourage the farmers to transfer their knowledge to other farmers by training farmers. Also, the extension agents and agricultural experts are required to establish contact with citrus growers and address their management problems in the center.

Based on the gardening management skills items ranking, it is proposed to agriculture organizations to establish training courses to estimate the cost of citrus and teach citrus growers to biologically combat weeds and to hold Farmer Field School (FFS) classes to teach soil drainage to improve soil quality and practically teach farmers on mechanized spraying and fertilizing, proper application of recommended fertilizers and applying pruning for aeration. Also, some training should be provided to farmers on observing health factors to prevent the diseases and their contamination and how to conserve water during practical training. Correspondingly, training classes should be provided to encourage farmers on doing the gardening inputs purchasing to reduce the cost, labor force recruitment, and products insurance.

Also, it is recommended to establish a farmers' production management system (collecting information about the price of seed, fertilizer and pesticide, recording the gardening affairs, estimating production costs and the income from production, maintaining financial and garden records, new data analysis for planning, and analyzing the products demand, supply, pricing and marketing) by applying ICT for structures and smart phone applications (such as farm manager) due to its lower cost and higher availability.

## ACKNOWLEDGEMENT

Authors would like to thank Sari Citrus farmers for their cooperation in this study and completing the questionnaires.

## REFERENCES

1- Allahyari, M. S., Saburi, M. S., & Keshavarz, F. (2011). Analyzing Farm Management Skills in Poultry Production Enterprises in Iran. *Life Science Journal,* 8(1), 61–67.

2- Asadi, A., Akbari, M., Sharifzadeh, A., & Alambeigi, A. (2009). *An Introduction to systemic approach in agriculture extension and development emphasizing on agricultural knowledge and information system.* Tehran, University Jihad.

3- Balasubramanian, V., Morales1, A. C., Cruz, R. T., & Abdulrachman, S. (1999). On-farm adaptation of knowledge-intensive nitrogen management technologies for rice systems. *Nutrient Cycling in Agro Ecosystems,* 53, 59–69.

4- Borimnejad, V. (2006). *Translation of farm management principles and strategies (written by Kenneth Olson).* First published in Tehran, TH: Aiezh Publication.

5- Carli, G., & Canavari, M. (2013). Introducing Direct Costing and Activity Based Costing in a Farm Management System: a conceptual model. *Procedia Technology,* (8), 397 – 405.

6- Dehghanian, S. (2002). *Translation of agricultural management (written by Hogo astin Hauser, Kai Langbhn and Peters).* 2nd. Ed., Mashhad: Mashhad University publication.

7- Dorward, P., Galpin, M., & Shepherd, D. (2003). Participatory farm management methods for assessing the suitability of potential innovations, a case study on green managing options for tomato producers in Ghana. *Journal of Agricultural Systems*, 75, 97–11.

8- FAO. (2003). World agriculture: Towards 2015/2030 – An FAO Perspective. Rome, FAO & London, Earth Scan Retrieved from: www.fao.org/3/a-y4252e.pdf.

9- Hasani, A. (2013). *Factors affecting the farm management skills among rice farmers in Savadkouh.* Agricultural Extension and Education Department Unpublished thesis. Sari, University of Sari. Iran.

10- Lantzosa, T., Koykoyrisa, G., & Salampasisb, M. (2013). FarmManager: An Android application for the management of small farms. *Procedia Technology,* 8, 587–592.

11- Mohammadi, D. (2001). *Analyzing the production management factors in top corn farmers' farms in Fars province.* Proceedings of the third Conference of Agricultural Economics, February, 17-19, (pp. 669-696). Mashhad: Iran Agricultural Economic Society.

12- Nuthall, P. L. (2006). Determining the important management skill competencies:The case of family farm business in New Zealand. *Agricultural Systems,* 88, 429–450.

13- Oenemaa, J., van Keulena, B., Schils, C., & Aartsa, H. (2011). Participatory farm management adaptations to reduce environmental impact on commercial pilot dairy farms in the Netherlands. NJAS - *Wageningen Journal of Life Sciences*, 58, 39–48.

14- Onyuma, S. E., Birachi, T., Cheruiyoit, K., & Icart, E. (2006). Effect of management in agricultural production: Evidence from Kenya. *African Journal of Business and Economics,* 1(1), 1-18.

15- Rolls, M. J. (2001). *Review of farm management in extension programs in central and Eastern European countries and agricultural organization of the United Nations.* Rome, Italy.

16- Shabanalifami, H. (2013). *Agricultural extension and education principles.* Tehran, TH: Payam Noor University publication.

17- Soltani, Gh., Najafi, B., & Torkamani, J. (1998). *Management of Agriculture unit (Third Edition),* Shiraz, Shiraz University.

18- Veisi, H., Mahmoudi, H., & Sharifimoghaddam, M. (2010). Explaining the farmers' behavior in pest integrated management technologies. *Research of Agricultural Economics and Development,* 4, 481-490.

13

# Measurement of Technical Efficiency and Production Risk in Zabol Dairy Farms, Iran

***Javad Shahraki [1], Mosayeb Pahlavani [2] and Nasrin Ohadi [3*]***

Abstract

Given entrepreneurs' special role in the economic growth and development, many developed and developing countries are attempting to motivate more number of people who have entrepreneurial characteristics towards entrepreneurship and entrepreneurial activities. The purpose of this research was to study the effectiveness of entrepreneurship training courses in Ilam Province, Iran. In this research a descriptive survey method was used. The statistical population consisted of 830 people that based on Cochran's formula and by using proportionate stratified random sampling method 130 people were selected as the sample. The research tool was a questionnaire which its validity was confirmed by a panel of experts' and its reliability was estimated using Cronbach's alpha as to be 0.83 proving its high reliability. Data were analyzed by SPSS Software Package. It was found that participants' inclination towards self-employment and entrepreneurship was increased after attending the courses. Also, significant differences were observed in participants' readiness to start a business, their familiarity with entrepreneurship concepts and their ability to make a business plan before and after attending the courses.

***Keywords:***
***dairy farms, risk production, technical efficiency, Zabol***

*[1,2] Associate Professor, Department of Economics, University of Sistan and Baluchestan, Zahedan, Iran*
*[3] Ph.D. Candidate of Agricultural Economics, Department of Economics, University of Sistan and Baluchestan, Zahedan, Iran*
** Corresponding author's email: nasrin.ohadi@yahoo.com*

## INTRODUCTION

Since ancient times, animal husbandry has been an important field of agricultural and economic activity. Sub-sector of animal husbandry and growing of poultry in Iran's agricultural sector was the second most important sub-sector after crop production. In this sub-sector, value-added agriculture has grown such that its share in agricultural sector ranged from 30 to 35 percent (Ministry of Agriculture, 2006).

Cattle breeding to produce milk and dairy products is one of the most important activities of agricultural sector in Iran. The number of dairy units (both active and inactive) was 24,659 for the year 2011 with total capacity of 2.7 million heads of cattle and their number and capacity have increased by 31 and 22.3 percentages as compared to 2008, respectively. Milk production in the dairy industry is reported to be more than 3.20 million tons in 2010, constituting about 48 percent of the country's total milk production. Milk production in the dairy industry has increased by about 33.2 percent in 2010 as compared to the statistics reported in 2007 (Statistical Center of Iran, 2011).

Population growth, particularly in developing countries and food poverty in other parts of the world, has led to the fact that adequate access to food for responsiveness to population's basic needs still remains high on the agenda of social-economic policy maker. Hence, it is essential to increase production of animal husbandry products with the same amounts of available inputs. Additionally, it is important to examine how much inputs are used to enhance production (Erilouzadeh and Saleh, 2010).

Mohammadi et al (2015) studied the efficiency of industrial dairy farms of Saqqez and Divandarreh counties by the super-efficiency approach. The super efficiency estimation showed that the mean of farms super-efficiency based on the assumption of variable return to scale for input-oriented approach is 1.01. About 58 percent of the studied dairy cattle farms were inefficient and the super-efficiency of about 42 percent of total farms got scores below the average. The amount of $\lambda^*k$ for all farms except for farms three and five was zero. Furthermore, these two farms in primitive super efficiency model became infeasible; therefore, they were considered as reference farms. The results of super-efficiency method and efficiency conventional Data Envelopment Analysis (DEA) method were compared together, and inefficient farms got the similar efficiency and super-efficiency scores. Efficient farms whose super-efficiency score was equal or greater than 1 stood over frontier production function in the conventional model.

Dourandish et al (2013) estimated multi-output technical efficiency for the dairy farms of the Northern Khorasan Province (applying stochastic production and stochastic distance frontiers). The Stochastic Production Frontiers (SPF), with and without the inclusion of inefficiency effects, estimated the average technical efficiency to be 0.96 and 0.93, respectively. In addition, the experience, targeting subsidies and main job of dairy farmers had significantly positive impact on technical efficiency. The results showed that the numbers of cattle, land size, and investment in equipments, labor, and feed costs have significantly positive impact on the income of milk production. The Stochastic Distance Frontiers (SDF), with and without the inclusion of inefficiency effects, estimated the average technical efficiency to be 0.91 and 0.94, respectively. The experience, targeting subsidies, and main job of dairy farmers had significantly positive impact on technical efficiency. The numbers of cattle, land size, and investment in equipment, labor, feed, and veterinary costs had significantly positive impact on the income of the milk production.

Mehrjou et al (2012) studied technical efficiency of dairy cattle cooperative companies. The results showed that the applied bran and salt were located in the first area of production function. Average technical efficiency of dairy farm units was approximately 94 percent that could be increased by 6 percent via improving management practices. Sing et al (2000) measured technical efficiency, allocative efficiency, and economic efficiency of cooperative dairy plants in Haryana and Punjab states of India by through the DEA method. The average technical efficiency, allocative efficiency, and economic efficiency were estimated to be 91.2, 73.1, and 66.7 percent, respectively. Alvarez and Arias (2004) investigated the relationship between technical efficiency and scale of milk production in North of Spain. The study lasted for 6 years (1993-1998). The results showed that farm size depended on technical efficiency, constant inputs, outputs price, as well as variable input and that there was a positive relationship between farm size and technical efficiency. USDA (2006) explored the influence of managerial factors on productivity of dairy cattle in America. The results showed that the number of animals in the farm, the genre of animals, the animals' age, the dairy farmers' age, the number of labor, the amount of capital, and size of farm had positive effects on milk production for non-commercial and commercial cattle. Mohammadi and Torkamani (2011) estimated technical efficiency and food rations of fattening calves by using production function. The results of this study showed that non-optimal use of food was the most important factor in non-profitability of the studied units in Fars. Zibaie and Soltani (1995) examined the technical efficiency of milk production units by linear programming methods, modified least squares, and maximum likelihood method. The results showed that the technical efficiency determined by maximum likelihood was higher than that determined by the other two methods.

In this study, the researchers evaluate all active dairy farms in Zabol County and measure technical efficiency for the studied units. Next, they examined the effect of various factors on the efficiency of dairy farms units. The results of this study can, then, be used to develop appropriate policies aiming to increase technical efficiency of units and to activate the inactive units.

## MATERIALS AND METHODS

So far, several methods of identifying and measuring technical efficiency are introduced by economists. Two main methods have been developed for measuring efficiency and productivity: the parametric (econometric) and non-parametric (mathematical programming) approach. These approaches have different strengths and weaknesses. The essential differences largely reflect the different maintained assumptions used in estimating the frontier. The main strength of the parametric approach stems from the fact that frontier is stochastic, and this allows the effects of noise to be separated from the effects of inefficiency. The non-parametric approach is free from the misspecification of functional form and other restrictions, but it does not account for statistical noise and is, therefore, vulnerable to outliers (Ferrier & Lovell, 1990; Fulginiti & Perrin, 1998; Kwon & Lee, 2004; Sharma et al., 1997).

### Stochastic frontier analysis (SFA)

Battese and Coelli (1995) considered the method of Stochastic Frontier Analysis (SFA) more suitable than Data Envelopment Analysis (DEA) for using in agriculture, particularly in developing countries. The main objective of these methods is to estimate frontier function and calculate the amount of this function for each firm's inputs as well as to measure the amount of each firm's outputs. The differences between actual production and frontier production for each firm lie in inefficiency. In the stochastic frontier model, the reason for the difference between actual production and frontier production is technical inefficiency and random factors. Hence, if the performance of a firm is less than frontier production, a part of inefficiency is created by technical and the other part is created by random factors. Yet, if firms can act higher than frontier production function, it is because of random factors. The weakness of stochastic frontier production function approach for calculation of technical efficiency scores is that computational technical efficiency is assumed to be uncorrelated with explanatory variables of stochastic frontier function. If the possibility of computational technical efficiency not correlating with production function parameters is not considered, the calculated values will be skewed and inconsistent (Battese & Coelli., 1995).

Aigner and Chu (1968) estimated parametric frontier production function by using Cobb – Douglas production function. The model is defined as follows:

$$Ln(y_i)=x_i\beta-u_i \qquad i=1,2,.....N \qquad (1)$$

$$Ln(yi)=xi\beta+vi-ui \qquad i=1,2,.....N \qquad (2)$$

where, $Y_i$ is scalar output of the ith farm, $X_i$ is a vector of input quantities and β is a vector of parameters to be estimated, $V_i$ is the disturbance term assumed to be independent and

symmetrically distributed $N(0,\sigma_V^2)$ and it captures the effects of random shocks outside the farmers control (e.g. weather, disease outbreaks, measurements errors, etc.), *Ui* is a non-negative random variable associated with technical inefficiency in production.

**Stochastic frontier production function model with risk of production inputs**

The nature of the stochastic frontier production function model with risk of production inputs was reviewed here (Khan et al., 2010; Tan et al., 2010).

$$Y_{it}=f(X_{it},\alpha)exp(\varepsilon_{it}) \qquad (3)$$

where, $Y_{it}$ is scalar output of the ith farm, $X_{it}$ is a $K\times 1$ vector of inputs quantities and explanatory variables, α is a $K\times 1$ vector of parameters to be estimated, i is the number of observations, t is the number of periods studied and $\varepsilon_{it}$ is compound error term that defined as follows (Khan et al., 2010; Tan et al., 2010).

$$\varepsilon_i=g(X_i;\beta)V_i-h(X_i;\delta)U_i \qquad (4)$$

where g $(Xi; \beta)Vi$ is risk of function and $h\ (Xi; \delta)$ $Ui$ is inefficiency of function. β and δ are vectors of parameters. The above model when the function $f(Xit,\alpha)$ was determined (eg, Cobb-Douglas, Transcendental or Translog) and with regard to the distribution assumptions for Uit (half normal) can be maximized by maximum likelihood. $V_{it}$ is the disturbance term assumed to be independent and symmetrically distributed $N(0,\sigma V_2)$ and it captures the effects of random shocks outside the farmers control. *Uit* is a non-negative random variable associated with technical inefficiency in production. Finally, technical efficiency is obtained from the following equation (Aigner et al., 1977):

$$EF_{it}=exp(-U_{it}) \qquad (5)$$

This index for farm which acts exactly on frontier production function and is absolutely efficient is equal to one. Otherwise, a computational number is obtained between one and zero. Frontier models can test various hypotheses (Khan et al., 2010; Villano et al., 2005).

Villano et al. (2005), according to the model proposed by Kumbhakar (2002), allowed positive or negative effects of inputs on production risk in comparison with the Just and Pope model. They determined error component by the following equation:

$$\varepsilon_i=g(X_i;\beta)[V_i-U_i] \qquad (6)$$

Assuming the equality g (Xi; β) Vi= h (Xi; δ) Ui can be written as:

$$Y_i=f(X_i;\alpha)+g(X_i;\beta)[V_i-U_i] \qquad (7)$$

The above equation is stochastic frontier production with flexible risk characteristics that was used by Batties and Coelli (1997). This state is reviewed to determine average and variance of product for the ith farmer, provided that there are quantities of inputs and technical inefficiency effects $(U_i)$:

$$E(Y_i\,|X_i,U_i) = f(X_i;\alpha) - g(X_i;\beta)U_i \qquad (8)$$

Variance of risk function is defined here as:

$$Var(Y_i|X_i,U_i) = g^2\,(X_i;\beta) \qquad (9)$$

Final product risk for the $j_{th}$ input is defined via the partial derivative of production variance by taking $X_j$ as positive or negative:

$$\partial Var(Y_i|\,X_i\,,U_i)\,/\partial X_{ij}\ \ >0 \text{ or } <0 \qquad (10)$$

Accordingly, technical efficiency for the $i_{th}$ farmer $(TE_i)$ is defined as follows:

$$TE_i=E(Y_i|X_i,U_i)/E(Y_i|X_i,U_i=0)=1-TI_i \qquad (11)$$

Where, $TI_i$ is technical inefficiency and is defined as potential missed product:

$$TI_i=(U_i.g(X_i,\beta)/(E_i|X_i,U_i=0)=(U_i.g(X_i,\beta)/(f(X_i;\alpha) \quad (12)$$

If the parameters of stochastic frontier production function are known and given, then the best criterion to predict $Ui$ is conditional expectation $TE_i$ that is determined for actual quantities of random variable $W_i = V_i - U_i$ (Villano et al., 2005).

A series of tests can be conducted to test the specification of the models. They are tested through imposing restrictions on the model and using the generalized likelihood ratio statistic to determine the significance of the restriction. The generalized likelihood ratio statistic (LR test) is given by Green (1997) as follows:

$$LR= -2(Loglikelihood\ H_0 - Loglikelihood\ H_1) \quad (13)$$

where, $L(H_o)$ and $L(H_1)$ are the values of the log-likelihood function under the null $(H_o)$ and alternative $(H_1)$ hypotheses, respectively. If the given null hypothesis is true, LR has approximately $\chi^2$ – distribution or mixed $\chi^2$ – distribution when the null hypothesis involves $\gamma$=0

**Data envelopment analysis**

Data envelopment analysis compares a set of homogeneous Decision Making Units (DMUs) relatively and assigns an efficiency score to each DMU by finding the distance of each unit with that of its peers on the best practice (frontier). Those units that lie on the frontier are recognized as efficient and those that do not as inefficient. Two basic DEA models are CCR (constant returns to scale) which was introduced by Charnes et al. (1978) and BCC (variable return to scale) which was introduced by Banker et al. (1993). The CCR model uses constant return to scale (CRS) concept to assess relative productive efficiencies of decision making units (DMUs) with multiple inputs and outputs. With the CCR model and assuming $m$ inputs, $s$ outputs and $n$ DMUs respectively, the linear programming problem for DMUk is represented as:

$$\text{Max} \quad h_x = \frac{\sum_{r=1}^{s} U_r Y_{rk}}{\sum_{i=1}^{m} V_i X_{ik}} \quad (14)$$

s.t.

$$h_x = \frac{\sum_{r=1}^{s} U_r X_{rk}}{\sum_{i=1}^{m} V_i X_{ik}} \leq 1$$

$Ur,\ Vi>0,\ r= 1,2,.....,s,\ \ i=1,2,...,m$

where, $hX$ is relative efficiency of the $kth$ DMU, Yrk is the rth outputs of the $kth$ DMU, $Xik$ is the ith inputs of the kth DMU, $Ur$ is the weight of the rth output and $Vi$ is the weight of the ith output.

According to above formulation, the maximum of a ratio of weighted outputs to weighted inputs are the relative efficiency scores of CCR model (Charnes et al., 1978).

The dual problem of CCR model can be written as:

$$\text{Min} \quad h_x = \theta - \varepsilon[\textstyle\sum_{r=1}^{s} S_x^+ + \sum_{i=1}^{m} S_i^-] \quad (15)$$

s.t.

$$\sum \lambda_j Y_{rj} - S_r^+ \geq Y_{rj}$$

$$\lambda_j \geq 0, S_r^+, S_r^- \geq \varepsilon \geq 0 \qquad \forall\ i,r,j$$

r=1,2,……,s, i=1,2,……,m, j=1,2,….,n

where, $\varepsilon$ is a small positive number, $\lambda_j$ is a weight of the $jth$ DMU, $S_r^+$ is a slack variable of the $rth$ output and $S_i^-$ is a slack variable of the ith input. Considering the convexity restriction ($\sum_i^n$ =1n λ j=1), Banker, Charnes and Cooper introduced BCC model and evaluated technical efficiency and scale efficiency of DMUs. The linear programming dual BCC model is represented by:

$$\text{Min} \quad h_x = \theta - \varepsilon[\textstyle\sum_{r=1}^{s} S_x^+ + \sum_{i=1}^{m} S_i^-] \ (16)$$

s.t.

$$\sum_{i=1}^{n} \lambda_j X_{ij} + S_i^- \leq \theta X_{ij}$$

$$\sum_{i=1}^{n} \lambda_j Y_{rj} - S_r^+ \geq Y_{ij}$$

$$\sum_{j=1}^{n} \lambda_j = 1$$

$\lambda_j \geq 0, S_r^+, S_i^- \geq \varepsilon \geq 0, \forall\ i,r,j$ i=1,2,..s, =1,2,…..,m, j=1,2,….,n

## Data and Variables

Statistical population included all active dairy farms of Zabol County; therefore, all active dairy farms of Zabol County were investigated by census method (30 dairy farms). The required data were obtained by interview and a questionnaire for the time span of 2012- 2013.

Variables of the study included production of milk in liters (Y), Forage consumption in kg ($X_1$), concentrate consumption in kg ($X_2$), bran consumption in kg ($X_3$), salt consumption in kg ($X_4$), the use of labor in day-person ($X_5$), and the number of cattle in person ($X_6$). Likewise, the explanatory variables of stochastic frontier analysis include: participation in educational classes ($Z_1$), membership in cooperative ($Z_2$), dairy farmers ($Z_3$), dairy farmer's educational level ($Z_4$), and duration of animal husbandry activity ($Z_5$).

## RESULTS ANS DISCUSSION

Table 1 shows summary of descriptive statistics of the variables used for Zabol County dairy farms. According to this table, average milk production is 1525 liters.

Table 2 shows summary of descriptive statistics of explanatory variable. According to this table, mean age for dairy farmer is 46 years and duration of animal husbandry activity is 19 years.

Table 3 reports the estimated results of production function and the choice of the best functional form. Based on the reported findings, Cobb-Douglas production function was selected and employed as the best production function form the conduct of the present study.

The results of stochastic frontier production function coefficients and factors affecting technical inefficiency were reported in Tables 4 and 5, respectively.

Table 1
*Descriptive Statistics of the Variables Used in the Model*

| Variable | Max | Min | Mean | SD |
|---|---|---|---|---|
| Milk production (liters) | 3840 | 500 | 1525 | 864 |
| Forage consumption (kg) | 2000 | 200 | 1080 | 6691 |
| Concentrate consumption(kg) | 4000 | 700 | 1073 | 1063 |
| Bran consumption (kg) | 2000 | 120 | 715 | 544 |
| Salt consumption (kg) | 80 | 12 | 38 | 23 |
| The use of labor (day-person) | 7 | 1 | 4 | 1 |
| The number of cattle (person) | 16 | 2 | 6 | 3.24 |

Table 2
*Descriptive Statistics of the Explanatory Variables Used in the Model*

| Variable | Max | Min | Mean | SD |
|---|---|---|---|---|
| Participation in educational classes | 1 | 0 | 0.86 | 0.34 |
| Membership in cooperative | 1 | 0 | 0.83 | 0.38 |
| Age of dairy farmers | 61 | 32 | 46 | 8.58 |
| Educational level of dairy farmers | 2 | 0 | 0.53 | 0.62 |
| Duration of animal husbandry activity | 45 | 6 | 19 | 10.36 |

Table 3
*Estimation of Production Functions Results and Choice of the Best Functional Form*

| The estimated model | SS | Significant level and degrees of freedom | F | Critical value | Result |
|---|---|---|---|---|---|
| Cobb-Douglas | 1.7 | $\alpha$ =5%,(6,18) | 1.35 | 2.66 | accepted |
| Transcendental | 1.5 | | | | |
| Cobb-Douglas | 1.7 | $\alpha$ =5%,(15,3) | 6.5 | 8.7 | accepted |
| Translog | 0.93 | | | | |

Table 4
*Coefficients of Stochastic Frontier Production Function Model*

| Independent Variables | Parameter | Coefficient | SD | t-value |
|---|---|---|---|---|
| C | $\beta_0$ | 1.44 | 0.74 | 1.94** |
| $LnX_1$ | $\beta_1$ | 0.13 | 0.13 | 1.01 |
| $LnX_2$ | $\beta_2$ | 0.08 | 0.1 | 0.81 |
| $LnX_3$ | $\beta_3$ | 0.21 | 0.17 | 1.22* |
| $LnX_4$ | $\beta_4$ | 0.09 | 0.16 | 0.57 |
| $LnX_5$ | $\beta_5$ | 0.1 | 0.29 | 0.34 |
| $LnX_6$ | $\beta_6$ | 0.67 | 0.22 | 2.92*** |

*** p<0.01, ** p<0.05, * p<0.1

Table 5
*Factors Affecting on Technical Inefficiency*

| Independent Variables | Parameter | Coefficient | SD | t-value |
|---|---|---|---|---|
| C | $\delta_0$ | 0.17 | 0.39 | 0.45 |
| $Z_1$ | $\delta_1$ | -0.18 | 0.3 | -0.6 |
| $Z_2$ | $\delta_2$ | 1.08 | 0.13 | 7.83*** |
| $Z_3$ | $\delta_3$ | -0.017 | 0.012 | -1.38* |
| $Z_4$ | $\delta_4$ | -0.095 | 0.092 | -1.03 |
| $Z_5$ | $\delta_5$ | 0.088 | 0.008 | 0.91 |
| Sigma-squared | $\sigma_s^2$ | 0.053 | 0.01 | 4.99* |
| Gamma | $\gamma$ | 0.99 | 0.002 | 414.09* |
| Log likelihood | - | 12.11 | - | - |
| LR test | - | 12.31 | - | - |

*** p<0.01, ** p<0.05, * p<0.1

According to Table 4, there is a direct relationship between bran and amount of output. Therefore, the amount of output can be increased considerably by increasing the consumption of this input. Additionally, there is a direct and significant relationship between the number of cattle and the amount of output such that more milk is produced by increasing the number of cattle.

Table 5 reported factors affecting technical inefficiency. Based on these results, there is a direct relationship between membership in cooperative and technical inefficiency, or there is an inverse relationship between technical efficiency and membership in the cooperative. Efficiency of members in cooperative dairy farms has declined because of the annoying laws and regulations in cooperatives. Furthermore, the results show a direct relationship between dairy farmers' age and technical inefficiency. Because young dairy farmers have more education and knowledge and more information as compared to old dairy farmers who use more traditional methods, they are more efficient.

Table 6 presents technical efficiency scores obtained via both SFA and DEA methods. Mean technical efficiency in SFA method was equal to 0.74.

The investigations under CRS assumptions shows that mean technical efficiency is 0.55 and mean efficiency score under VRS assumption is 0.63. Comparison of score efficiency calculated by both SFA and DEA methods shows that mean technical efficiency in SFA method is more than that in the DEA method. Moreover, the mean ef-

Table 6
*SFA and DEA Technical Efficiency Scores*

| | SFA | DEA(CCR) | DEA(BCC) |
|---|---|---|---|
| Max | 0.99 | 1 | 1 |
| Min | 0.43 | 0.21 | 0.49 |
| Mean | 0.74 | 0.55 | 0.63 |

Table 7
*The Estimation Results of Inputs Risk Function*

| Input | C | t-value |
|---|---|---|
| Participation in educational classes | -0.89* | -2.67 |
| membership in cooperative | 0.53* | 1.87 |
| Age of dairy farmers | -0.54 | -0.78 |
| Educational level of dairy farmers | -0.12* | -2.25 |
| Duration of animal husbandry activity | -0.13 | -0.45 |
| The coefficient of determination | $R^2$=0.18 | |

ficiency score under VRS assumption is higher than that under CRS assumptions. It shows that dairy farms are able to convert their inputs into output efficiently, but their lower technical efficiency is due to their disadvantageous size.

The estimation results of inputs risk function are summarized in Table 7.

In order to investigate the influence of the applied inputs on production risk, production risk function was estimated in a linear form. In fact, in order to estimate inputs marginal production risk, remaining logarithm of production function definitive component upon inputs logarithm in the model was regressed by using ordinary least squares method. In Table 7, estimated coefficients show the type of effect inputs have on production risk and $R^2$ represents what percentage of production risk is related to inputs.

Based on the results shown in Table 7, $R^2$ has low value, showing that a low percentage of production risk is related to inputs. Saha (2001) and Villano and Fleming (2006) ported similar results.

## CONCLUSION AND RECOMMENDATIONS

Based on the results of the study, the mean technical efficiency of dairy farms is 0.74 percent as reported by the SFA method, and is 0.55 and 0.63 as reported by the DEA model under CRS and VRS assumptions, respectively. The factors affecting technical efficiency are dairy farmers' age and membership in the cooperative dairy farms. There is a direct relationship between bran, number of dairy farmers, and amount of output. Based on these results, the following suggestions are provided for improving dairy farms in Zabol County:

Given the difference between efficient units and inefficient units, it is recommended that training related to optimal use of inputs should be provided and proper management be performed in an attempt to increase the efficiency of dairy farms.

Based on results, there is a direct relationship between the used bran, number of cattle, and the amount of output and these inputs use in the first area of production function; therefore, it is recommended to use more amount of bran and number of cattle can increase the production.

Since there is an inverse relationship between membership in cooperative and technical efficiency, proper actions need to be performed so as to eliminate dairy farms' problems and government as an affecting factor must increase its protection from cooperatives.

The results showed that young dairy farmers are more efficient than older dairy farmers due to their higher educational level and knowledge of animal husbandry. Consequently, they need to cooperate someway so that dairy farmers could be able to transfer their information and experiences to the older ones.

## ACKNOWLEDGEMENTS

Authors would like to thank Zabol dairy farmers for their cooperation in this study and their taking time completing the questionnaires.

## REFERENCES

1- Aigner, D.J., & Chu, SF. (1968). On estimating the industry production function, *American Economic Review, 58,* 826-839.

2- Aigner, D., Lovell, C., & Schmidt, P. (1977). Formulation estimation of stochastic frontier production function models. *Journal of Econometrics, 6,* 21-37

3- Alvarez, A., & Arias, Co. (2004). Technical efficiency farm size a conditional analysis. *Agricultural Economics, 30,* 241-250.

4- Banker, R., & Gadh, V., & Gorr, W. (1993). A

Monte Carlo comparison of two production frontier estimation methods, Corrected ordinary least squares and data envelopment analysis, *European Journal of Operational Research, 67,* 332-343

5- Battese, G. E., & Coelli, T.G. (1995). A model for technical inefficiency effect in a stochastic frontier production function for panel data. *Journal of Empirical Economics, 20*, 325-332.

6- Battese, G.E., & Coelli, T.J. (1997). Guest editor's introduction to the special issue on papers presented at the New England conference on efficiency and productivity. *Journal of Productivity Analysis, 8*, 323-324.

7- Charnes, A., & Cooper, W. W., & Rhodes, E. (1978). Measuring the efficiency of decision making units. *European Journal of Operational Research, 2,* 429-444.

8- Dourandish, A., Nikoukar, A., Hosseinzadeh, M., & Lavshabi, A. (2013). Estimating multi-output technical efficiency for the dairy farms of the Northern Khorasan Province (Applying stochastic production and stochastic distance frontiers). *Agricultural Economics and Development, 27*(2), 114-122.

9- Erilouzadeh, R., & Saleh, A. (2010). The economic evaluation of dairies, dairy farm case study. *Journal of Agricultural Development Economics, 17*, 1 - 41.

10- Ferrier, G. D., & Lovell, C. A. K. (1990). Measuring cost efficiency in banking: econometric and linear programming evidence, *Journal of Econometric, 46*, 229-245.

11- Fulginiti, L. E., & Perrin, R. K. (1998). Agricultural productivity in developing countries. *Agricultural Economics, 19*, 45-51.

12- Green, W. (1997). Frontier Production Function, *in Handbook of Applied Econometrics (Volum 2: Microeconomics), Hashem Pesaran, M., Schmidt, P., Blackwell Pub. Ltd., USA.*

13- Khan, A., Huda, A.Z., & Alma, A. (2010). Farm household technical efficiency, as study on rice producers in selected Area of Jamalpur district in Bangladesh. *Europen Journal of Social Sciences, 14,* 262-271.

14- Kwon, O. S., & Lee, H. (2004). Productivity improvement in Korean rice farming: parametric and nonparametric analysis, *The Australian Journal of Agricultural and Resource Economics, 48*(2), 323-346.

15- Kumbhakar, S. C. (2002). Specification and estimation of production risk, risk preferences and technical efficiency. *American Journal of Agricultural Economics, 84*, 386-390.

16- Mehrjou, A., Moghadasi, R., Mirsalehpoor, M., & Mehjou, N. (2012). Examine the technical efficiency of the dairy cooperative Kohgiluyeh Boyer Ahmad. *Cooperative, 6,* 21-40.

17- Ministry of Agriculture Jihad. (2006). Statistical annuals. Retrieved from http://maj.ir.

18- Mohammadi, S.J., Hosseini Yekani, S.A., & Ghaderzadeh, H. (2015). Studying the efficiency of industrial dairy farms of Saqqez and Divandarreh Cities: Using super-efficiency approach. *Agricultural Economics and Development, 29*(2), 159-168.

19- Mohammadi, H., & Torkamani, J. (2011). *Economic analysis determination of broiler diets on calf fattening units In Fars Province.* Proceedings of the third Conference of Agricultural Economics, February, 17-19, Mashhad: Iran Agricultural Economic Society.

20- Saha, A. (2001). Risk in Hive traditional rice cultivation: A- enquiry in west Bengal agricultural, Indian. *Journal of Agricultural Economic, 56,* 57-70.

21- Sharma, K. R., & Leung, P., & Zaleski, H. M. (1997). Productive efficiency of the swine industry in Hawaii: stochastic frontier vs. data envelopment analysis. *Journal of Productivity Analysis, 8,* 447-459.

22- Sing, So., Fleming, E., & Coelli, T. (2000). Efficiency productivity analysis of cooperative dairy plants In Haryana Punjab States of India. Working Papers Series in Agricultural Resource Economics Issn 1442 1909, University of New Engl. Armidale. Australia.

23- Tan, S., Herein, N., Kuyvenhoven, A., & Quick, F. (2010). Impact of Land fragmentation on rice producers' technical efficiency in South-East China, *Wageningen Journal of Life Sciences, 57*, 117–123.

24- Statistical Center of Iran. (2011). Statistical Yearbook. Retrieved from http://amar.org.ir

25- United States Departments of Agriculture. (2006). Independence of managerial practices expected financial efficiency. USDA Economic Research Service. Retrieved from http://usda.gov/wps /portal/usdahome Villano, R., Donnel, C., & Battese, G. (2005). *An investigation of production risk, risk preferences technical fficiency: Evidence from rain fed lowland rice farms in the Philippines*. Working paper Series in Agricultural Resource Economics, University of New England.1,12-30.

26- Villano, R., & Fleming, E. (2006). Technical inefficiency and production risk in rice farming: evidence from Central Luzon Philippines. *Asian Economic Journal, 20,* 29-46.
27- Zibaei, M., Soltani, GH. (1995). Different methods for estimating the frontier production function and technical efficiency of milk pro-

# Impact of Peripheral Factors on Reducing the Consequences of the Targeted Subsidies from the Greenhouse Owners' View Point: The Case of Semnan Province, Iran

*Mohammad Sadegh Sabouri [1*], Siavash Noori [1] and Mostafa Karbasioun [2]*

Abstract

The purpose of this study was to investigate the peripheral factors on reducing the consequences of the targeted subsidies from the view point of greenhouse owners of Semnan Province, Iran. The population of the study comprised greenhouse owners of Semnan Province. The sample size was calculated using Cochran formula and encompassed a number of 287 greenhouse owners. The main research tool used in this study was a questionnaire. SPSS v21was used to analyze the data obtained through the questionnaires. Based on hypotheses and theories, questions were designed, and after testing validity and reliability (based on the dependent variable Cronbach's alpha was calculated equal with 87.8) indices, modifications were performed. To examine the hypotheses of this research, the correlation coefficient was estimated, and a multiple regressions analysis was run. The results showed that current expenses, using the proper structures for constructing greenhouse, method of fuel use, the perfect place of heaters, and the use of skilled workers in the greenhouse had a key role in reducing the impact of targeted subsidies plan (dependent variable) on greenhouse owners. These variables explained 40.2 per cent of the variance in the dependent variable.

***Keywords:*** ***greenhouse, peripheral factors, subsidies, Semnan Province***

[1] *Faculty of Agriculture, Department of Agricultural Extension and Education, Garmsar Branch, Islamic Azad University, Garmsar, Iran*

[2] *Faculty of Agriculture, Department of Agricultural Development, Shahrekord University, Shahrekord, Iran*

* *Corresponding author's email: Sabouri5413@yahoo.com*

## INTRODUCTION

A greenhouse is a building in which plants are grown. Nowadays, greenhouse products are considered as one of the most important elements of people's daily consumption. This is more important when we consider the fact that in Iran, farmers are forced to produce the crops in intensive areas due to the shortage of water resources, land quality and drastic climate changes. However, targeted subsidies plan had some harmful effects on greenhouse production in the last couple of years (Jozwik; 1992, Arde & Ghazvi, 2010; Mohammadi et al., 2013).

The Iranian targeted subsidy plan also known as the subsidy reform plan was passed by the Iranian Parliament on January 5, 2010. The goal this plan is to replace subsidies allocated to food and energy (80% of the total) with targeted social assistance, in accordance with Five Year Economic Development Plan and moving towards free market prices in a five-year period. According to the government reports, approximately 100 billion dollars is spent on subsidizing energy prices per year in which 45 billion dollars is paid for the prices of fuel alone and also for many consumable goods including bread, sugar, rice, cooking oil and medicine (Kazemi et al., 2011).

As a consequence of the implementation of targeted subsidies plan, many greenhouse owners have declared to be bankrupt due to increasing fuel prices (Mohammadi, 2013). The main problem of greenhouse owners is not only provision of different material for production such as, fuel, seeds, plastics, fertilizer, pesticide; but also marketing and supplying human resources whose prices have dramatically increased over past years (Dehghani et al., 2007). On the other hand, many researchers have emphasized that the performance of greenhouses in Iran is not desirable at all in comparison with other countries due to the lack of proper technical knowledge of the owners (Mergen, 2011). Such poor performance of these owners of greenhouses has made them more fragile against the shocks produced by changes of energy costs.

Although, most studies have underscored educational needs as the most important factor (Mattson, 2008; Sabouri & Minaee, 2009), it seems that in Iran, identifying peripheral factors is a very crucial issue that needs to be seriously taken into account.

*The overall objective of this study was to investigate the impact of peripheral factors on dependent variable which is the consequences of the targeted subsidies from the view point of greenhouse owners in Semnan Province.*

## MATERIALS AND METHODS

An applied research with causal-relative method was used. The study was carried out in Semnan Province, Iran. Based on Iranian General Census (2011), there are nearly 700 official active greenhouses in this province.

The statistical population comprised all greenhouse owners of Semnan Province, which cultivate at least 1,000 square meters (0.1 hectare) of greenhouses in each period of planting. Based on Cochran formula, the sample group was determined 287 greenhouse owners. In this study, random sampling was used. The main research tool used in this research was questionnaire. Based on hypotheses and theories, questions were designed and after testing validity and reliability (based on the dependent variable Cronbach's alpha was calculated 87.8), modifications were performed. Field study was considered for filling the questionnaires. Descriptive statistics data analyses were executed by version 21 of SPSS software. In addition, correlation and multiple regressions analysis were applied for inferential statistics.

## RESULTS

The research findings showed that 12.5% of greenhouse owner's were below the age of 30 and 1.4% of them were above the age of 60. In terms of age, the largest population of greenhouse owners (92 individuals) was between 31 to 40 years old (Table1).

The level of education of 0.7% greenhouse owners was familiarity with reading and writing, and 24.7% had the B.A. degree and above. In addition, the majority of individuals were high school graduates (34.1%). A number of 98 on average had high school diploma and 29.3%

(84 persons) of greenhouse owners had the technician graduation (Table 1).

Based on the results, 57.1% of the target group declared the amount of their greenhouse lands is below 0.5; whereas, 30.7% declared that the amount of their greenhouse lands is between 0.5 to 1 ha. In addition, 12.2% (the minimum frequency) indicated that their lands are above 1 ha

Descriptive statistics of variables are shown in Table 2. According to that, "using the suitable

Table 1

*The Personal Characteristics of the Target Group: Greenhouse Owners' Age, Level of Education and Amount of Their Greenhouse Lands (n=287)*

| Individual characteristics | Group | Frequency | Percentage |
|---|---|---|---|
| **Age (year)** | Up to 30 years | 36 | 12.5 |
| Min: 24 | 31-40 | 92 | 32.1 |
| Max:65 | 41-50 | 64 | 22.3 |
| SD:8.653 | 51-60 | 91 | 31.7 |
| Mean:43 | more than 60 years | 4 | 1.4 |
| **Sex** | Female | 24 | 8.4 |
| | Male | 263 | 91.6 |
| **Education level:** | Reading & writing level | 2 | 0.7 |
| | Eight class | 32 | 11.1 |
| | High school diploma | 98 | 34.1 |
| | Associated diploma | 84 | 29.3 |
| | Bachelor's Degree or higher | 71 | 24.7 |
| **Amount of lands** | To 0.5 Ha | 164 | 57.1 |
| The lowest:1000.000 | 0.5-1 | 88 | 30.7 |
| The highest:13500.000 | Over 1 Ha | 35 | 12.2 |
| SD:3400.77331 | | | |
| Average:5404.5296 | | 32 | 11.1 |
| **Fixed expenses (IRR[1])** | To 30 Million | 6 | 2.1 |
| Min: 30 million | 31-60 | 249 | 86.8 |
| Max: 85 million | 61-90 | | |
| SD: 15.20383 | | | |
| Mean: 67.8780 | | | |
| **Current expense (IRR)** | 2-4 Million | 34 | 11.8 |
| | 4.1-6 | 67 | 23.3 |
| | 6.1-8 | 35 | 12.2 |
| | Over 8 | 151 | 52.6 |

Table 2

*Descriptive Statistics of Economic Variables*

| Variable | Mean | SD |
|---|---|---|
| Reducing the interest rate of loan received by greenhouses | 4.74 | 0.459 |
| Pricing the products expertly | 4.56 | 0.471 |
| Direct payment of subsidy to the greenhouse owner by government | 4.49 | 0.466 |
| Using the proper structures for constructing greenhouse | 4.33 | 0.471 |
| Use of skilled workers in the greenhouse | 4.54 | 0.500 |
| Using the suitable productive soil | 4.79 | 0.405 |
| Using the suitable seeds | 5.00 | 0.00 |
| Using pesticides and fertilizer with optimum performance | 4.89 | 0.311 |
| The assessment of time of cultivation as reduces the fuel consumption | 4.11 | 0.583 |
| Doubling walls in greenhouse | 2.41 | 1.083 |
| Controlling greenhouse' heat energy | 4.01 | 0.454 |
| The perfect place Heaters | 4.46 | 0.500 |
| Method of Fuel use | 4.56 | 0.497 |

Scale: 1 = very low, 2 = low, 3 = average, 4 = high, 5 = very high.

[1] *IRR: Iranian Currency ($1= 35000 IRR)*

Table 3
*Correlation Coefficient between Variables*

| Variable | Coefficient | p-value |
|---|---|---|
| Amount of greenhouse lands | -0.021 | 0.727 |
| Revenue | -0.079 | 0.184 |
| Fixed expenses | -0.021 | 0.719 |
| Current expenses | 0.180** | 0.002 |
| Using the proper structures for constructing greenhouse | 0.173** | 0.003 |
| Use of skilled workers in the greenhouse | 0.156* | 0.048 |
| Using the suitable and productive soil | -0.002 | 0.977 |
| Using pesticides and fertilizers with optimum quality. | 0.021 | 0.722 |
| The assessment of time of cultivation as reduces the fuel consumption | 0.042 | 0.479 |
| Doubling walls in greenhouse | 0.011 | 0.849 |
| Controlling greenhouse' heat energy | -0.030 | 0.608 |
| The perfect place Heaters | 0.125* | 0.035 |
| Method of Fuel use | 0.162** | 0.006 |

**p<0.01 *p<0.05

Table 4
*The Regression Analysis Summary*

| Model | R | $R^2$ | Adj$R^2$ |
|---|---|---|---|
| Enter | 0.634 | 0.402 | 0.386 |

Table 5
*The Regression Coefficient*

| Variable | B | Beta | SE | t | p-value |
|---|---|---|---|---|---|
| Constant | 12.324 | - | 1.893 | 8.0047 | 0.000 |
| Current expenses | 0.667 | 0.835 | 0.091 | 7. 365 | 0.000 |
| Using the proper structures for constructing greenhouse | 0.463 | 0.717 | 0.070 | 6.609 | 0.000 |
| Method of Fuel use | 0.204 | 0.303 | 0.025 | 4.260 | 0.000 |
| The perfect place Heaters | 0.183 | 0.270 | 0.062 | 2.972 | 0.003 |
| Use of skilled workers in the greenhouse | 0.105 | 0.243 | 0.054 | 3.745 | 0.009 |

seeds" and "using pesticides and fertilizers with optimum quality" were in first and second order, respectively.

In order to study the relationship of research variables, the researcher calculated the Pearson correlation coefficient. As shown by Table 3, there was a positive and meaningful correlation between reducing the consequences of the plan is targeted subsidies with current expenses, using the proper structures for constructing greenhouse, use of skilled workers in the greenhouse, the perfect place heaters and method of fuel use.

To determine the role of variables were used regression Inter method. Variables entered into the model the explanation 40.2% of the variance in the dependent variable (Table 4).

Table 5 shows the finding of the regression analysis of independent variables. According to Table 5, there was a linear regression between reducing the consequences of the plan is targeted subsidies (dependent variable) and current expenses, using the proper structures in constructing greenhouse, method of fuel use, the perfect place heaters and use of skilled workers in the greenhouse.

## DISSCUSION

Production of greenhouse is one of the manifestations at Industrial agriculture, however, a great number of greenhouse owners were bankrupt from the beginning of the target subsidies plan. This survey was done with the purpose of regression analysis of effective factors on reducing the consequences of targeted subsidies from the viewpoint of greenhouse owners of Semnan Province of Iran.

This study showed that there is a positive and meaningful role between reducing expenses with current expenses, using the proper structures in constructing greenhouse, method of fuel use, the perfect place heaters, and use of skilled workers in the greenhouse.

The finding of regression analysis showed reducing of current expenses can play an important role in reducing expenses. Current expenses reduction can reduce the effects of targeted subsidies to the greenhouses. This finding is consistent with the findings of Mattson (2008) and Latimer et al. (2002).

In addition, it was found that that there is a relationship between reducing the consequences of targeted subsidy (reducing expenses) and using the proper structures in constructing greenhouse. Use of a suitable structure plays a role in reducing energy consumption and consequently reduces the impact of targeted subsidies. This finding is consistent with Suvedi et al. (2010) and Ford (2005) results.

Method of fuel use plays role in reducing the consequences of the plan is targeted subsidies. These findings were in accordance with Ford (2005) and Arde and Ghazvi (2010).

The perfect place heaters and the use of skilled workers in the greenhouse can contribute to reducing the consequences of targeted subsidies. Eventually, the results of the research verified studies done by Mattson (2008) and Sabouri and Minaee (2009).

Therefore, the peripheral factors mentioned above can be considered as best methods for reducing the consequences of targeted subsidies. The study findings appear; when the consequences of the plan targeted subsidies will be reduced current expenses, appropriate structures to be designed, the use of skilled labor and heating appliances with appropriate design.

## ACKNOWLEDGEMENT

Authors would like to acknowledge supports offered from the Islamic Azad University of Garmsar Branch. Also, they appreciate the helps of English proofreader and the reviewers of the manuscript which enriched the article and made it publishable.

## REFERENCES

1- Arde M.J., & Ghazvi, M. (2010). *Principles of biological control in greenhouse crops*. University Connecticut. Research Institute of Medical Plant. USA: Central Maryland Research and Education Center.

2- Dehghani Sanij, H., Zarei, G.H., & Heydary, N. (2007). *The first technical workshop of water consumption by planting the greenhouse crops*. Tehran: University of Agriculture Engineering.

3- Ford, C. (2005). Educational priorities of small farmers in West Tennessee.*Journal of Agricultural Education, 36* (1), 31-37.

4- Iranian General Census (2011). *Agricultural statistics*. Tehran: Statistical Center of Iran.

5- Jozwik, F.X. (1992). *The Greenhouse and Nursery Handbook: A Complete Guide to Growing and Selling Ornamental Container Plants*. Mills, WY: Andmar.

6- Kazemi, A. Mehregan, M.R., & Shakouri Ganjavi, H. (2011). Presentation of a linear multi-purpose programming model for allocating optimal energy resources of Iran. *Perspective of Industrial Management, 3*(1), 43-65.

7- Latimer, J., Scoggins, H., Barden, V., & Lambur, M. (2002). *Need's assessment survey of the Virginia greenhouse industry*. Virginia: Polytechnic Institute and State University. Mattson, N. (2008). Assessment survey of the New York state greenhouse industry. Department of horticulture plant science building .Cornell University Ithaca. New York.

8- Mergen, I. (2011). Expensive trials and errors for farmers. *Journal of Green World, 3*(4), 75- 83.

9- Mohammadi, A. (2013). I*nvestigation of problems in process of production in level of Shahrood's greenhouses*. Retrieved from http:// agricul tureenterpreneurship. Persianblog. ir

10- Mohammadi, M., Abbaspourfard, M.H., Goldani, M., & Emadi, B. (2013). The effect of greenhouse area and vegetation percentage on the efficiency of ground to air exchanger in cooling greenhouse. *Journal of Agricultural Machinery, 2*(3), 16-23.

11- Sabouri, M.S, Minaee, A.H. (2009). *Classification of educational needs of greenhouse owner from the perspective of agriculture experts and greenhouse owners* (2thed.). Agriculture education and promotion researches. Islamic Azad University, Garmsar Branch, Iran

12- Suvedi, M., Jeong, E., & Coombs, J. (2010). Education needs of Michigan farmers. *Journal of Extension, 48*(3), 1-11.

# 15

# Analysis of Factors Affecting Canola Plantation Development in Tabriz and Marand Counties, Iran

***Ghader Dashti, Babollah Hayati, Noushin Bakhshy * and Mohammad Ghahremanzadeh***

Abstract

This study identifies and analyzes factors influencing canola plantation development in Tabriz and Marand Counties. The Censored Model was used to analyze cross-sectional data collected from 372 farmers using a questionnaire. Due to the weakness of the Tobit model in separating factors affecting the adoption decision of farmers and factors affecting the rate of adoption, the Heckman Model was employed to separate the contributions made by these factors. The results of estimated Probit model in the first stage of the Heckman Approach showed that machinery ownership had an important effect on canola adoption, as a 1% increase in machinery ownership had led to 0.158% increase in canola adoption probability. Contact with extension agents, farm income proportion, education, and farmers' experience influenced canola plantation probability positively, and the age and number of fragmentations had a negative impact on it. The significance of inverse Mill's ratio indicates that the factors affecting the decision to start planting and the amount of canola plantation are not the same. The Heckman's second step estimation results indicated that the loan amount, canola relative benefit, and family labor had a positive effect, and that machinery cost and farm distance from the road had a negative effect on canola acreage. Relative benefit was the most effective element, as 1% increase in relative benefit results in a 0.342% increase in canola plantation.

***Keywords:***
***Canola Adoption, Censored Model, Heckman two step procedure***

*University of Tabriz, Faculty of Agriculture, Department of Agricultural Economics*
** Corresponding author's email: bakhshyecon@yahoo.com*

## INTRODUCTION

Canola is the third most important source of vegetable oil in the world after soybean and palm. Rapeseed contains about 48% oil and has an appropriate fatty acid composition (Taheri et al., 2010). Per capita consumption of oil and fat is 17.4 kg in Iran and 90% of Iran's population oil need (2610 thousand tones) is imported from Brazil, Canada and Swiss (Mostofi, 2008; Youssefi et al., 2010). Given that domestic production of edible oils is insufficient to meet increasing demands due to population growth, the policy approach to Iran's agriculture has been increasing strategic agricultural products (especially oilseeds) to meet domestic consumption and to reduce dependence on food imports. In this regard, area expansion and yield improvement have been envisaged as the most significant solutions (Abyar, 2002).

Canola contains a high percent of oil. Characteristics and compatibility of canola with different climate conditions have increased its importance as hopes for the supply of edible oil in Iran. Meanwhile, despite the attempt made to increase canola production, East Azerbaijan Province has experienced considerable variation in canola cultivation to the extent that the number of canola cultivars has dropped from 1957 in 2007 to approximately 867 in 2010; it means 55% of canola cultivators during these years were decreased. Tabriz and Marand counties have an important role in whole province canola acreage determination. About one-third of province's canola cultivated area was assigned to these two counties.

Study of trend of canola plantation shows farmers' primary interest in canola plantation and in their turning away from this product. Farmers' attitudes toward canola are not stable; they may plant canola one year and give it up next year. Therefore, it must be investigated why farmers adopt different behaviors toward canola (cultivation continuity, or stop or deciding to start plantation for the first time).

Canola adoption and its plantation development can be a solution for oil import dependence problems. Whereas canola is an almost new crop in Iran, adoption models can be employed for this study. Studies of adoption of new crops by farmers have been often framed within the traditional adoption-diffusion model of innovation. Rogers (1962) developed a model of diffusion, which has become widely established in the marketing literature. The diffusion process consists of four key elements: an innovation, the social system on which the innovation impacts, the communication channels of that social system, and time (Wright & Charlet, 1995). Notwithstanding, this model has some weaknesses, because it ignores individual farmer characteristics, social relations in decision-making and noneconomic goals. The other model is a farm structure model. This model looks at the economic constraints imposed by farm size and generally argues that larger and more economically viable farms are more capable of assuming the risks of new behaviors. Finally, researchers combined two theoretical explanations of farmer adoption behavior, the diffusion-innovation model and farm structure model (Lubell, 2004).

The decision to adopt a new crop or technology has been widely documented throughout the literature. Shapiro et al. (1992) used the Tobit Model to study the adoption of double cropping soybeans and wheat in the Midwest. The results showed that farmers start double cropping in an attempt to boost revenue and lower the risks. As Shapiro shows higher income encourages farmers to adopt new thing, so we can consider income and benefit as factors affecting canola cultivation adoption. Foster and Rosenzweig (1995) studied adoption and profitability of high-yielding seed varieties (HYV) associated with the Green Revolution of rural Indian households. They used a model incorporating learning by doing and learning spillovers. The estimates indicated that imperfect knowledge about the management of the new seeds was a significant barrier to adoption; this barrier diminished as farmers' experience with the new technologies increased. As they stated own experience and neighbors' experience with HYVs significantly increased HYV profitability, therefore experience and participation in extension classes could be an important factor influencing canola cultivation. Munshi (2004) studied technology diffusion in the Indian Green revolution. As they stated, information flows are weaker in a heterogeneous population when the performance of a new technology is sensitive to unobserved individual characteristics, preventing individuals from learning from neighbors' experiences. This characterization of social learning is tested with wheat and rice. The rice-growing regions display greater heterogeneity in growing conditions and the new rice varieties were also sensitive to unobserved farm characteristics. Wheat growers respond strongly to neighbors' experiences, while rice growers do not. Oladele (2005) applied the Tobit Model to study the discontinued adoption of agricultural technology among farmers in southwestern Nigeria. The results indicated that the lack of extension visits and lack of input required were the most important factors on discontinuing adoption of improved maize. Accordingly, it could be deducted that inputs like machinery ownership, credits and so on, can influence canola cultivation. Amao and Awoyemi (2008) studied adoption of improved cassava varieties and its welfare effect on households in Osogbo. The results indicated that age, access to extension agents, crop yield, marital status, labor, production input, and education influenced the adoption positively, and family, size, had a negative effect, and that poverty was higher among the non-adopters. Their results show the importance of farmer characteristics on adoption. Oyekale and Idjesa (2009) studied adoption of improved maize seeds and production efficiency in Rivers state, Nigeria. The findings showed that education, farming experience, mono-cropping, minimum tillage, and use of fertilizer significantly influences adoption. It can be conclude that some characteristics could be very important on adoption and cultivation of new crops like canola. Using Tobit regression, Kheil et al. (2009) studied maize boom in the uplands of northern Vietnam and showed that the framers' area allocation to maize was mainly determined by education, wealth, sex, race, and distance to paved roads. Scandizzo and Savastano (2010) modeled adoption and diffusion decisions of farmers toward genetically modified crops under a real option framework. Both adoption and diffusion depend on internal and external factors, such as information and prices. The results emphasize the role of profitability in expanding new crops. It is an important factor that encourages farmers to plant new crops. A number of studies have been conducted in Iran. Salami and Einallahi Ahmad Abadi (2001) applied Tobit and two-stage Heckman Models to specify factors affecting sugar beet production in Khorasan, Iran. The results showed that the factors influencing the farmers' decision to cultivate sugar beet are not the same as those affecting the area under cultivation. Abyar (2002) conducted a survey of factors influencing soybean area expansion in Golestan province of Iran. Using the Tobit method, the study found out that farm size, farmers experience, access to farm machinery, and the type of irrigation water source were the most effective factors in the soybean area expansion. Shafiei (2007) identified factors affecting olive plantation development in Kerman province, using Logit Model. The results showed that variables including education, children above 14 years old, contact with extension agents, and acreage of garden can affect olive adoption. As these studies indicate, some factors can influence adoption and other can influence cultivated area acreage. Those which can affect the adoption (first stage) are machinery ownership, income proportion, age, education, experience, fragments, and contact with extension agents. These variables are mostly characteristic variables. On the other hand, some variables can affect canola acreage like family labor, machinery cost, relative benefit, credit, and distance. These variables are mostly economic elements that can change farmers' profit.

In Iran, due to the strategic importance of oilseed crops, identification and analysis of factors affecting canola plantation development can help economic specialists in the process of policy and strategy formulation to achieve canola growth. In recent years, special attention has been paid to canola in East Azerbaijan province of Iran. Despite support programs such as guarantees and providing bank credits for canola cultivars, considerable success has not been achieved in canola area development. The results of this study can provide useful information that could help planners and public officials to make decisions and policies more rational and effective. Analysis of factors affecting canola plantation development can provide valuable information for policymakers with insight into how proposed agricultural policies might affect canola producers. The policy planners face the challenge of formulating suitable agricultural policy by which the desired growth rate in canola output may be achieved. But intervention and incentives are not constant across farmers and groups of farmers have different responses to these policies. For this reason, policymakers should examine the source of the change in farmers' behavior in addition to the aggregate change. The evaluation of a policy with censored data reveals that farmers react to interventions and incentives by changing their behavior with a move to / from the limit or by changing their behavior away from the limit. Such reactions are distinct and researchers should isolate the separate effects. Understanding the separate effects provides a more precise perception of the effects of a given policy. Understanding this effect can provide a significant contribution to the assessment of a policy designed to change participation rates among farmers. For instance, if the analysis of a policy demonstrates that the majority of the change in non-adopter behavior was attributable to prior participants rather than an increase in the adoption rate, such analysis could conceivably result in the search for a policy that has a stronger effect on farmers' participation rates.

## MATERIALS AND METHODS

### Methods of data analysis

Generally, there are two main methods to study the behavior of farmers: parametric and nonparametric approaches. In the frame of parametric methods, single equation estimations are widely used for agricultural studies. In econometric perspective, two types of errors could occur when such a model is used for agricultural product's acreage response. The first error is sample selection bias, Sample selection bias means the sample includes only those farmers who plant canola, and those farmers who do not plant canola are eliminated which could be potentially canola cultivators. The second error is to consider equal factors influencing on canola adoption decisions and factors affecting acreage of canola cultivation. While these two groups of factors can be different. Tobin (1958) introduced truncated or censored models to eliminate first error (Salami & Einallahi Ahmad Abadi, 2001). In censored models it is assumed that the dependent variable has been clustered at a limiting value, usually zero. If we use only the observations above the limit value, then we'll face bias estimation. The Tobit Model was designed to deal with biases introduced by censoring. Hence it uses all observations, both those at the limit and those above the limit, the estimations provide more information than commonly realized (McDonald & Moffit, 1982). Although Tobin could eliminate the first error, but the other problem is not yet fixed.

We can think of the canola plantation as a two stage process. First, farmers have to decide whether to adopt canola or not. Then, if they decide to adopt, they have to choose how much to plant. Decisions of whether to adopt and how much to adopt, maybe considered as joint or separate decisions. When the decisions are considered to be joint decisions, the Tobit Model is appropriate for analyzing the factors affecting the decision (Greene, 1990). But when we consider the decision separated, the Heckman procedure is appropriate.

We consider two different models. The Tobit Model estimates the likelihood of adoption and the extent of adoption, which could be expressed in the following way:

$$Y_i^* = \beta' X_i + U_i \qquad \begin{cases} Y_i = \beta' X_i + U_i & \text{if} \quad Y_i^* > 0[1] \\ Y_i = 0 & \text{if} \quad Y_i^* \leq 0 \end{cases} \tag{1}$$

where the $Y_i^*$ is the latent variable, $Y_i$ is observed variable, $\beta$ is a (k x 1) vector of parameters, and $U_i$ is a random error term that is normally distributed with mean zero and variance $\sigma^2$ (Tobin, 1958).

The estimator described in (1) does not alone indicate the effect of a change in X on Y. Following Tobin (1958) it can be shown that:

$$E(y) = x\beta F(z) + \sigma f(z) \tag{2}$$

Where $z = X\beta/\sigma$, $\sigma f(z)$ is the unit normal density and $F(z)$ is the cumulative normal distribution function. Furthermore, the expected value of Y for observations above the limit represented by $Y_i^*$ is given by?:

$$E(Y^*) = x\beta + \sigma f(z)/F(z) \tag{3}$$

Hence, the basic relationship between the expected of all observations, $E(Y)$, the expected value conditional upon being above the limit, $(Y^*)$ and the probability of being above the limit, $F(z)$, is:

$$E(Y) = F(z)E(Y^*) \tag{4}$$

The effect of a change in explanatory variable, X on the dependent variable, Y, can be disaggregated into: (i) the change in Y if it is above zero (adopters) weighted by the probability of being above zero [F(z)]; and (ii) the change in the probability of being above the limit weighted by the expected value of Y if above zero. That is (Gebremedhin et al., 2003):

$$\frac{\partial E(Y)}{\partial X_i} = F(z)\left(\frac{\partial E(Y^*)}{\partial X_i}\right) + E(Y^*)\left(\frac{\partial F(z)}{\partial X_i}\right) \tag{5}$$

Whereas the Tobit Model was designed to deal with estimation bias associated with censoring, the Heckman Model is a response to sample selection bias (Sigleman &Zeng, 1999). The choice to participate or be selected to participate in any program may not necessarily be random. Consequently, selection bias may exist (Kinuthia et al., 2011).

The extent of canola adoption is conditional on first adopting canola, and therefore, there is need to control for the factors that affect adoption before assessing determinants of extent of adoption. Single-equation approaches to these types of problems fail to capture the logical two-step decision process that potential participants undertake (Lohr & Park, 1995; Ngwira et al., 2014). The decision of the farmer will be formulated based on two interrelated choices. The first choice is whether to adopt or not. If the decision to adopt is positive, then the second choice is how many acres will be allocated to canola.

A farmer decision to adopt or not is guided by the perceived utility that will be derived out of engagement in that activity rather than in any another activity (Kinuthia et al., 2011). It is assumed that the farmers adopt new technologies only when the perceived utility or net benefits from using such a technology is significantly greater than that of the existing technology. Although the utility that is derived from the agricultural choices is not directly observable, differences among farmers in the non-observable underlying utility function can be modeled through socio-economic and agronomic variables. The ith farmer will choose to plant canola if the utility of canola cultivation, UiA, is greater than utility of other crop cultivation,UiB, i.e. UiA>UiB. Otherwise, the farmer will not adopt canola plantation (Adesina & Zinnah, 1993; Caviglia & Khan, 2001; Ngwira et al., 2014). The index function used to estimate the adoption of canola can be expressed as (Ibrahim et al., 2012):

$$i=1,2,\ldots,N \qquad [6]\gamma' W_i + V_i Z^*_i = $$
$$\begin{cases} Z_i = 1 & \text{if} \quad Z_i^* > 0 \\ Z_i = 0 & \text{if} \quad Z_i^* \leq 0 \end{cases} \tag{6}$$

where $Z^*_i$ is a latent variable denoting the difference between utility from planting canola UiA and the utility from not planting canola UiB. The farmer will cultivate canola if $Z^*_i = U_{iA} - U_{iB} > 0$. The term $\gamma'$ $W_i$ provides an estimate of the difference in utility from adopting canola (UiA-UiB), using socio-economic and agronomic characteristics, $W_i$, as explanatory variables, while $V_i$ is an error term (Ibrahim et al., 2012).

When a farmer adopts canola, he has self-selected to participate instead of a random assignment. After this stage decides about how many acres that allocates to canola. We state the determinants of the canola cultivated area as a linear function of the vector of explanatory variables ($W_i$) and an adoption dummy variable ($Z_i$). The linear regression can be specified as (Greene, 1990):

$$Y_i = \beta' W_i + \delta Z_i + e_i \quad (7)$$

Where $Y_i$ is the land area devoted to canola, $W_i$ is a vector of independent variables with coefficients $\beta'$, $e_i$ is the error term (normally distributed with zero-mean) and $Z_i$ is a dummy variable; $Z_i=1$ if the farmer plant canola and $Z_i=0$ otherwise.

In estimating equations (6) and (7), it needs to be noted that the relationship between canola adoption and acreage could be interdependent, thus estimating equation (7) using the ordinary least squares (OLS) will lead to biased estimates. To address this problem, a two-step Heckman's procedure was employed to analyze factors affecting canola plantation. The model is appropriate because it addresses simultaneity problems (Kinuthia et al., 2011; Ibrahim et al., 2012; Ngwira et al., 2014).

According to Heckman (1979), for the estimated parameters of equation (7) to be efficient, there should be no correlation between the two error terms ($e_i$ and $V_i$). However, sample selection bias results in a nonzero correlation between the two errors. To correct for this selection bias, the Heckman Model first estimates the first stage (6) to obtain a sample selection indicator called Inverse Mills Ratio (IMR). The IMR measures the covariance between the two errors and is an indicator of whether there is significant sample selection bias or not. The predicted errors and the IMR in the first stage are there entered into the second stage together with the Wi vector of regressors (Ngwira et al., 2014).

The Mackinnon non nest test can be applied for choosing a functional form of second stage. First the next equation must be estimated

$$Log\, y_t = b_0 + \Sigma b_i log X_{it} + \theta v_t + e_t$$

Then the coefficient of $v_t$ must be tested with Wald test. The model is linear if the coefficient was the significant model and logarithmic vice versa.

From the Probit equation the inverse of the Mill's ratio, LAMBDA ($\lambda$), which is the ratio of the ordinate of a standard normal to the tail area of the distribution, can be computed. The IMR reflects the probability that an observation belongs to the selected sample and is obtained as follows (Ben-Houassa, 2011):

$$\lambda_i = \frac{\phi\left(\frac{\gamma' W_i}{\sigma}\right)}{\Phi\left(\frac{\gamma' W_i}{\sigma}\right)} \quad (8)$$

This variable can remove the heteroscedasticity of the model, and as indicated earlier, the IMR parameter estimate is used to test for sample selection bias. The null hypothesis for sample selection bias is that the coefficient of IMR is zero, i.e., the IMR collapses to zero hence there is no sample selection bias. Therefore, when the IMR is significant, the null hypothesis is rejected (i.e. there is a significant sample selection bias). On the other hand, when the IMR is not significant, the null hypothesis of sample selection bias is not rejected implying that a single-equation ordinary least squares (OLS) estimation of the equation (7) would yield efficient estimates (Ngwira et al., 2014).

**Source of data**

The study was carried out in Tabriz and Marand counties, which are situated in the northwest of Iran. Tabriz is the capital of East Azerbaijan province which is situated in an area of about 1780km$^2$. Marand is a city in East Azerbaijan province, northwest of Tabriz. Marand has an area about 3340km$^2$.

East Azerbaijan has experienced considerable volatility in canola cultivation. Tabriz and Marand counties have an important role in whole province canola acreage determination. In 2008 approximately half of province's canola production was devoted to these two counties. Table 1 presents canola statistics in Iran.

Comparing figures 1 and 2 shows the same trend of canola cultivars' behavior in East Azerbaijan Province and Marand and Tabriz counties. Farmers initially adopt canola plantation and then turn away from this product.

The cross-sectional data for the analysis were obtained from primary sources, a combination of a written questionnaire, and an oral face-to-face interview. Structured questionnaires were conducted to obtain data on the socio-economic and agronomic characteristics of the respondents such as age, family size, level of formal education,

Table 1
*Canola Statistics of Iran*

| Year | Cultivated area (hectare) | Growth rate (percent) | Production (ton) | Growth rate (percent) | Yield (kg per hectare) | Growth rate (percent) | Price (IRR) | Growth rate (percent) |
|---|---|---|---|---|---|---|---|---|
| 1999 | 5067 | -37 | 5221 | -21 | 1030 | 26 | 1200 | 14 |
| 2000 | 17852 | 252 | 13507 | 159 | 757 | -26 | 1400 | 17 |
| 2001 | 26387 | 48 | 27591 | 104 | 1046 | 38 | 1710 | 21 |
| 2002 | 49000 | 86 | 69250 | 151 | 1413 | 35 | 2050 | 20 |
| 2003 | 72000 | 47 | 112164 | 62 | 1557 | 11 | 2500 | 22 |
| 2004 | 73521 | 2 | 105929 | -5 | 1441 | -7 | 2830 | 13 |
| 2005 | 127463 | 73 | 210199 | 98 | 1650 | 14 | 3110 | 10 |
| 2006 | 174500 | 37 | 296000 | 41 | 1696 | 28 | 3420 | 10 |
| 2007 | 140825 | -19 | 219000 | -26 | 1555 | -8 | 3700 | 8 |
| 2008 | 120000 | -15 | 150000 | -31 | 1250 | -20 | 3700 | 0 |
| 2009 | 105000 | -12 | 160000 | 7 | 1500 | 20 | 4000 | 8 |
| 2010 | 110000 | 5 | 185000 | 16 | 1650 | 10 | 6200 | 55 |

East Azerbaijan Province Organization of Jihad-e-Agriculture (2010)

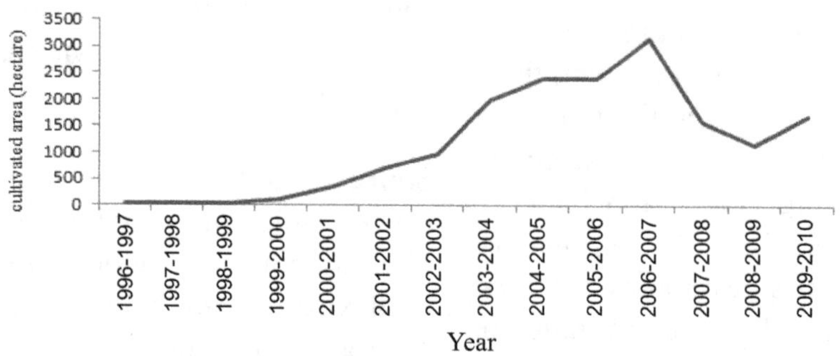


*Figure 1*. Changes in canola cultivated area in east Azerbaijan province
Source: East Azerbaijan Province Organization of Jihad-e-Agriculture (2010)

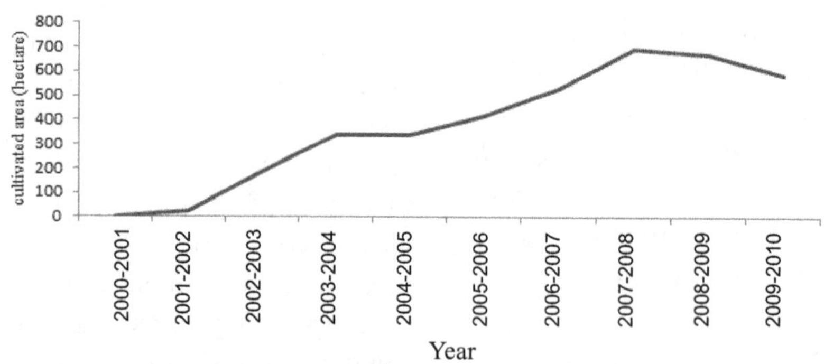


*Figure 2*. Changes in canola cultivated area in Tabriz and Marand counties
Source: East Azerbaijan Province Organization of Agriculture (2010)

canola farming experience, farm size, input quantities, input costs, output incomes, farm distance from the main road and farmland patch number. Cochran formula was applied to determine the sample size, which resulted in n = 372. Then this sample size was devoted to each county proportionally (200 for Marand and 172 for Tabriz) and the socio-economic factors and agronomic characters were gathered in 2010.

## RESULTS AND DISCUSSION

Required data derived from a sample of 372 farmers, of whom255 (69%) planted canola and 117 (31%) planted other seeds. Whereas canola guaranteed price by Iran's government is the same for all farmers, we cannot use it as a variable, because it has insufficient variation to

Table 2
*Results of estimating Tobit Model of Canola Plantation Acreage*

| Variable | Coefficient | t statistic | Elasticity of index | Elasticity of E(y) |
|---|---|---|---|---|
| Education | 0.038*** | 4.399 | 0.159 | 0.252 |
| Proportion | 0.020*** | 7.560 | 0.488 | 0.771 |
| Credit | 0.035*** | 8.341 | 0.207 | 0.327 |
| Real price | 0.005*** | 4.173 | 1.411 | 2.230 |
| Distance | -0.103*** | -2.756 | -0.133 | -0.210 |
| Relative benefit | 0.339*** | 3.072 | 0.249 | 0.394 |
| Fragment | -0.089* | -1.918 | -0.123 | -0.194 |
| Machinery cost | -0.236*** | -3.769 | -0.297 | -0.469 |
| Intercept | -1.718*** | -2.855 | - | - |

log likelihood=-182.12 Number of Obs.=372
square correlation between observed and expected values=0.86
wald chi-square statistic= 288.80 with 8 d.f. p-value= 0.000

***, ** , *= significant at 10% , 5% and 1% levels respectively

enter the model. Therefore, we used another variable that demonstrates the real price received by farmers. For generating such a variable, we calculate the price of canola after reducing moisture drop. Product's moisture content (M.C.) can influence storage and may sometimes lead to spoilage. When the government buys the product, it determines the maximum moisture content level. If product's moisture exceeds that level, a penalty is charged. The amount of the penalty is determined by the amount of moisture content above the acceptable level. The government reduces the product price based on the canola moisture percent and then pays farmers. Now the researches can use real prices that have enough variance to be interpreted. Finally, the Tobit regression was estimated and the results were summarized in Table 2.

Among the variables, real price and farm income proportion had the most impact on canola growing. Whereas a 10% increase in real price and farm income will result in an increase in canola cultivated acreage by 22.3% and 7.71%, respectively, and a 10% increase in real price and farm income will result in a canola adoption probability, by 14.11% and 4.48%, relative benefit of canola producing, compared with other crops, had a positive impact on the cultivated area under canola. It is logical that farmers would devote their limited land resources to that crop whose relative benefit tends to be favorable. The received credit and education had a positive impact and machinery costs per hectare had a negative impact on canola plantation.

Table 3
*Results of Estimating Probit Model*

| Variable | Coefficient | t statistic | Weighted aggregate elasticity | Marginal effect |
|---|---|---|---|---|
| Machinery ownership | 0.772*** | 3.317 | - | 0.158 |
| Farm income proportion | 0.060*** | 5.144 | 0.403 | 0.012 |
| Age | -0.026*** | -2.867 | -0.339 | -0.005 |
| Education | 0.047* | 1.796 | 0.057 | 0.009 |
| Contact with extension | 0.366** | 1.994 | - | 0.075 |
| agents | 0.410*** | 7.853 | 0.298 | 0.084 |
| Experience | -0.558*** | -4.134 | -0.268 | -0.114 |
| Fragment | -0.711 | 1.351 | -0.191 | - |
| Intercept | | | | |

Likelihood ratio =201.763 with 7 d.f. P¬¬-value=0.000percentage of right predictions=0.85
number of right predictions=318 Number of Obs.=372
Estrella r-square= 0.512, Maddala r-square= 0.418, Cragg-Uhler r-square= 0.595, McFadden r-square= 0.447

***, ** , *= significant at 10% , 5% and 1% levels respectively

Table 4
*Results of Estimating OLS*

| Variable | Coefficient | t statistic | Elasticity at means |
|---|---|---|---|
| Inverse of Mill's ratio | -0.407*** | -3.678 | - |
| Credit | 0.049*** | 11.610 | 0.266 |
| Distance | -0.112*** | -2.641 | -0.098 |
| Relative benefit | 0.464*** | 3.635 | 0.342 |
| Machinery cost | -0.185*** | -2.549 | -0.159 |
| Family labour | 0.064* | 1.824 | 0.056 |
| Intercept | 1.149*** | 5.011 | 0.655 |
| D.W=1.78, $\overline{R}^2$=0.57 F=6.49 | P-value=0.01 | Number of Obs.=255 | |

***, ** , *= significant at 10% , 5% and 1% levels respectively

Number of fragments per farm also influenced canola cultivation negatively. Fragmentation of land implies that farmers have to transport inputs to several isolated plots in different locations. Moreover fragmentation can lead to decreasing plot size. Farmland distance from main road had a negative impact on canola plantation.

Although the results of the Tobit Model show the importance of the variables, it cannot separate the factors affecting the adoption decision and the factors affecting the amount of adoption. In this context, the Heckman Model was estimated, and Table 3 and 4 present the first and second step of Heckman Model estimation, respectively.

Probit regression was estimated using maximum likelihood. The model LM2 statistic is 7.8 which show that the null hypothesis of homoscedasticity could not be rejected. Variance inflation factor for all variables were less than 10 which shows that there is not collinearity. As Table 3 shows 1% increase in farm income proportion (farm income relative to total income, including employment earnings, other business activities and…) leads to 0.403% increase in canola adoption probability. Experience of canola production influences canola plantation positively. This experience will help the farmer assess the worthiness of the new crop in meeting the farmer's own objective, better than new entrants. Farmer's formal education had a positive effect on canola. Contact with extension agents, also increases canola adoption, whereas the adoption probability of canola will increase by 0.075%. Machinery ownership is the last factor has a positive impact on the adoption probability of canola by 0.158%. Obtaining machinery at the right time is a big matter for farmers, if not results in untimely harvest of canola and increases risk of canola shattering and production loss. So owning machinery eliminates such a problem. Furthermore, increasing landowner's age decreases canola adoption probability of 0.339%, suggesting that older farmers are likely to adopt canola less than young farmers. Number of fragments per farm had a negative impact on canola adoption.

Results from the first stage of the Heckman method (Probit Model) expresses factors effective in probability of canola adoption, and results from the second stage of this method (OLS) show factors affecting the amount of canola acreage. The inverse Mill's ratio, which is calculated from the first stage enters the OLS estimation as an additional explanatory variable. The results are shown in Table 4. The significance of Mill's ratio shows that factors effective in decision making about canola adoption are not identical with factors determining the amount of land devoted to canola. Accordingly, using Heckman two-stage is appropriate for this study. Among all variables, the relative benefit of canola had the most impact on canola area. A 1% increase in relative benefit results in 0.342% increase on canola plantation. Farm machinery cost per hectare had a negative effect on canola acreage and farm distance from main road is another variable that influences canola plantation negatively. A 1% increase in the distance leads to 0.098% decrease in the canola cultivated area. Family labor numbers had a positive effect on canola plantation. With increasing 1 person in the family labor, the canola acreage will increase 0.064 hectares. Moreover, received credit amount had a positive impact on canola area.

## CONCLUSION

Identification of factors affecting canola plantation development can provide insight into the decision-making process farmers engage in when deciding to adopt and/or retain the rapeseed cultivation. The Tobit and Heckman's two-stage Models were applied to analyze the cross-sectional data collected from a sample of 255 canola planters and 117 other farmers of Tabriz and Marand Counties of Iran.

This study found that the real price is an important factor motivating canola adoption. Accordingly, the government should have to increase certified price for canola and decrease production drop percent determined in purchasing time. Machinery ownership is an effective element on canola plantation acreage. The government can establish some machinery leasing companies rent farm equipment at lower rental rates or can help farmers obtain credits for equipment buyers. Fragmentation is a serious problem in adoption decisions. Overcoming this problem would be possible by organizing agricultural production cooperatives. Hence education and contact with extension agents had a positive impact on canola production; it is recommended that more extension classes be conducted by agents. These agents can increase farmers' knowledge and assist them in applying this information for on-farm use. It is noteworthy that none of the proposed solutions will work alone. Whereas the farmer's behavior is likely to be influenced by different factors, policy makers must consider all factors to design a policy package for canola plantation development.

## ACKNOWLEDGMENT

Authors would like to thank Tabriz and Marand Counties' farmers for their cooperation in this study and their time completing the questionnaires.

## REFERENCES

1- Adesina, A. A., &Zinnah, M. M. (1993). Technology characteristics, farmers' perceptions and adoption decisions: A TobitModel application in Sierra Leone. *Agricultural Economics*, *9*(4), 297-311.

2- Abyar, N.(2002). A study of factors influencing soy bean area expansion in Golestan Province. *Journal of Agricultural Economics and Development*, *10*(38), 67-82.

3- Amao, J. O.,&Awoyemi, T. T. (2008). Adoption of improved cassava varieties and its welfare effect on producing households in Osogbo Adp zone of Osun State. *Gene Conserve, 7*(29), 1-11

4- Ben-Houassa, K. E. (2011). Adoption and levels of demand of fertilizer in cocoa farming in Cote D'ivoire: does risk aversion matters? *Paper selected for presentation at the CSAE conference, economic development in Africa.*

5- Caviglia, J. L., & Khan, J. R. (2001). Diffusion of sustainable agriculture in the Brazilian tropical rain forests: a discrete choice analysis. *Economic Development and Cultural Change, 49*(2), 311-333.

6- East Azerbaijan Province Organization of Jihad-Agriculture. (2010). *Agricultural Statistics Year Book*, Department of Planning and Economics, Tabriz, Iran.

7- Foster, A. D., & Rosenzweig, M. R. (1995). Learning by doing and learning from others: human capital and technical change in agriculture. *Journal of Political Economy, 103*(6), 1176- 1209.

8- Gebremedhin, B., Ahmed, M., & Ehui, S. K. (2003). Determinants of adoption of improved forage technologies in crop-livestock mixed systems: evidence from the highlands of Ethiopia. *Tropical Grasslands, 37*, 262-273.

9- Greene, W.H. (1990). *Econometric analysis* (2nd-Eds). New York: Macmillan publishing company, USA.

10- Heckman, J. J. (1979). Sample selection bias as a spesification error. *Econometrica, 47*(1), 153-161.

11- Ibrahim, M., Florkowski, W. J., & Kolavalli, S. (2012). The determinants of farmer adoption of improved peanut varieties and their impact on farm income: evidence from Northern Ghana. Selected peper prepared for presentation at the agricultural and applied economics association annual meeting, Seattle,WA.

12- Kheil, A., Saint-Macary, C., & Zeller, M. (2009). Maize boom in the uplands of northern Vietnam, economic importanceand environmental implication. In: *International research on food security, natural resource management and rural development conference*, PP. 1-6.

13- Kinuthia, E. K., Owuor, G., Nguyo, W., Kalio, A. M., & Kinambuga, D. (2011). Factors influencing participation and acreage allocation in tree planting program: a case of Nyeri district, Kenya. *Agricultural Science Research Journal,1*(6), 129-133.

14- Lohr, L., & Park, T.A. (1995). Utility-consistent

discrete continuous choices in soil conservation. *Land Economics, 71*(4), 474-490.

15- Lubell, M. (2004). Collaborative watershed management: a view from the grassroots. *Policy Studies Journal, 32*(3), 341-361.

16- Mostofi, S. (2008). Investigation of oilseed and its products. Agricultural Planning, Economic and Rural Development Research Institute Tehran,1-50.

17- Munshi, K. (2004). Social learning in a heterogeneous population: technology diffusion in the Indian Green Revolution. *Journal of Development Economics, 73*(1), 185-213.

18- McDonald, J., & Moffit, R. (1982). The uses of Tobit analysis. *The Review of Economics and Statistics, 62*(2), 318-321.

19- Ngwira, A., Johnsen, F.H., Aune, J. B., Mekuria, M., & Thierfelder, C. (2014). Adoption and extent of conservation agriculture practices among smallholder farmers in Malawi. *Journal of Soil and Water Conservation, 69*(2), 107-119.

20- Oladele, O.I. (2005). A Tobit analysis of propensity to discontinue adoption of agricultural technology among farmers in southwestern Nigeria. *Journal of Central European Agriculture, 6*(3), 249-254.

21- Oyekale, A. S. &Idjesa, E. (2009). Adoption of improved maize seeds and production efficiency in Rivers state, Nigeria. *Academic Journal of Plant Sciences, 2*(1), 44-50.

22- Rogers, E. (1962). *Diffusion of innovation* (5thEds). New York: Free Press.

23- Salami, H., & Einallahi Ahmad Abadi, M. (2001). Application of Tobit econometric model and the two-stage Heckman method in determining factors affecting sugar beet production in Khorasan province. *Iranian Agricultural Sciences, 32*(2), 433-445.

24- Scandizzo, P. L., &Savastano, S. (2010). The adoption and diffusion of GM crops in United States: a real option approach. *AgBio Forum, 13*(2), 142-157.

25- Shafiei, L. (2007). Studying effective factors on olive cultivation in Kerman. *Journal of Agricultural Economics and Development, 15*(58), 1-22.

26- Shapiro, B.I., Brorsen, B.W. & Doster, D.H. (1992). Adoption of double-cropping soybeans and wheat. *Southern Journal of Agricultural Economics, 24*(02), 33-40.

27- Sigleman, L. & Zeng, L. (1999). Analysing censored and sample selected data with tobit and Heckit models. *Political Analysis, 8*(2), 167-82.

28- Taheri, F., Mousavi, S.N., & Rezayie, M.R. (2010). The impact of removing energy subsidies on rapeseed production costs in Marvdasht County. *Journal of Agricultural Economics Research, 2*(3), 77-89.

29- Tobin, J. (1958). Estimation of relationships for limited dependent variables. *Econometrica, 26*(1), 24-36.

30- Wright, M., & Charlert, D. (1995). New product diffusion models in marketing: an assessment of two approaches. *Marketing Bulletin, 6*(4), 32-41.

31- Youssefi, A., Nshauian, A., & Azizi, M. (2010). Yield differences of brassica oilseed rape cultivars based on physiological parameters. *American- Eurasian Journal of Agricultural and Environmental Sciences, 9*(4), 436-439.

# 16

# Rice Farmers' Behaviors and Attitudes Toward Agricultural Extension Programs in Healthy Food Production (Case of Lahijan and Amlash Counties, Northern Iran)

*Fatemeh Askary Bozayeh[1*], Fathollah Keshavarz Shal [2], Zahra Mojib Haghghadam [3] and Parisa Shahinrokhsar [4]*

Abstract

This study aimed to explain the behavior and attitudes of farmers in the adoption of Integrated Pest Management (IPM) technologies in rice fields. The main instrument of research was a questionnaire whose validity was confirmed by a panel of professors and executive experts, and its reliability was confirmed by apreliminary test. The sample employed in the research study comprised farmers who were participating in extension programs of integrated pest management held in Lahijan and Amlash during 2012-2014. The findings showed that the predominant activity of farmers was the use of combined methods (chemical and non-chemical), leaving crop residue on the field, and using biological control, with the rank average of 9.27, 9.25, and 9.12 from the total average of 6.5, respectively. In addition to, private extension cooperatives, rice supervisors, agricultural extension staff, educators, and TV programs with the average of 11.14, 10.87, 10.80, and 10.71 of the total average of 7.49 had the highest role in informing the farmers. Based on such findings, some suggestions were also given, such as concentration on special aspects of integrated pest management, attention to the education of rural women, and employing young and educated members of farmers' families as arms of agriculture extension in transferring innovation of integrated pest management to families.

***Keywords:***
***Amlash, healthy food production, Integrated Pest Management (IPM), Lahijan***

[1] Ph.D. Student of Geography and Rural Planning at Kharazmi University, Master of Research in Guilan Agricultural and Natural Resources Research and Education Center

[2] Ph.D.of Human Geography, Director of Planning and Development Department of Agriculture in Gilan Agricultural and atural Resources Research and Education Center

[3] Ph.D. Studentof Entomology at Guilan University, Master of Research in Guilan Agricultural and Natural Resources Research and Education Center

[4] Ph.D. Student of Irrigation and Drainage at Ferdosi University, Scientific Member of Guilan Agricultural and Natural Resources Research and Education Center

* Corresponding author's email: fa868_bozayeh@yahoo.com

## INTRODUCTION

One of the greatest challenges of human society that has been consideredas a new global movement is the problem of food security of the increasing population. The need for food supply, in spite of the two crises of population growth and limitation of primary resources, has fueled the movement toward the use of industry and technology in the process of production. Agriculture, since 1950s, accompanied by the green revolution, decades of entering of technology including chemical input, high-yielding varieties, new methods of watering, new machineries, and so on. Only two decades after the arrival of technology, indiscriminate and inappropriate application of chemical inputs has exposed the production process in agriculture to serious problems and crisis whose consequence has beenmuch worse than that of the lack of access to technology. Due to the dangers and crisis of chemical input application in agriculture sector that has created different problems in addition to production quality and health, a movement has been initiated toward sustainable and organic agriculture in recent decades, especially since the 1980s, and tools, activities, and researches by agriculture specialtiesare directed towardthis process (Office of Promotion Affairs and the Improvement of Extensional Systems, 2007).

Agricultural protection technology, which includes a wide range of activities such as integrated pest management, conservative tillage, biologic control, and integrated nutrition of the soil, is a fundamental basis of sustainable agriculture. Among these technologies, integrated pest management has a special place in activities and agriculture planning due to its high effectiveness on health of agriculture, ecosystem, as well asproduction of healthier products (Veisi et al., 2010).

Each year, despite using toxin and chemical materials in agriculture, not only pests and diseases have not become less, but also the level of pollution has increased, and the process of production has faced a serious problem. In the third millennium, the most important challenges for governments are food security and energy that achieving it was not possible,if the current situationin the production system continues, so, reform the management methods of production are essential (Sharifi Moghaddam & Ezatian, 2008).

Integrated pest management that was proposed in response to this need is a system which consists of various parameters and its promotion and facilitation needs considering regionally specific social, economic, and ecological features. In fact, integrated pest management is based on regional feature and ecological conditions as well aslocal socio-economicfeatures (Sharifi et al., 2009). Introduction and facilitation of technologies is one main duty of promotion sector, as it is responsible for the development and growth of agriculture and has gained importance due to the recent needs for knowledge and information about the adoption of sustainable agriculture technology (Veisi et al., 2009).

According to researches, although many attempts have been made for the education of villagers in recent decades, their knowledge and awareness is not sufficient about the integrated pest management, suggesting the inefficiency of current activities (IPM education and extension). The success of integrated pest management practices depends on farmers' motivations, skills, knowledge, as well as the participation of local groups and societies (Ettehadi et al., 2011). Effective promotion of integrated pest management needs identification of the main parameters of the integrated pest management for policy-making and practical planning with participation of agriculture (Noori et al., 2011).

Accordingly, the Office of Extension Management and Improvement of Farming System in Guilan Province has held educational and extensional courses about healthy production in the counties of Lahijan and Amlash with an emphasis on rice cultivation, In this program, farmers were trained on how to produce a healthy product. In this research, the degree of effectiveness of these extension activities is analyzed from the perspectives of educated farmers (Table 1).

Table 1
*Summary of Related Literature*

| Title | Year | Authors | Findings |
|---|---|---|---|
| Identify requirements for using integrated pest management in horticultural crops. | 2013 | Moradi and Omidi Najafabadi | The result of factor analysis divided necessaries into five factors of educational-extensional, economic planning, policy-making, technical supervision and planning that totally explains 66% of the total variance of variables. |
| Affective factors in accepting gardeners of Shoushtar-county in the field of integrated pest management. | 2012 | Shoushtari and Amani | A positive, significant relationship was found between level of education, product income, total owned land, amount of water and dry land, mechanization level, social position and degree of accessing communicative channel with level of acceptance of gardeners at the level of 0.01%. |
| Considering effective factors on the degree of knowledge, skill and level of tendency of participating gardeners of the Dalaho county in integrated pest management programm | 2012 | Ghorbani Piralidehi et al. | The result of stepwise regression showed that participation in classes, job record, degree of production, interpersonal reliance and social position are effective on knowledge of integrated pest management of gardeners. |
| Considering extensional and educational factors affecting on applying IPM of the products by farmers of Karaj county. | 2012 | Faiz Arbabi et al. | The result of multiple regressions denotes that variables participating in educational-extensional classes of IPM of the product, effect of educational extensional methods onincreasingknowledge of farmers, contact with agriculture promotion, effect of educational extensional methods forincreasing farmers' skills and area under cultivation has explained the positive function at applying IPM and 67% of changes of dependent variables. |
| Affective factors in accepting IPM of wheat pest by relying on approach of farmer fieldschool. | 2011 | Noori et al. | Two variables of extensive contacts and communicative channels denote changes of variance of the dependent variable to the degree of 37%(degree of acceptance of IPM). |
| Effectiveness of training of IPM from the perspectives of cotton farmers of Garmsarcounty | 2009 | Ghane et al. | There is a significant relationship between personality variables and the variables of knowledge, attitude and skill and variables of participation in educational classes, features of the educational period with variables of attitude and skill. |
| Effect of information resources and communicative channels in accepting IPM of rice | 2011 | Hejazi and Sharifi | Calculating correlation coefficient denotes the existence of a positive relationship between the variables of the area under cultivation, education, being an excellent farmer; agree with reducing herbicides and pesticides and the possibility of controlling pests of rice without using herbicides by benefiting resources and information channels. |
| Limitations at accepting IPM, case study 240 rice farmers participating in 14 IPM educational plans in Jamo region | 2012 | Kumari | Lack of knowledge and skill among farmers, complicated nature of the technology of IPM, inaccessibility to inputs and tools for implementing the plan of IPM, the small size of farm, lack of information about recent strategies of IPM, promotion, services, involvement of experts of IPM andparticipation of rural society such as the main limitation for accepting IPM among farmers is the limitations for the adoption of this technology. |
| Accepting IPM among farmers, case study of Delta of Nigeria | 2008 | Sing et al. | There was a significant, negative and positive relationship between the size of the farm and the degree of adoption of products of rice and cotton. |
| Accepting technology of IPM, case study of potato farmers in the region of Karachi, Ecuador 2005 | 2005 | Mauceri et al. | Hours of training and educational publication are effective mechanism for transferring the technology of IPM to farmers and its acceptance. Hours of training and educational publication, although are somewhat lower cost,but they have considerable effect on this field. |

## MATERIALS AND METHODS

The present study was a survey conducted by a descriptive method. The study tool was a questionnaire composed of open and closed questions in which the main questions were prepared on the basis of the research paradigm. The statistical society was composed of farmers participating in educational-extension training courses of integrated pest management in Lahijan and Amlash. Theyhave been under the education of IPM in the form of farmer field school (FFS) since 2012-2014. Given the limitedsize of the statistical society, almost all educated people (140 people, including 90 people in Lahijan and 50 people in Amlash) were studied.

Data collected by questionnaire were entered into Statistical Package for the Social Sciences (SPSS) after being codedand processed. They were descriptively studied with a central index and were analyzed with a correlation test and stepwise regression analysis.

## RESULTS

From among 140 respondents, 119 were male and 21 were female (i.e.,85 percent of beneficiaries were male and 15 percent were female). Since rural women play a significant role in cultivating rice and in investment in education of rural women as an important part of the rural population, and given the fact that cost-benefit analyses of world bank shows that investment in education of women has the highest degree of return among the investments (Moridsadat et al., 2010), it is necessary to pay attention to the education of women in this field.

Amongthe participants, 0.7 percent were younger than 25, 9.3% were between 26-35, 12.9% were between 36-45, and 77.1% were older than 45 years old. The level of education of respondents shows that 23.6% were illiterate, 37.9% hadlimitedliteracy of reading and writing, 29.3% haddiploma, and 7.9% hadliteracy of higher than diploma (in fact it can be said that 62.9% of beneficiaries had not benefited from the advantages of literacy). The main job of 127 respondents was farming among whom 37 people had secondary jobs. The respondents' job experience showed that the minimum and maximum limits ofjob experience were2 and 70 years, respectively. In addition, the average time of occupation was 32.7 years. Furthermore, 3.6 percent had work experience of less than 10 years, 13.6% between 10-19 years, 15.7% between 20-29 years, 23.6% between 30-39 years, 42.1% over 40 years of work experience, and 57.9 percent had more than 30 years of farming experience. This suggests that adolescents have lower tendency toward farming. 27.1% of respondents had personal-leased lands, 67.9% had personal lands and 5% had leased land. The minimum and maximum limits of the cultivation area of the studied population were 0.1 and 4 hectares, respectively. Average cultivation area is 1.14 hectares, showing a low cultivation area.

Table 2

*Prioritizing the Use of Ecological and Non-Ecological Features at Rice Cultivation Using the Friedman Test (n=140)*

| Variable | Friedman test | |
|---|---|---|
| | **Mean rank** | **Priority** |
| Using integrated methods | 9.27 | 1 |
| Releasing plant residues | 9.25 | 2 |
| Using biologic control | 9.12 | 3 |
| Using aresistance variety | 7.96 | 4 |
| Using manure in farms | 7.66 | 5 |
| Using crop rotation | 7.17 | 6 |
| Using animals and birds | 5.35 | 7 |
| Second cultivation | 5.10 | 8 |
| Pure use of chemical toxin | 5.08 | 9 |
| Using green fertilizer | 4.86 | 10 |
| Using legume cultivation | 4.11 | 11 |
| Burning plant residues in farm area | 3.07 | 12 |
| $\chi^2$=628.342 | Sig=0.000 | df=11 |

For the use of ecological features about rice cultivation, 12 items were designed and farmers were asked to indicatethe degree of their use of available optionsby rating them on a Likert type scale. In Table 2, it is obvious that using the integrated method, releasing plant residues at thefarm area, and using biological control with little differences are the options that farmers use in rice production, while all the other items are given lower priorities (Table 2).

## Attitude of farmerstowardproducing healthy products

To discover farmers' attitudes toward producing healthy products of rice,nine statements were-designed, and farmers were asked to show their viewpoint about each item by rating them on aLikert type scale. According to Table 3, it is obvious that the role of government in encouraging and guaranteed purchase of products, the negative effect of indiscriminate consumption of fertilizer and chemical pesticides on water, soil and foods, the effect of farmers' poverty and weak financial power on incorrect applicationof inputs, and the preference of mechanical operations for weed control were placed in the first to fourth priorities, with the means of 6.68,

Table 3
*Prioritization of Farmers' Attitudes towards Producing Healthy Products Using Friedman Test*

| Variable | Friedman test | |
|---|---|---|
| | Mean rank | Priority |
| Government can have a more effective role in proceeding plans of IPM by encouraging farmers and guaranteed purchase of rice produced through this method | 6.68 | 1 |
| Indiscriminate consumption of fertilizer and chemical toxins pollutes the atmosphere and foods | 6.63 | 2 |
| Farmers' poverty and financial weakness are effective in the incorrect application of agriculture inputs | 6.38 | 3 |
| Control weeds by mechanical operation like better weeding by chemical herbicides | 6.16 | 4 |
| Species rotation causes fertility of soil | 4.93 | 5 |
| The main reason of farmers' inattention for sustainable agriculture and production of healthy product is the lack of correct awareness | 4.47 | 6 |
| Farms should cultivate so as to sustain their ability for long-term production even if it | 4.28 | 7 |
| means less production and profit | 3.45 | 8 |
| Consecutive cultivation of a species of rice in a land causes a pest outbreak | 1.03 | 9 |
| Cultivation of legume decreases the degree of consuming chemical fertilizers | Sig=0.000 | df=8 |
| $\chi^2$=523.263 | | |

Table 4
*Prioritization of the Effect of Educational-Extension Methods of Producing Healthy Products from the Farmers' Viewpoint Using Friedman Test*

| Variable | Friedman test | |
|---|---|---|
| | Mean rank | Priority |
| Group training courses | 7.24 | 1 |
| Visiting of educational farms | 6.40 | 2 |
| Implementing of farmer's, field school projects | 6.32 | 3 |
| Establishing demonstration farms | 6.01 | 4 |
| Personal contact | 5.66 | 5 |
| Distributing extension publications | 5.19 | 6 |
| Using TV programm | 5.12 | 7 |
| Displaying educational films | 5.06 | 8 |
| Using radio programm | 4.56 | 9 |
| Holding an exhibition of agricultural products | 3.44 | 10 |
| *X=296.84* | Sig=0.000 | df=9 |

Table 4
*Prioritization of the Degree of the Studied Farmers' Use of Information Resources Using Friedman Test*

| Variable | Friedman test | |
|---|---|---|
| | Mean rank | Priority |
| Private Cooperatives of extension | 11.14 | 1 |
| Rice supervisors | 10.87 | 2 |
| Extension workers | 10.80 | 3 |
| TV | 10.71 | 4 |
| Pioneering farmers | 8.09 | 5 |
| Extension publications | 7.33 | 6 |
| Radio | 6.85 | 7 |
| Agriculture input shops | 6.82 | 8 |
| Educational film | 6.62 | 9 |
| Internet | 5.61 | 10 |
| Educational and research centers | 5.36 | 11 |
| Newspaper | 4.97 | 12 |
| Book | 4.95 | 13 |
| Journal | 4.85 | 14 |
| *X=788.393* | Sig=0.000 | df=13 |

6.63, 6.38, and 6.16, respectively. The other items were placed in lower priorities (Table 3).

With respect to farmers'viewpoints about the effect of educational-extensional methods on producing healthy products of rice, 10 items were designed, and farmers were asked to express their viewpoints about each item by using a Likert type scale. Regarding Table 4, it is obvious that group training courses, visiting of educational farms, implementing of farmer's, field school projects, andestablishing demonstration farms with a mean of 7.24, 6.40, 6.32, and 6.01 are methods that studied farmers have emphasized more than the planners, and implementers of educational-extension plans should pay due attention to them (Table 4).

**Prioritization of the degree of farmers' use of information resources about producing healthy products**

With respect to the degree of farmers' use of information resources about producing healthy products, 14 items were designed, and farmers were asked to show their viewpoints about each item by using a Likert type scale. According to Table 5, it is obvious that private cooperatives of extension, supervisors of rice, extension workers, and TV programs have made the greatest contributions to informing farmers with a mean of 11.14, 10.87, 10.80, and 10.71, respectively. The other items were placed in lower ranks (Table 5).

**The relationship between respondents' personal characteristics and the effectiveness of extensional activities in producing healthy products**

Pearson and Spearman correlation coefficients were used to test the relationship between personal features and degree of effectiveness of extensional activities at producing healthy prod-

Table 6
*The Relationship between Personal Features and Degree of Effectiveness of Extensional Activities in Producing Healthy Products of Rice*

| Independent variable | Correlation coefficient | r | p-value |
|---|---|---|---|
| Gender | Pearson | -0.069 ns | 0.417 |
| Age | Pearson | -0.176* | 0.038 |
| Level of education | Spearman | 0.222** | 0.008 |
| Years of occupations | Pearson | -0.084ns | 0.321 |
| Main or side job | Pearson | -0.057ns | 0.504 |
| Farm ownership | Pearson | -09.194* | 0.021 |

*p<0.05 **p<0.01 ns: Non-significant

Table 7
*The Relationship between Economic Features and The Effectiveness of Extensional Activities In Producing Healthy Rice Products*

| Independent variable | Correlation coefficient | r | p-value |
|---|---|---|---|
| Area under cultivation | Spearman | 0.406** | 0.000 |
| Amount of production | Spearman | 0.428** | 0.000 |
| Using agricultureloan | Pearson | -0.142ns | 0.095 |

**p<0˙01   ns: Non-significant

ucts of rice, and as it is observed in Table 6, the effectiveness of extension activities is significantly related to the age, educational level, and farm ownership at the level of 0.05, 0.01, and 0.05% level of significance, whereas it did not highly correlate with gender, years of occupations, and the job type. In fact, as farmers grew older, the effectiveness of extension activities would bereduced. Higher educational level was related to higher effectiveness, and leasing farm was accompanied with lower effectiveness (Table 6).

### The relationship between economic features and effectiveness of extensional activitiesin producing healthy products

The relationship between economic activities and the effectiveness of extensional activities in producing healthy rice product was tested by Pearson and Spearman correlation coefficients. As can be seen in Table 7, there is a direct significant relationship between cultivation area and the level of production, as well as the effectiveness of extensional activities of producing healthy products of rice at the level of 0.01%, whereas no significant relationship was found between the use of agricultural loans and the effectiveness of the extension activities (Table 7).

### The relationship between farmers' knowledge and attitude and the effectiveness of extensional activitiesin producing healthy products

A correlation coefficient of Kendall Tau-b and Spearman were used to test the relationship between farmers' knowledge and attitude and the

Table 8
*The Relationship between Farmers' Knowledge and Attitude about Index of Producing Healthy Products and Degree of Effectiveness*

| | Independent variable | Dependent variable | Correlation coefficient | r-value | p-value |
|---|---|---|---|---|---|
| Attitude | Cultivation of legume reduces the degree of using fertilizer | Effectiveness of extension activities | Kendal tau b | 0.192** | 0.002 |
| | Consecutive cultivation of a species of rice in a piece of land over some years causesa pest outbreak | | | | |
| | Weeds are better to be controlled by mechanical operations than by chemical herbicides | | | | |
| | Species rotation of rice cause fertility of soil | | | | |
| | Indiscriminate consumption of fertilizer and chemical toxins pollutes the atmosphere and foods | | | | |
| | Financial weakness of farmers is effective in the incorrect application of inputs | | | | |
| | Farms should be cultivated so as to save their long-term ability of production | | Spearman | 0.265** | 0.002 |
| | Government can play more effective role in proceeding plans of IPM by encouraging farmers to produce healthy products and guaranteed purchase of rice produced through this method | | | | |
| | The main reason for the inattention of application of sustainable agriculture and production of healthy product is the lack of correct awareness in this field | | | | |
| Knowledge | Familiarity with sustainable agriculture | | Kendal tau b | 0.230** | 0.001 |
| | | | Spearman | 0.276** | 0.001 |

**p<0.01

Table 9
*Summary of Multivariable Regression*

| Step | R | $R_{Sq.}$ | $R_{Ad}$ | Std.Error of |
|---|---|---|---|---|
| 1 | 0.249 | 0.062 | 0.055 | 5.289 |
| 2 | 0.309 | 0.096 | 0.083 | 5.212 |

Table 10
*Standard and Non-Standard Coefficients of The Effect of Information Resources on Applying IPM*

| | Variable | B | Standard error | Beta | t | p-value |
|---|---|---|---|---|---|---|
| $X_1$ | Agriculture promoter and teachers | 0.759 | 0.297 | 0.212 | 2.557 | 0.012 |
| $X_2$ | Radio | 0.887 | 0.393 | 0.187 | 2.254 | 0.026 |

effectiveness of extensional activitiesin producing healthy rice product. Table 8 shows a direct and positive significant relationship between farmers' knowledge and attitude, as well asthe effectiveness of extension activities. In other words, as farmers have a more positive attitude toward producing healthy products; it affects the effectiveness of extensional activities to the same extent. In addition, as they obtain more and more information about sustainable agriculture,the effectiveness of extensional activities of producing healthy productsis affected to the same degree (Table 8).

**Considering the effect of information resources on applying integrated pest management**

Stepwise multivariable regression was used to predict the changes of dependent variable through independent variables. The goal of regression analysis is to consider the effect of each independent variable of research on the dependent variable (application of training of integrated pest management) and predicting the changes of the dependent variable, and determining the share of each independent variable in explaining the variance of the dependent variable. The results show that only 0.083% of the dependent variable was accounted for by thevariables listed in Table 9 through two steps into the regression.

In addition, given the existing coefficient in Table 10, standard equation of the regression line in the second step is as below:

$Y=27.50+0.75(X_1)+0.88(X_2)$

According to the result of Tables 9 and 10, it can be said that the effect of information resources on applying education of integrated pest management (0.083) was insignificant. It can be explained bytwo reasons: either these resources-lacked sufficient information for presenting them to the farmers, or the farmers could not use these resources due to various reasons. It seems that the first reason (lack of presenting sufficient information) is more tenable.

**CONCLUSION AND RECOMMENDATION**

The findings of this research, which are inconsistent with those of many other studies conducted across these counties or other counties, suggest that integrated pest management is a combination of different actions that overlap and are in close relationship. It undoubtedly requires planning and implementing educational and promotional trainings at extended levels for beneficiaries, along with supplying economicandsocio-cultural facilities. This finding is similar to the result reported by Pretty (1997), Mauceri et al. (2005), Ghane et al. (2009), Kumari (2011), Moradi and Omidi Najafabadi (2013), (see Table 1), according to which it is necessary to change and improve the procedures of extensional activities in the promotion of sustainable agricultural technologies such as integrated management.

Overall, the results, given the overall goal and objectives of the study, were obtained from the following sources:

Interviewing with farmers and collectioninformation through a questionnaire containing open and closed questions polling their opinions about their attitude and behavior, the details of which are given in the corresponding tables. Aging of over 45 years and an illiteracy coefficient of 65% of respondents were a limiting factor in information retrieveal that was reduced by the accuracy of interviewers to much extent.

Field observation and informal interview with authorities (e.g., private educators, experts of Jihad Agriculturein Lahijan and Amlash) is the second source of data for this research. It shows that this innovation, like any another new idea, needs capacity building, providing facilities, continuation and improvement of training courses, and, above all, using of material and moral incentives, which is in agreement with Fuglie and Kascak (2001), emphasizing the roles of policy support, regional factors, as well asincome levels on the acceptance of innovation. In addition improving knowledge of farmers through extensional trainings and developed social communications are in agreement with De Harreraand Sain (1999), under scoring the roles of communication and information channels inthe adoption of integrated pest management.

Findings of research about implementing pest integrated management among rice farms indicated that among the identified actions, farmers mostly apply 'combined methods', 'leaving crop residue in the field' and 'using biological control', therefore, as a necessity, it is suggested that these items should be emphasizedin educational-promotional plans about integrated management. Moreover, considering the effectiveness of promotional educations and degree of achievement, and behavior changing of farmers, it is suggested that department of extension affairs, before holding educational trainings, should do a primary assessment of the society to assess the level of difference of learning of learners after education and apply it in future planning. Another important point is senility and the erosion of man power in the farmers' community that calls for injecting the spirit of happiness and creating change in their viewpoints and behavior. It is not possible by current condition and requires economic and social infrastructure. Therefore, it is suggested to choose an educated member from the farmer's family andto train him/her to act as arms for promoting the agriculture in the field of pest-integrated management in the family and todistribute this idea as innovators at the level of society.

## ACKNOWLEDMENT

Funding for this research wasprovided by the Guilan Agriculture Organization (Jihad Agriculture of Guilan), which is highly appreciated.

## REFERENCES

1- De Harrera, A. P. & Sain, G. (1999). *Adoption of maize conservation tillage in Azuero, Panama. Economics*. Working Paper, CIMMYT. 99-01

2- Ettehadi, M., Rousta, K., & Mohammadgholinia,J. (2011). Factors influencing knowledge, attitudes and skills of participants in farmer field school approach towards the development of integrated pest management technology. *Journal of Agricultural Extension and Education, 4*(4), 13-28.

3- Faiz Arbabi, S., Mirdamadi, S.M., & Omidin Najafabadi, M. (2012). Considering extensional and educational factors effecting on applying IPM of the products by farmers of Karaj County. *Journal of Agricultural Extension and Education, 5*(1), 1-14.

4- Fuglie, O. K., & Kascak, C. A. (2001). Adoption and Diffusion of Natural Conserving Agricultural Technology. *Review of Agricultural Economics, 23*(2), 386-403.

5- Ghane, F., Namdar, R. & Chizari, M. (2009). Effectiveness of training of IPM from the perspectives of cotton farmers of Garmsar County. *Journal of Agricultural Training Management Research, 3*(8), 59-67.

6- Ghorbani Piralidehi, F., Zarafshani, K., & Jalilian, F. (2012). Considering effective factors on the degree of knowledge, skill and level of tendency of participating gardeners of Dalaho County in plan of integrated pest management. *Journal of Environmental Science, 9*(3), 41-58.

7- Hejazi, S.Y., & Sharifi, M. (2011). The effect of information resource and communicative channel in accepting IPM of rice, case study of Droudzan County in Fars province. *Journal of Cultivation, 24*(4), 48-56.

8- Kabir. M.H., & Rainis, R. (2013). Sustainable Development Strategies and Challenges for Promotion of Integrated Pest Management (IPM) Program in Bangladesh Agriculture. *American-Eurasian Journal of Agricultural and Environmental Sciences, 13*(7), 988-995.

9- Kumari, G. (2012). Constraints in Adoption of Integrated Pest Management (IPM) Practices by rice growing farmers of Jammu Division. *Indian Research Journal of Extension Education,* Special Issue, 2, 15-17.

10- Mauceri, M., Alwang, J., Norton,G., & Barrera,V. (2005). *Adoption of Integrated Pest Management Technologies: A Case Study of Potato Farmers in Carchi, Ecuador*. Selected Paper prepared for presentation at the American Agricultural Economics Association Annual Meeting. Providence, Rhode Island, July 24-27.

11- Moradi, P., & Omidi Najafabadi, M. (2013). Identifing the required factors in applying integrated pest management in the horticulture crops in the view of experts, Case study of Karaj County. *Journal of Agricultural Extension and Education, 6*(2),73-84.

12- MoridSadat, P., Rezvanfar, A., Shabanali Fami,H., & Pezeshki Rad, G. (2010). Considering proper content of organizing promotion for rural women of Khozestan province. *Journal of Social-psychological Studies of Women, 8*(1), 31-55.

13- Noori, S., Lashkarara, F., & Shojaei, M. (2011). Factors affecting adoption of integrated pest management in wheat production with an emphasis on farmer field school approach in Kermanshah province. *Journal of Agricultural Extension and Education, 4*(3), 15-29.

14- Office of Promotion Affairs and Improvement of Extensional Systems (2007). *Instruction of IPM/FFs of operational approach of producing healthy and sustainable agriculture products (concepts, principles and administrative process at operation level)*. Ministry of Agriculture.

15- Ofuoka, A.U., Egho, E.O., & Enujeke, E.C. (2009). Integrated Pest Management (IPM) Adoption among Farmers in Central Agro-Ecological Zone of Delta State, Nigeria. *Advances in Biological Research, 3* (1-2), 29-33.

16- Pretty, J. N. (1997). *Regenerating agriculture: policies and practice for sustainability and self-reliance*, Joseph Henry press, Washington,D.C.

17- Sharifi Moghaddam, M., & Ezatian, R. (2008). *Content principles of the model project of comprehensive management of agriculture health (healthy farmer, healthy agriculture in healthy village)*. Joint projects of ministry of agriculture and health & medical education.

18- Sharifi, M., Sharifzadeh, A., & Mahboubi, M. (2009). Identify requirements for using integrated pest management in horticultural crops (county of Karaj). *Journal of Agricultural Sciences and Natural Resources, 16*(2), 1-12.

19- Shoushtari, M., & Amani, A. (2012). Effective factors on gardeners accepting of Shoushtar ounty in the field of integrated pest management. *Journal of Agricultural Extension and Education, 5*(2), 45-55.

20- Sing, A., Vasisht, A.K., Kumar, R., & Das, D.K. (2008). Adoption of integrated pest management practices in Paddy and Cotton: A Case Studyin Haryana and Punjab. *Agricultural Economics Research Review, 21*(2), 221-226.

21- Veisi, H., Mahdavi Damghani, A., Liaghati, H., & Sabahi, H. (2009). Analysis of the reasons of non-acceptance of technologies of IPM among farmers of Guilan and Mazandaran Province. *Journal of Environmental Science, 7*(1), 45-56.

22- Veisi, H., Mahmoodi, H., & Sharifi Moghaddam, M. (2010). Explaining behavior of farmers at accepting technologies of IPM. *Iranian Journal of Agricultural Economics and Development, 41* (4),481-490.

# 17

# Empirical Assessment of Environmental and Health Risks in Intensive Poultry Production in Kwara State, Nigeria

*Amolegbe Khadijat Busola* *, Usman Mutiat Abimbola and Animashaun Jubril Olayinka*

*Keywords:*
*adoption, environment, health, poultry, risk*

Abstract

The study was designed to assess the environmental and health risks arising from intensive poultry production in Kwara State, Nigeria. A total of 120 poultry farmers in Kwara State under the Poultry Association of Nigeria (PAN) were selected randomly for the study, and structured questionnaires were administered. The data collected were analyzed using descriptive statistics, chi-square test, and the linear regression. The study revealed that about 79.5% of the poultry farmers were literate, having completed at least a tertiary educational level. In addition, the results showed that the environmental and health issues were significant during production and processing. It was further revealed that the type of battery cage used and educational level of the farmers were positively significant, while the year of establishment was negatively significant when it comes to the adoption practices. This implies that the higher the level of education, the more likely farmers adopt improved practices. Moreover, it was observed that the majority of the poultry farmers do not abide by the environmental and public health laws and regulations; therefore, it is recommended that government should enforce strict supervisory agencies of the sanitation/health policies aiming at environmental preservation and protection.

*Department of Agricultural Economics and Farm Management P.M.B 1515, University of Ilorin, Ilorin, Nigeria*
** Corresponding author's email: busolatinwol@gmail.com*

## INTRODUCTION

According to the Animal Production and Health Division, Food and Agriculture Organization of the United Nations, FAO (2007), recently, the poultry industry in the world has made tremendous changes to meet the increasing demand for inexpensive and safe supply of meat and egg. This increasing demand has been accompanied by structural changes within the sector, which is characterized by the emergence and growth in commercial and industrial farming establishments as well asthe intensification of poultry operations (FAO, 2007; Shashank, 2013). Poultry production in Nigeria consists of local and exotic breeds, and these local or backyard breeds of poultry constitutes about 84% of total poultry production. Notwithstanding, the exotic or foreign breeds contribute about 14% to the total poultry production in Nigeria (Adene & Oguntade, 2006). Obioha (1996) observed that of all the poultry species, the fowls (chicken) are the most common and important in the tropics. These are reared under the free-range systems and the intensive or semi-intensive systems. Intensive poultry management systems are usually found in urban areas, where there are markets for egg and meat, and this system recommends standard practices such as disease control methods, housing, and feeding depending on breed of choice. (Abubakar et al., 2007; Tadelle, 1996; Alabi et al., 2014).

The poultry sector is very important; it supplies food and raw materials, generates employment, and serves as a basis for research works. Poultry meat and eggs are a good source of vitamins and minerals. They supply rich protein and are palatable and generally acceptable. Poultry birds mature earlier than most breeds of livestock (cattle, swine, goats, etc.) and can bring economic returns within about 10–12 weeks, which is relatively short compared to other livestock. Poultry production systems are influenced by some factors. These factors are: housing (Natukunda et al., 2011), feed source (Byarugaba et al., 2002), health and disease (Simainga et al., 2011), and environment. Intensive poultry farming is a highly efficient system which saves land, feed, labor and other resources, and increases production. In this system, the poultry farm environment is very well controlled by the farmer. Therefore, production takes place all year round and does not depend on the seasons. According to the World Watch Institute (2006), about 74% of the global total poultry meat and 68% of total poultry eggs worldwide are produced from intensive poultry farming system.

As with any production process, the production of poultry products results in wastes such as the manure from bird excrement, hatchery wastes, litter from bedding materials such as sawdust, wood shavings, and so on, offal, various kinds of feed, medications and pesticide packages, cleaning materials, used ventilation filters,as well ason-farm mortalities (FAO, 2008; Akanni & Benson, 2014). If recycled and managed properly, most of these byproducts can provide valuable organic and inorganic nutrients. Yet, they also give rise to potential environmental and human health concerns and act as vectors for insects and vermin, and pathogenic microorganisms (FAO, 2008; Akanni et al., 2014; Hossen, Hoque & Nahar, 2015). Specific concerns that are well documented in different studies include degradation of surface water and/or nearby groundwater, owing to increasing nutrients such as nitrogen and phosphorus (and potassium in some locations). Also of concern is the issue of air quality affected by dust particles, hydrogen sulfide, ammonia and other volatile organic substances discharged from poultry production facilities. Due to the global concern aboutclimate change and the health effects, the effect of greenhouse gas emissions is now a major concern (FAO, 2008; Akanni et al., 2014).

According to Alabi et al. (2014), a study by the Environmental Protection Agency in 2007 reported that chicken droppings usually contaminate the litter spread in poultry houses and poses great environmental threats during the process of disposing the litter. This is because improper disposal leads to air pollution from unpleasant odors, breeding of flies, and water pollution. To manage the nuisance of odor created by the poultry industry, studies (Environmental Protection Agency, 2007) suggest that poultry farms are to be located at least 500 meters away from human settlements. Accordingly, pollution associated with poultry production has been of great concern, and many authorities and several studies have suggested a thorough pollution control measure by states, where non-compliant poultry farmers are fined or even jailed (Alabi et al., 2014; Anosike, 2007).

The planning, construction, and operation of the poultry installation of any size should consider the issues involved in storing, managing, and using waste byproducts. This is because of the effects of these wastes on the human health and the environment. On a global scale, much research (FAO, 2008; Hossen et al., 2015) has been conducted on intensive poultry management systems and its environmental impact; and ways to reduce, manage, and use these poultry wastes. Such as the use as manure, animal feed components and for fuel energy. However, many poultry farmers do not comply with the environmental laws and public health laws. The majority of these poultry farmers do not recycle their by-products, and because of not doing this, some discharge into the river, bury, and burn the poultry wastes, thus causing environmental and health hazards to the people and community around. Therefore, this study seeks to answer the following questions: What are the types of environmental concerns arising from intensive poultry production? What are the patterns of public health issues arising from intensive poultry production? and, what are the determinants of adoption of the improved practices by the farm for mitigating the environmental and health risks?

The main objective of this study was to assess the environmental and health risks in intensive poultry production in Kwara State, Nigeria. The specific objectives were to: identify the type of environmental concerns arising from intensive poultry production; assess the patterns of public health issues arising from intensive poultry production; and examine the determinants of adoption of the improved practices by the farm for mitigating the environmental and health risks.

**Hypothesis:**

**$H_0$**: There is no significant difference between the expected frequencies and observed frequencies in each category.

**$H_1$**: There is a significant difference between expected frequencies and observed frequencies in each category.

Decision rule: Reject $H_0$ if significant value is less than $\alpha=0.05$.

## MATERIALS AND METHODS

### Study area

The study area was Kwara State. Kwara State

*Figure 1.* Map Highlighted Section: Kwara State, Nigeria
https://guardian.ng/news/group-cautions-can-against-divisive-politics-in-kwara/

was created on 27th May 1967. Kwara State is located in the North-central part of Nigeria. It lies in the middle belt of Nigeria. The State is situated between latitudes 8° and 10° North and longitude 3° and 6°. It covers an area of about 32,500sq/km and has River Niger as its natural boundary along its Northern and Eastern margins. It shares a common boundary with Niger State in the North, Kogi State in the East, Oyo, Ekiti and Osun States in the South, and an international boundary with the Republic of Benin in the West (Fig. 1). The climate of the state is characterized by almost equal wet and dry seasons (i.e., each lasting 6 months). The total annual rainfall ranges from 800mm to 1,200mm in the Northwest and 1000mm to 1500mm in the southeast part of the state. The common rivers are Oshin, Awon, Asa, and Moro. The state has a mean temperature of 30° C to 35°C. Kwara State consists of 16 Local Government Areas. They are: Asa, Baruten, Edu, Ekiti, Ifelodun, Ilorin East, Ilorin South, Ilorin West, Irepodun, Isin, Kaiama, Moro, Offa, Oke Ero, Oyun, and Patigi.

**Sampling techniques, methods of data collection and analysis**

A total of 120 poultry farmers in the state registered under the Poultry Association of Nigeria were selected randomly for the study. The Local Governments Area in which the poultry farms are located is: Asa, Ilorin East, Ilorin South, Ilorin West, Moro, Ekiti, Ifelodun, Isin, Offa, Oke-ero and Oyun. Primary data were obtained with the use of structured questionnaires from the sample farms. Descriptive analysis, chi-square, and linear regression were employed for the study. Descriptive statistics such as frequency distribution, percentage, tables, mode, and mean were used to identify the type of environmental concerns arising from intensive poultry production and assess the patterns of health risks arising from intensive poultry production.

**Chi-square test**

The chi-square test was used to determine whether there is a significant difference between the expected frequencies and observed frequencies in one or more categories.

**Linear regression**

The linear regression was used to model the relationship between the dependent and independent variables by fitting a linear equation to the observed data. The regression was used to examine the determinants of adoption of the improved practices by the farms in mitigating the environmental and health hazards.

$$Y = \alpha+\beta_0+\beta_1X_1+\beta_2X_2+\beta_3X_3+\beta_4X_4+\beta_5X_5+\beta_6X_6+\beta_7X_7+\beta_8X_8+\varepsilon$$

$\alpha$ = Intercept
$\beta_1$ - $\beta 8$ = Parameters (Coefficients)
Y = Sum of adoption practices
$X_1$= Farm size
$X_2$ = Number of flocks
$X_3$ = Total cost of production
$X_4$ = Educational level
$X_5$ = Year of establishment
$X_6$ = Type of poultry
$X_7$ = Farm produce
$X_8$ = Type of battery cage
$\varepsilon$ = Error term

## RESULTS AND DISCUSSIONS

**Socio-economic features of the poultry farmers**

Table 1 shows the socio-economic features of the poultry farmers; these includethe year of establishment, educational level, farm size, number of flocks, type of poultry and type of intensive system etc. of the poultry farmers.

From Table 1, it can be observed that the majority of the farms wereestablished in 2011-2015. About 0.9% of the farmers had primary education, 16.1% had secondary education, and 79.5% had tertiary education, while 3.6% had no formal education. This indicates a high level of literacy among the poultry farmers. Importance of education in managing businesses more productively with a tendency of adopting new technology and innovation cannot be over-emphasized. Education leads to acquisition of new skills and efficient allocation of limited resources (Awosanya, 2002). Most of the poultry farmers do not fully utilize their farm land. Most of them only make use of a fraction of their farm land, while the other part is left unused or used for another purpose. About 50% of the farmers

Table 1
*Socio Economic Features of the Poultry Farmers*

| Items | Frequency (%) |
|---|---|
| **Year of establishment** | |
| 1994-1999 | 8(6.7) |
| 2000-2005 | 27(22.9) |
| 2006-2010 | 31(26.2) |
| 2011-2015 | 54(44.1) |
| **Educational level** | |
| Primary | 1(0.9) |
| Secondary | 22(16.1) |
| Tertiary | 93(79.5) |
| **No formal education** | 4(3.6) |
| Farm size | |
| 0.06 hectares | 5(0.96) |
| 0.25 hectares | 8(4.8) |
| 0.5% | 58(50) |
| 1 hectare | 39(34.6) |
| 2 hectares | 7(6.73) |
| 3 hectares | 3(2.88) |
| **Number of flocks** | |
| < 2000 | 46(38.4) |
| >2000 | 74(61.5) |
| **Types of flock** | |
| Layers | 46(38.3) |
| Broilers | 3(2.5) |
| Cockerels and broilers | 5(4.2) |
| Cockerels and layers | 3(2.5) |
| Broilers and layers | 18(15.0) |
| Cockerels, broilers and layers | 45(37.5) |
| **Types of poultry** | |
| Chicken | 85(73.0) |
| Chicken and Turkey | 29(25.2) |
| Chicken and Guinea fowl | 3(0.9) |
| Turkey and Guinea fowl | 3(0.9) |
| **Types of produce** | |
| Eggs only | 32(26.7) |
| Meat only | 7(5.8) |
| Eggs and meat only | 23(19.2) |
| Eggs and point of lay | 2(1.7) |
| Eggs and day old chicks | 10(8.3) |
| Meat and day old chicks | 4(3.3) |
| Egg production, meat production and point of lay | 24(20) |
| Egg production, meat production and day old chicks | 15(12.5) |
| Egg production, meat production, day old chicks and point of lay | 3(2.5) |
| **Types of intensive system** | |
| Deep litter system | 19(15.0) |
| Battery cage system | 60(48.6) |
| Deep litter and cage system | 41(36.4) |
| **Type of cage** | |
| California design | 2(2.2) |
| Multiple deck system | 77(84.6) |
| Single deck system | 11(12.1) |
| Multiple and single deck system | 1(1.1) |

operate on a farm size of 0.5 hectares, while 34.6% operate on farm size of 1 hectare. The modal flock size of the farmers is 2000 birds. 38.4% of the farmers have a flock size less than 2000, while 61.5% of the farmers have a flock size of 2000 and above. The mean flock size is 2199 birds. About 38.3% and 2.5% of the farmers has a flock type of layers and broilers, respectively. 4.2% have cockerels and broilers, while 2.5% had cockerels and layers. 15.0% had broilers

Table 2

*Perceived environmental concerns associated with poultry during production*

| Environmental concerns | High Frequency (%) | Medium Frequency (%) | Low Frequency (%) | Negligible Frequency (%) | Chi-square (df) | p-value |
|---|---|---|---|---|---|---|
| Fecal waste | 53 (44.2) | 47(39.2) | 14(11.7) | 6(5.0) | 55.000(3) | 0.001 |
| Water waste | | 6(5.0) | 69(58.0) | 44(37.0) | 50.739(2) | 0.001 |
| Feed waste | | 3(2.5) | 65(54.6) | 51(42.9) | 53.311(2) | 0.001 |
| Carcass waste | 8 (6.7) | 31(26.1) | 70(58.8) | 10(8.4) | 83.521(3) | 0.001 |
| Cracked/stale egg waste | | 8(6.9) | 88(75.9) | 20(17.2) | 96.276(2) | 0.001 |
| Dust | 14 (12.5) | 28(25.0) | 44(39.3) | 25(22.3) | 46.125(4) | 0.001 |

and layers, while 37.5% had all the three types of birds. A high level of both broilers and layers are being raised by the poultry farmers.

Most of the poultry farmers are into egg production, less is involved in point of lay and day old chicks. About 26.7% of the farmers produce eggs only, 5.8% produces meat only, 19.2% produces both eggs, and meat, 1.7% produces eggs and point of lay. In addition, about 8.3% of the farmers produce eggs and day old chicks, 3.3% produces meat and day old chicks, 20% produces eggs, meat, and point of lay, 12.5% produces eggs, meat and day old chicks, while 2.5% produces eggs, meat, day old chicks, and point of lay. The majority of the farmer in the study area adopted the use of the battery cage system. About 15.0% use deep litter system and 48.6% indicated battery cage system only, while 36.4% stated both deep litter and battery cage systems. The predominance of the battery cage system in the study area maybe due to the fact that it is easy to manage and also reduce the number of egg cracks. About 84.6% use the multiple deck design as the type of the battery cage system for their poultry production.

**Environmental concerns arising from poultry during production**

These are wastes generated on the farms during poultry production. These wastes include fecal, water, feed, carcass, disposal of cracked egg, and so on.

From Table 2, a high level of fecal waste (44.2%) during production was observed. About 39.2% were rated medium, 11.7% were rated low, and 5.0% were rated as being a negligible environmental problem during poultry production of the various farms. Therefore, mismanagement of the fecal waste can lead to the growth of worms, offensive smell, and contamination of surface water and ground water through runoff into the water body, when not properly disposed. About 58.0% and 37.0% of the water waste has low and negligible environmental problem during poultry production as stated by the farmers. This indicates that there is a low level of water waste during production and this doesn't pose much threat to the poultry farmers. Feed waste has about 54.6% as low and 42.9% as a negligible environmental problem during poultry production. The carcass has 6.7% as high, 26.1% as medium,

Table 3

*Perceived Environmental Concerns Arising from Poultry During Processing*

| Environmental concerns | High Frequency (%) | Medium Frequency (%) | Low Frequency (%) | Negligible Frequency (%) | Chi-square (df) | p-value |
|---|---|---|---|---|---|---|
| Feed waste | 3(3.3) | 10(10.9) | 48(52.2) | 31(33.7) | 54.696(3) | 0.0009 |
| Carcass waste | 6(6.5) | 27(29.3) | 49(53.3) | 10(10.9) | 50.000(3) | 0.0005 |
| Feathers and organs of slaughtered birds | 40(43.0) | 29(31.2) | 10(10.8) | 14(15.1) | 24.720(3) | 0.0078 |
| Water waste | 1(1.1) | 11(11.8) | 44(47.3) | 37(39.8) | 54.398(3) | 0.0006 |
| Cracked/stale egg waste | | 3(3.3) | 60(66.7) | 27(30.0) | 54.600(2) | 0.0004 |
| Dust | 3(3.6) | 4(4.8) | 33(39.3) | 44(52.4) | 61.238(3) | 0.0006 |

Table 4
*Recycling and Non-recycling Methods for the Environmental Problems Generated on the Farm*

| Recycling methods | Not used at all Frequency (%) | Used occasionally Frequency (%) | Used all time Frequency (%) | Non recycling methods | Not used at all Frequency (%) | Used occasionally Frequency (%) | Used all time Frequency (%) |
|---|---|---|---|---|---|---|---|
| Conversion into biofuel | 92(98.9) | 1(1.1) | | Sell | 16(14.4) | 81(73.0) | 14(12.6) |
| Composting as manure | 32(33.3) | 58(60.4) | 6(6.3) | Discharge into water | 84(83.2) | 17(16.8) | |
| Process as animal feed component | 71(77.2) | 19(20.7) | 2(2.2) | Bury | 27(24.1) | 58(51.8) | 27(24.1) |
| Use as fertilizer | 31(33.0) | 36(38.3) | 27(28.7) | Burn | 31(27.7) | 52(46.4) | 29(25.9) |
| | | | | Dump | 1(4.0) | 12(48.0) | 12(48.0) |

58.8% as low, and 8.4% as being a negligible environmental problem during poultry production. Accordingly, both feed and carcass waste has a high percentage of environmental concerns and hence were rated low during poultry production, which indicates that they can be easily managed and disposed properly.

About 75.9% of cracked/stale egg waste was rated low and 17.2% as being negligible environmental concerns during poultry production. Dust has 12.5% as high, 25.0% as medium, 39.3% as low, and 22.3% as being a negligible environmental problem during poultry production. This indicates that cracked/stale egg waste and dust during poultry production can be easily maintained. Also from table 2, it can be gleaned that the p-value for fecal waste, water, waste, feed waste, carcass waste, cracked/stale egg waste, and dust during production are all equal to 0.001, which is less than the α-level (α=0.05); accordingly, the null hypothesis can be rejected. Therefore, the environmental problems during poultry production are all significant at the 0.05 (5%) level of significance. This indicates that the variations in farmers that faces high level, medium level, low level, and negligible fecal waste and other wastes during poultry production are not the same. Most farmers that rated these wastes as being a low environmental problem does not mean the wastes are not having an effect on the environment. For example, a wet fecal waste if not managed properly gives rise to the growth of worms and maggots and also ammonia gas from the wet waste, which can lead to a chronic respiratory problem.

**Environmental concerns arising from poultry during processing**

These are the wastes generated on the farms during poultry processing. These wastes include feed, carcass, feather, and organ of slaughtered birds, cracked/stale egg etc.

From Table 3, it can be observed that the majority of the farmers rated their feed waste as low with 52.2% and 33.7% as being negligible environmental problems during poultry processing. About 53.3% of the farmers rated carcass waste as low and 10.9% of them as being a negligible environmental problem during poultry processing. The majority of the farmers stated that feathers and organs of slaughtered birds have 43.0% higher and 31.2% medium environmental problems during poultry processing. In addition, about 47.3% and 39.8% of the farmers stated that water waste has low and negligible environmental problems during poultry processing. About 66.7% and 39.3% indicated cracked/stale egg waste has low and negligible environmental problem during processing. Finally, more than 52.4% stated dust as being negligible during processing.

All of these figures indicated that environmental problems (wastes) of the poultry farms during processing are either low or negligible, which means that they can be easily maintained and managed properly. Also, from table 3, the p-value for fecal waste, carcass waste, feather and organs of slaughtered birds, water, waste, cracked or stale egg waste and dust during poultry processing are 0.0009, 0.0005, 0.0078, 0.0006, 0.0004, 0.0006, respectively which are all less than the α-level (α=0.05). Therefore, the null hypothesiscan be rejected. Therefore, the environmental concerns during processing are all significant at the 0.05 (5%). Most farmers that rated these wastes as being a low environmental problem does not mean the wastes do not have any effect on the environment.

**Recycling and non-recycling methods (managements) of the environmental concerns generated on the farm**

These are the management of the poultry waste which can include conversion into biofuel, composting as manure, and process inanimal feed component. The non-recycling management strategies include selling, burning, burying, dumping, and discharging into water.

Most of the poultry farms have devised a series of management strategies to be able toget rid of their poultry wastes so as to mitigate their effects on the environment. From Table 4, it can be observed that the majority of the farmers do not practice the recycling method of disposing their poultry waste; instead, they practice the non-recycling method by dumping, burning, swelling, and even burying the poultry wastes. It was discovered that 51.8% of the farmers bury their waste. This may, however, lead to groundwater contamination and thus constitutes a source of risk to human life according to Carr, 1994. About 46.4% burns their poultry waste, which could cause atmospheric pollution that might pose some danger to human and animals' lives. Also about 48.0% dump the wastes generated on their farms. This may, however, be washed off into nearby pits, streams, and rivers, and thus cause a damaging effect to both the human and aquatic lives.

Only about 60.4% of the farmers' compost comprised manure, which could also lead to an environmental issue when the manure is applied to the land in excess of the receiving crop threshold level and the ability to utilize the nutrients as justified by Charles, 2008.

**Health problems associated with poultry during production**

These are the health hazards that do occur during poultry production. They include accidental hazards, for example, sprains, strains, burns, etc., biological hazards which include antibiotic resistance, bacterial infection, etc.

From Table 5, it can be observed that about 50.8% of the farmers rated biological health hazard during poultry production as low, while 34.2% said it is at a medium level. About 45.8% of them stated chemical health hazard as medium and 45.0% as a low health hazard.

About 43.2% and 47.5% of the farmers rated respiratory hazard as medium and low health hazards, respectively during poultry production. About 48.3% of them stated that physical hazard is low during poultry production, while 35.8%

Table 5
*Perceived Health Hazards Associated with Poultry during Production*

| Environmental hazards | High Frequency (%) | Medium Frequency (%) | Low Frequency (%) | Negligible Frequency (%) | Chi-square (df) | p-value |
|---|---|---|---|---|---|---|
| Biological (diseases transmitted from birds to humans, antibiotic resistant, bacteria etc.) | 1(0.8) | 41(34.2) | 61(50.8) | 17(14.2) | 68.102(3) | 0.0005 |
| Chemical (respiratory problems resulting from dust, skin and eye diseases due to exposure to gases) | 1(0.8) | 55(45.8) | 54(45.0) | 10(8.3) | 79.492(3) | 0.0003 |
| Respiratory problem (chronic respiratory diseases and phlegm) | 2(1.7) | 51(43.2) | 56(47.5) | 9(7.6) | 79.356(3) | 0.0001 |
| Physical (noise, exposure to heat and cold, musculo-skeletal problem etc.) | 5(4.2) | 43(35.8) | 58(48.3) | 14(11.7) | 62.000(3) | 0.0006 |
| Accidental (sprains, strains, eye and skin irritation, burns etc.) | 1(0.8) | 42(35.0) | 69(57.5) | 8(6.7) | 97.932(3) | 0.0004 |

said it isat a medium level. The accidental hazard is indicated as low by 57.5% of the farmers, while 35.0% of them said it's at a medium level.

Also, from Table 5, the p-value for biological, chemical, respiratory, physical, and accidental hazards are 0.0005, 0.0003, 0.0001, 0.0006, and 0.0004, respectively are all less than the α-level (α = 0.05). Hence, the null hypothesis can be rejected. Therefore, the health hazards during production are all significant at the 0.05 (5%) alpha level. This shows that most farmers that rated these hazards as being a low health problem does not mean they do not have an effect on the environment.

## Health hazards associated with poultry during processing

These include zoonotic disease such as Avian Tuberculosis, Avian influenza and so on, exposure to ammonia gases, musculoskeletal problem, and so on.

From Table 6, it can be observed that the majority of the farmers,that is, 53.0% and 21.4%, rated zoonotic disease as low and negligible health hazards during poultry processing. About 50.4% of them stated exposure to gases such ammonia is at a low level. 58.0% of the farmers stated that chronic respiratory problem and phlegm arealso at a low level during poultry processing; about 72.9% stated musculoskeletal problem as low; and 48.7% stated sprains, strains, and eye irritation also as low during poultry processing. In a broad sense, all the health hazards occurring during processing can be easily managed and maintained. Also from Table 6, the p-value for zoonotic, exposure to gases, chronic respiratory problem and phlegm, muscles and sprains, strains, eyes, and skin irritation hazards are 0.0002, 0.0005, 0.0001, 0.0003, 0.0004, respectivelyand are all less than the α-level (α=0.05), hence we reject the null hypothesis. Therefore, the health hazards during processing are significant at the 0.05 (5%) level of significance. This implies that most farmers that rated these hazards as being a low health problem does not mean they do not haveanyeffect on the environment. A low incidence of Avian Tuberculosis if not treated accordingly can lead to death of the infected individual.

## Improved practices used for mitigating environmental and health issues of the poultry farms

These include practices such as the use of disinfectant, giving of antibiotics, timely vaccination and sanitation, isolation or slaughtering of infected birds, the use of protective masks, gloves and so on. for maintaining the environmental and health issues of the poultry.

From Table 7, it can be observed that about 95.8% of the farmers' practice proper and timely vaccination all time while only 4.2% practice occasionally. About 90.0% of the farmers give antibiotics to their birds while about 82.2% isolate the infected birds. More 40.0% of the farmers practice slaughtering of infected birds while 21.7% do not. About 57.1% of them make use of protective masks and nose guards occasionally, while only 10.1% do not. About 33.9% of the farmers put on gloves all time, and 57.6% use them occasionally, while 8.5% do not use them at all. More than 56.3% of the farmers use farm clothes and 80.8% of them make use of

Table 6
*Health Hazards Associated with Poultry during Processing*

| Health hazards | High Frequency (%) | Medium Frequency (%) | Low Frequency (%) | Negligible Frequency (%) | Chi-square (df) | p-value |
|---|---|---|---|---|---|---|
| Zoonotic diseases | | 30(25.6) | 62(53.0) | 25(21.4) | 21.586(2) | 0.0002 |
| Exposure to gases | | 47(39.5) | 60(50.4) | 12(10.1) | 28.879(2) | 0.0005 |
| Chronic respiratory diseases and phlegm | 1(0.8) | 43(36.1) | 69(58.0) | 6(5.0) | 99.241(3) | 0.0001 |
| Musculo-skeletal problem | | 21(17.8) | 86(72.9) | 11(9.3) | 81.017(2) | 0.0003 |
| Sprains, strains, eye and skin irritation and burns | 2(1.7) | 43(36.1) | 58(48.7) | 16(13.4) | 61.034(3) | 0.0004 |

Table 7
*Improved Practice Used for Mitigating Environmental and Health Issues of the Poultry Farms*

| Improved Practices | Not used at all Frequency (%) | Used occasionally Frequency (%) | Used all time Frequency (%) |
|---|---|---|---|
| Proper and timely vaccination | | 5(4.2) | 115(95.8) |
| Giving of antibiotics | | 12(10.0) | 108(90.0) |
| Isolation of infected birds | 2(1.7) | 19(16.1) | 97(82.2) |
| Slaughter of the infected birds | 26(21.7) | 46(38.3) | 48(40.0) |
| Wearing of protective mask and nose guards | 12(10.1) | 68(57.1) | 39(32.8) |
| Regularly putting on of gloves | 10(8.5) | 68(57.6) | 40(33.9) |
| Always putting on overall and farm clothes | 1(0.8) | 50(42.0) | 67(56.3) |
| Proper and timely sanitation | | 7(5.8) | 113(94.2) |
| Use of disinfectant | | 23(19.2) | 97(80.8) |

Table 8
*Linear Regression Result of Determinants of Adoption of the Improved Practices by the Farm for Mitigating Environmental and Health Risks of the Poultry Farms*

| Variables | Coefficients | Standard error | t-value | Significance value (p≤10%) |
|---|---|---|---|---|
| Constant | 294.107 | 112.994 | 2.603 | 0.01 |
| Farm size | -0.035 | 0.142 | -0.256 | 0.79 |
| Number of flocks | -0.009 | 0.000 | -0.066 | 0.94 |
| Total cost of production | -0.008 | 0.000 | -0.063 | 0.95 |
| Educational level | 0.338*** | 0.418 | 2.618 | 0.01 |
| Year of establishment | -0.314** | 0.056 | -2.513 | 0.02 |
| Type of poultry | 0.007 | 0.209 | 0.058 | 0.95 |
| Farm produce | 0.122 | 0.070 | 0.963 | 0.34 |
| Type of battery cage | 0.222* | 0.626 | 1.885 | 0.06 |
| $R^2$ = 27% | | | | |

***p<0.01, **p<0.0 and *p<0.1

disinfectant in their various poultry. This implies that the majority of the farmers actually make use of these practices in their various farms in mitigating the environmental and health issues concerning their farms.

**Determinants of adoption of the improved practices by the farm for mitigating environmental and health risks of the poultry farms**

From Table 8, the result shows that educational level, year of establishment, and type of battery cage are significant at 1%, 5%, and 10% level of significance, respectively. The educational level is positive and significant at 1%, which implies that the higher the level of educational attainment, the more the farmers will adopt the practices. Education is very crucial in managing businesses more productively with a tendency of adopting new technology and innovation. According to Awosanya (2002), education leads to acquisition of new skills and efficient allocation of limited resources. Year of establishment is negative and significant at 5% due to the fact that the farms that were recently established adopted the improved practices less than those established earlier. Type of cage is positive and significant at 10%, which implies that a higher percentage of the farmers use the multiple cage system (84.6%), which is a more sophisticated cage than the other type of cage and the fact that the constant variable is also positive and significant at 5% and 10% implies that there are other explanatory variables that influence the adoption practices which are not captured in the model. Such variables may be too subjective to be captured in this econometric model.

The ANOVA (Table 9) was also used to test for the significance level of the regression model. The null hypothesis was rejected.The model is significant at the 0.01 of alpha levels. Furthermore, the result of $R^2$ value (27.4%) indicates that 27 percent of the variation expressed

Table 9
*ANOVA Table for the Linear Regression Result*

| Model | SS | df | Mean squares | F-value | p-value |
|---|---|---|---|---|---|
| Regression | 102.541 | 8 | 12.818 | 2.788 | 0.01 |
| Residual | 271.268 | 59 | 4.598 | | |
| Total | 373.809 | 67 | | | |

in the model is explained by the explanatory variables and about 72.6% are left unexplained.

## CONCLUSION AND RECOMMENDATIONS

From the results obtained, both the environmental and health issues of the poultry farms were significant, and the improved practices adopted by the farms were used by the poultry farmers. Accordingly, the study concludes that the environmental and health issues are affecting the production and processing of poultry farms. From the results obtained, it was observed that the majority of the poultry farmers does not abide by the environmental and public health laws and regulations;therefore, it is recommended that the poultry farmers should employ the recycling method in the management of their waste products by convertingthem into biofuel, composting them into manure, using themas fertilizer, processing them into animal feed component, and so on. These are of great benefits to the population at large. Government should, then, enforce strict supervisory agencies of the sanitation/health policies aimingat environmental preservation and protection.

## REFERENCES

1- Abubakar, M.B., Ambali, A.G & Tamjdo, T. (2007). Rural chicken production: Effects of gender on ownership, and management responsibilities in some parts of Nigeria and Cameroon. *International Journal of Poultry Science, 6* (6), 413-416.

2- Adene D.F., & Oguntade A.E. (2006). Poultry Sector Country Review. FAO Animal Production and Health Division FAO, Italy, Rome; 2006.

3- Akanni, K. A., & Benson, O. B. (2014). Poultry Wastes Management Strategies and Environmental Implications on Human Health in Ogun State of Nigeria. *Advances in Economics and Business, 2*(4), 164-171.

4- Alabi A., Aghimien C.I., Osasogie D.I. & Erie O.G. (2014). Environmental Effects of Poultry Production in Edo State, Nigeria. American *Journal of Experimental Agriculture 4*(12), 1773-1782.

5- Awosanya E.O, (2002). An Economic Analysis of Egg Production in Kwara State of Nigeria. M.Sc Project Work of the Department of Agricultural Economics and Farm Management, University of Ilorin.

6- Byarugaba, D.K., Olsen, J.E., & Katunguka-Rwakishaya E. (2002). Production, Management and Marketing Dynamics of the Rural Scavenging Poultry in Uganda. Second FAO/INFPD Electronic Conference on Family Poultry 2002 on Bangladesh Model Retrievd April 5, 2012, from http://www.fao.org/ag/ againfo/ themes/fr/infpd/documents/econf_bang/add _paper9.html

7- Environmental Protection Agency (FEPA) (2007): A Synopsis of the Laws and Regulations on the Environment in Nigeria, FEPA Office, Abuja, Nigeria; 2007.

8- FAO (2007): Poultry production and the environment – a review P. Gerber, C. Opio and H. Steinfeld Animal Production and Health Division, Food and Agriculture Organization of the United Nations, Viale delle Terme di Caracalla, 00153 Rome, Italy

9- FAO 2008: Poultry waste management in developing countries Charles Michael Williams, North Carolina State University, Department of Poultry Science, Raleigh NC, United States of America

10- Hossen, M. S., Hoque, Z., & Nahar, B.S. (2015). Assessment of poultry waste management in Trishal upazila, Mymensingh. *Research in Agriculture Livestock and Fisheries, 2*(2), 293-300.

11- Natukunda K, Kugonza D R and Kyarisiima C C 2011: Indigenous chickens of the Kamuli Plains in Uganda: I. Production system and flock dynamics. *Livestock Research for Rural Development. Volume 23, Article #220*. Retrieved February 19, 2017, from http://www.lrrd.org/lrrd23/10/natu23220.htm

12- Obioha FC. A Guide to Poultry Production in the Tropics, Enugu: Africana Educational Publishers Limited; 1996.

13- Maheshwari, S. (2013): Environmental Impacts of Poultry Production. *Poultry, Fisheries & Wildlife Sciences*. 1: 101

14- Simainga S, Moreki J C, Band F and Sakuya

N 2011: Socioeconomic study of family poultry in Mongu and Kalabo Districts of Zambia. *Livestock Research for Rural Development. Volume 23, Article #31*. Retrieved February 19, 2017, from http://www.lrrd.org/lrrd23/2/ sima23031.htm

15- Tadelle, D. (1996). *Studies on village poultry production systems in the central highlands of Ethiopia.*Unpublished dissertations, Swedish University of Agricultural Sciences, Uppsala, Sweden.

16- World Watch Institute (2006). *State of the World* Watch Institute, p. 26

17- Anosike, V. (2007). Health Implications of Poultry Waste Utilization for Urban Agriculture: An Ecohealth Approach. *Epidemiology, 18*(5), pp 53.

# 18

# On Different Characteristics of Iranian Farmer Women in Rural Development Activities Based on Animal Husbandry

*Hossein Bevrani* [1*] *and Simin Naseri* [2]

Abstract

Rural women play a significant role in domestic and socio-economic life of the society so that national development is not possible without developing this important and substantial segment of our society. Given the importance of women in agriculture, this study was conducted to examine the role of Iranian women in the farm management decision-making processed concerning animal husbandry. The sample of the study comprised 200 farm women, who were recruited by the help of the proportionate stratified sampling technique. Data were collected in East Azerbaijan in the Northwest of Iran. Women's contribution to agriculture is thus fundamental for the socio-economic development of rural areas. They are extensively involved in agriculture as female farm heads, co-owners, family farm workers, and also employees. In order to find out the relationship between different characteristics of rural women and participation in animal husbandry, the GLM multivariate method was employed as a robust statistical method. As the final results showed, women age, family size, level of aspiration, extension contact, perceived role overloads, and inter-spouse communication correlated highly with work participation in animal husbandry.

***Keywords:***
***animal husbandry, General Linear Models, rural development***

[1] *Professor, Department of Statistics, Faculty of Mathematical Sciences, University of Tabriz, Tabriz, Iran*

[2] *PhD Graduate in Agricultural Extension from UAS, Bangalore, India*

* *Corresponding author's email: bevrani@gmail.com*

## INTRODUCTION

In today's Iranian society, women have demonstrated their capabilities and potential talent, which, under proper conditions can greatly contribute to national wealth and economic development. The awareness and voluntary participation of Iranian women in a wide range of economic fields have greatly increased since they have been provided with more technical and scientific information. Therefore, enhancement of women's participation in economic activities is one of the most important objectives of many governmental organizations. Given the importance of women in agriculture, this study was conducted to examine the role of Iranian women in animal husbandry-based farm management decision-making process.

Ommani and Chiizari (2006) found that Iranian farm women workforce is used as gratis family labours and therefore, the real value of the women's roles is unknown in surplus economic production of peasant families. Since women have a significant role in agriculture workforce in Iran, it is very important to identify this role if we seek to improve the rural economy and support food production.

Rahmani et al. (2007) found that in Guilan females account for 76% of the labour input for rice and 80% of tea, in Mazandaran 50% for rice and in Gorgan 40% for cotton and 90% of summer crops. Women provide labour input for fruits, animal husbandry, dairy products, and preservation and processing activities. Sarada (2007) found that participation of women through the provision of training in management skills, appropriate technology and resources must receive priority in agricultural development programs if such programs are to significantly contribute to economic development and the reduction of poverty.

In a study on participation level of farm women in entrepreneurship activities under self-help groups in Kadur Taluk of Chikmagalur district, Bhagyavathi et al. (2008) found that the majority of the farm women (52.5%) belonged to the medium participation category, followed by high (30.83%) and low (16.67%) participation categories, respectively.

Anonymous (2008) mentioned that the challenge for rural development programs is to identify the constraints on women's full participation in economic and public life at local level, and to develop specific initiatives in their favour. Women were also seeking a better balance in the division of labour in the domestic household, needing encouragement for their personal and professional development and more support in their bid to achieve financial independence, and to participate fully in decision-making.

Behzadnasab (2008) found that rural female literacy was only 1% of the total female population of ten years and older in 1956. This rate was 60% in 1996, and has risen since then. Mechanization is another factor that has contributed to the decline in female labour participation. With some exceptions, girls are not trained to learn, and they do not show an interest in learning to use mechanized inputs or to manage farms.

Karbasioun et al. (2008) mentioned that agricultural and other rural products produced by women were sold in the market. But women often participated as unpaid family workers. Therefore, their work may be perceived as a part of household labour. Thus, women were active in households as well as cash-earning activities for their families. But neither men, nor women made a distinction between purely household related activities, and those related to the market.

Aref (2010) outlined some of the barriers to participation of farm women in agricultural activities. These barriers included lack of knowledge, lack of ability of individuals to participate, lack of effective and strong government institutions, inadequate focus on human resource development, and dependency on government and lack of authority in communities.

Naseri and Bevrani, (2013; 2015) reported that Iranian farm women had low levels of extension participation. Their results showed a high level of mass media exposure, in case of marginal and small farm families. They revealed that the majority of farm women of all the three groups had medium and high level of inter-spousal communication, and that the majority of farm women belonged to low level of cosmopolitan.

The reviewed studies revealed that farm women have an important and significant role in Iran agriculture and economy, which could be supported by proper and special program for farm women. This study was conducted to examine the role of Iranian women in the farm management decision-making process based on animal husbandry.

## MATERIAL AND METHODS

The study was conducted in Tabriz, East Azerbaijan Province, Iran. It comprises two districts of Markazi and Khosroshar, six Taluk and seventy-five villages. As was pointed out, the study was conducted in Tabriz area to know the participation of farm women in different agricultural activities. However, Khososhahr and Markazi districts were purposively selected. Out of a list of the total villages in these districts, and other farm women population based on the census data, 200 farm women selected from the age group of older than 18. This age group is considered as the active population in agriculture.

The total sample size of the study was 200 farm women, which were obtained through the proportionate stratified sampling technique. Cooper and Schindler (1999) defined this technique as a process by which each stratum is properly responded, so the sample drawn from it is proportionate to the stratum's share in total population. This approach is more appropriate than any other stratified sampling procedure and will generally have a higher statically efficient than the simple random sample. The details are given in Table 1.

The variables selected and empirical measurements used in the study are presented in Table 2.

The statistical design explains what observations to make and how to make. It also indicates what types of statistical test are to be applied for analyzing the data. In the present study, the ex-post facto research design was used.

According to Gay (1992), content validity is the degree to which a test measures an intended content area and is determined by expert judges. Relevancy weightage is defined as the ratio of actual score obtained by an item to maximum possible score of that item. The relevancy weighted for each statement was calculated using the formula:

The responses of 40 judges were obtained in a four-point scale, including most relevant, relevant, undecided, least relevant given the scores of four, three, two, and one, respectively. After collection of the judge opinions, the relevant percentage and weight age scores were obtained.

Considering the magnitude of obtaining relevancy scores and other criteria like coverage of diverse dimensions, 39 items from animal husbandry component were accepted

The primary scale was administered to a group of 30 respondents randomly selected in a non-sample area. The respondents were selected from a different socio-economic background in order to cover women from different categories. The Cronbach's Alpha was estimated at 0.895, which was highly significant at one per cent level indicating the high reliability of the scale.

In order to find out the relationship between different characteristics of rural women with the participation in animal husbandry, the GLM

Table 1
*Selection of Respondents Based on Proportionate Stratified Sampling*

| Districts | Name of Taluk | Popu. | % of Popu. | No. of village | % of village | Selected villages number % | | Number of respondent |
|---|---|---|---|---|---|---|---|---|
| Markazi | Ajichai | 11864 | 26 | 10 | 14.3 | 8 | 14.5 | 52 |
| | Espiran | 2955 | 6.6 | 16 | 22.9 | 13 | 23.6 | 13 |
| | Maydanchai | 18319 | 40.2 | 23 | 32.9 | 18 | 32.8 | 80 |
| Khosroshahı | Sarvsahra | 4502 | 10 | 7 | 10 | 5 | 9.1 | 20 |
| | Lahijam | 3193 | 7 | 6 | 8.5 | 5 | 9.1 | 14 |
| | Tazekand | 4871 | 10.2 | 8 | 11.4 | 6 | 10.9 | 21 |
| Total | | 45704 | 100 | 70 | 100 | 55 | 100 | 200 |

Table 2
*Operationalization and Measurement of Independent Variables*

| Variables | Empirical measurement |
|---|---|
| **Dependent variable** | |
| Participation of farm women in animal husbandry | Scale developed for the study |
| **Independent variables** | |
| Age | A structured schedule |
| Education | A procedure developed by Trivedi (1963) |
| Annual income | A structured schedule |
| Landholding status | A structured schedule |
| Family size | A structured schedule |
| Animal possession | A structured schedule |
| Marketing infrastructure | A structured schedule |
| Innovativeness | A procedure developed by Moulik and Rao (1965) |
| Level of aspiration | A procedure developed by Seema (1986) |
| Perceived role overload | A structured schedule |
| Extension participation | A structured schedule |
| Extension contact | A structured schedule |
| Mass media exposure | A procedure developed by Trivedi (1963) |
| Inter-spouse communication | A structured schedule |
| Cosmo-politeness | A procedure developed by Desai (1981) |

(general linear model) multivariate method was applied as strong statistical method. GLM is a general procedure for analysis of variance and covariance (ANCOVA) as well as regression. It can be used for univariate, multivariate, and repeated measures designs. The GLM Multivariate procedure allows you to model the values of multiple dependent quantitative variables based on their relationships to categorical and scale predictors (MANCOVA).

## RESUTS AND DISCUSSION

The present research was designed to study the participation of farm women in agricultural development in Iran. Fifty percent of total farm women belong to 18- 35 age group. A majority (81.3 %) of marginal farm women and 85.7 per cent of small farm women, and 72.5 per cent of big farm women were less than 45 years age group. A great majority (94.50 %) of farm women of Markazi and Khosroshar, were literate, received education up to the middle school level and high school level. A majority of farm women in both districts belonged to the medium level of income. 58.60 per cent of the big farm families had a larger number of farm women with the medium level of income. 73.60 percent of the marginal respondents had low level family income, whereas 71.40 per cent of small farm families had a medium level of family income. Women exhibited a high level of decision-making participation in many aspects of agricultural activities, particularly in animal husbandry activities.

### Relationship between independent variables and animal husbandry participation in agricultural activities

A look at Table 3 reveals that age, family size, level of aspiration, perceived role overload, extension contact, and inter-spouse communication had a positive relationship with work participation. Other variables, including annual income, education, and landholding marketing infrastructure, innovativeness, mass media exposure, and cosmo politeness had a negative relationship with work participation. Animal possession had relationship with work participation of farm women in animal husbandry at five per cent level. Perceived role overload and inter-spouse communication were related to work participation at 1% level of significant. As shwon in Table 3, the independent variables of age, education, marketing infrastructure, innovativeness, perceived role overload, marketing infrastructure, cosmopolitans, and mass media exposure had a positive rela-

Table 3
*Relationship between Independent Variables with Animal Husbandry Participation in Agricultural Activities*

| Independent variables | Animal husbandry activities (n=200) | | | | | |
|---|---|---|---|---|---|---|
| | Work participation | | Decision-making participation | | Overall participation | |
| Age | 0.033 | (0.324) | 0.051 | (0.238) | 0.012 | (0.436) |
| Education | -0.061 | (0.195) | 0.039 | (0.290) | 0.003 | (0.481) |
| Annual income | -0.042 | (0.276) | -0.194** | (0.003) | -0.002 | (0.487) |
| Landholding status | -0.070 | (0.161) | -0.197** | (0.003) | -0.049 | (0.249) |
| Family size | 0.108 | (0.064) | -0.057 | (0.210) | 0.074 | (0.150) |
| Animal possession | 0.116* | (0.049) | -0.085 | (0.115) | 0.109 | (0.063) |
| Marketing infrastructure | -0.006 | (0.468) | 0.078 | (0.136) | -0.014 | (0.424) |
| Innovativeness | -0.044 | (0.270) | 0.039 | (0.293) | -0.017 | (0.404) |
| Level of aspiration | 0.043 | (0.274) | -0.028 | (0.346) | 0.032 | (0.329) |
| Perceived role overload | 0.428** | (0.000) | 0.078 | (0.139) | 0.410** | (0.000) |
| Extension contact | 0.043 | (0.272) | -0.077 | (0.139) | -0.006 | (0.468) |
| Mass media exposure | -0.047 | (0.255) | 0.050 | (0.240) | -0.101 | (0.078) |
| Inter-spouse communication | 0.497** | (0.000) | -0.054 | (0.244) | 0.488** | (0.000) |
| Cosmo politeness | -0.043 | | 0.008 | | -0.017 | |
| | (0.272) | | (0.454) | | (0.405) | |

**p<0.01 * p<0.05

tionship with decision–making. Other variables such as annual income, land holding, extension contact, animal possession, family size, animal possession, and level of aspiration had a negative relationship with participation in decision-making in animal husbandry. Annual income and land holding had a relationship with decision-making participation in animal husbandry at one per cent level.

Table 3 indicates that the independent variables such as age, education, family size, and animal possession, level of aspiration, inter-spouse communication, and perceived role overload had a positive relationship with overall participation. It shows that other variables had a negative relationship with overall participation in animal husbandry. Moreover, inter-spouse communication and perceived role overload had a relationship with overall participation in animal husbandry at one per cent level.

GLM multivariate tests based on animal husbandry activities by the independent variables

GLM (general linear model) is a general procedure for analysis of variance and covariance

Table 4
*Tests of Between-Subjects in GLM Multivariate Tests Based On Animal Husbandry Activities by Independent Variables (n=200)*

| Source | Dependent Variable | Type III Sum of Squares | df | Mean Square | F | p-value |
|---|---|---|---|---|---|---|
| GLM Mult. Model | Overall | 15264.816[a] | 28 | 545.172 | 4.136 | .000* |
| | Work | 14565.246[b] | 28 | 520.187 | 5.106 | .000* |
| | Decision | 388.231[c] | 28 | 13.865 | .878 | .646 |
| Intercept in the GLM model | Overall | 5342.564 | 1 | 5342.564 | 40.533 | .000* |
| | Work | 2232.891 | 1 | 2232.891 | 21.919 | .000* |
| | Decision | 656.517 | 1 | 656.517 | 41.556 | .000* |
| Error | Overall | 21879.830 | 166 | 131.806 | | |
| | Work | 16910.672 | 166 | 101.872 | | |
| | Decision | 2622.518 | 166 | 15.798 | | |
| Corrected Total | Overall | 37144.646 | 194 | | | |
| | Work | 31475.918 | 194 | | | |
| | Decision | 3010.749 | 194 | | | |

a. R Squared = 0.411 b. R Squared = 0.463 c. R Squared = 0.129

(ANCOVA) as well as regression. It can be used for univariate, multivariate, and repeated measures designs.

The GLM Multivariate procedure allows you to model the values of multiple dependent scalar variables based on their relationships to categorical and scale predictors (MANCOVA).

Reasons to use GLM as a statistical tool for this particular research can be summarized as:

1. The presence of nominal, ordinal and scale types of independent variables. For example, marital status and districts are nominal, and level of aspiration and education is ordinal. Age and annual income are scalar variables.

2. The presence of more than one dependent variable; for example, animal husbandry activities have three variables, including overall, work and decision making.

3. In this specific research, it was not suitable to use regression, even multivariate regression because the regression method will only predict one dependent variable.

Table 4 shows the effects of independent variables on animal husbandry on three different aspects including overall, work, decision making at the same time.

The omnibus or overall F test, tests the null hypothesis that there is no difference in the means of the dependent variables for the different groups formed by categories of the independent variables. The multivariate formula for F is based not only on the sum of squares between and within groups, as in ANOVA, but also on the sum of cross products, that takes covariance into account as well as group means. Significance tests for multiple dependents all follow the F distribution and so an F value and corresponding significance level are printed out for each of these tests.

As the results showed, there was a significant relationship within independent variables in models of overall, work animal husbandry participation, and independent variables did not significantly affect participation in decision-making. Also about the individual effects of independent variables on dependent variables, this variable had significant effects at 5 per cent level.

1. For overall animal husbandry activities: perceived role overload, inter-spouse communication.

2. For work participation: perceived role overload, inter-spouse communication, family size.

3. For decision making activities: education, innovativeness.

Results of the GLM method were clearly shown in Figure 1. Also, Figure 2 shows the results of GLM and significance test in that the dotted lines show insignificant relations and solid lines shows significant relations between animal husbandry activities and independent variables.

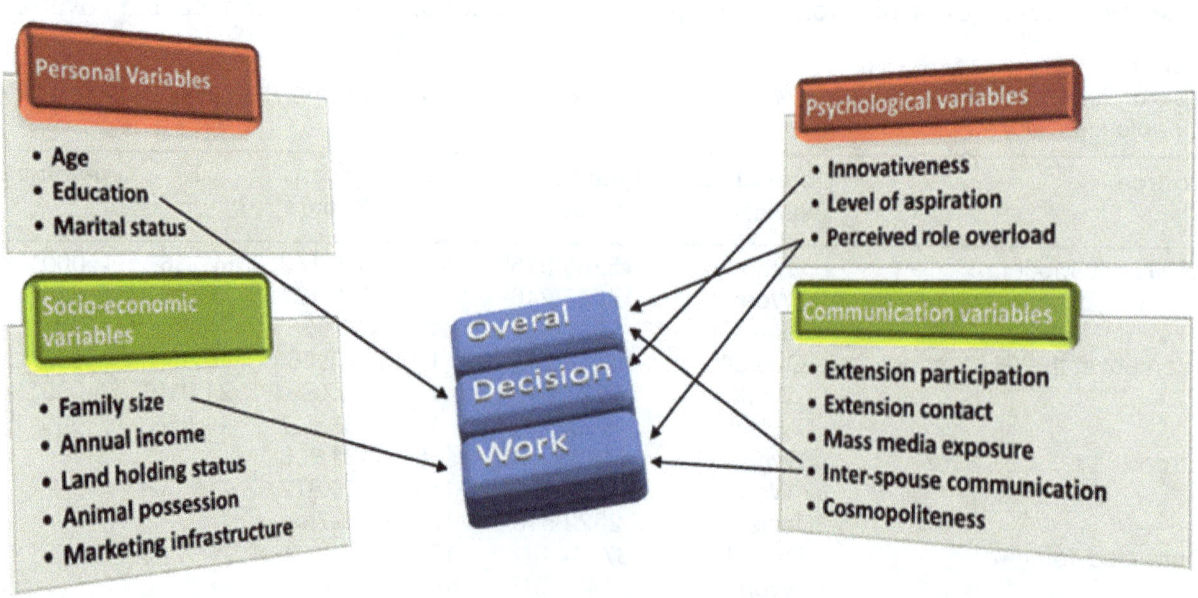


*Figure 1*. Result of GLM multivariate tests based on animal husbandry activities with indipendent variables

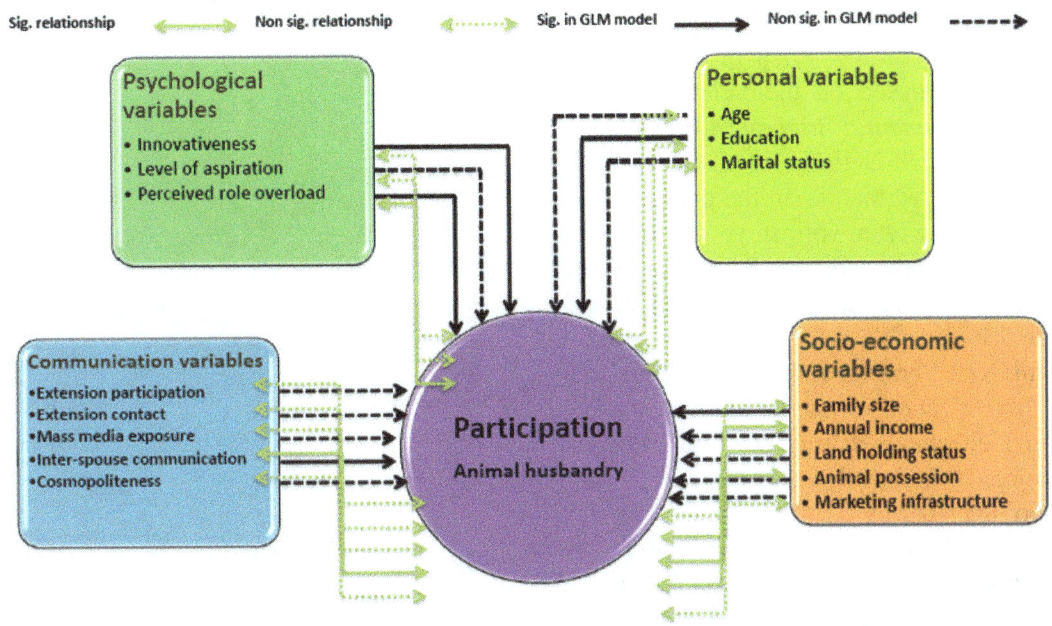


*Figure 2.* Emprical model of the study based on animal husbandry activities with indipendent variables

## CONCLUSION

In Iranian farms, women play an important role in achieving rural development goals. In Iran's rural society, every family can be counted as a production unit. Results showed that age, education, family size, animal possession, level of aspiration, inter-spouse communication, and perceived role overload had a positive relationship with overall participation. Examination of the effects of independent variables on animal husbandry at three different parts including overall participation, work participation, and decision-making participation showed a significant relationship within independent variables in models and overall participation and participation in animal husbandry, and none of the independent variables significantly influenced participation in decision-making.

## ACKNOWLEDGMENT

The authors would like to thank the Editor and referees for the constructive comments and illustrative suggestions, which helped greatly improve the quality and presentation of the manuscript.

## REFERENCES

1- Anonymous, D. (2008). *United Nations women participation in development*: emerging trends in community development, Oxford University Press, New York, USA.

2- Aref, F. (2010). Barriers of agricultural development in Iran: A case study of Fars province. *Journal of American Science, 6* (11), 1034-1037.

3- Behzadnasab, Z. (2008). Women in agriculture. *Jihad Magazine Education, 29* (30), 85-89.

4- Bhagyavathi M., Venugopal L., Naresha, S., & Nagabhushanam, K. (2008). Participation level of farm women in entrepreneurship activities under self Help groups. *Mysore Journal of Agricultural Sciences, 42* (1), 159-163.

5- Cooper, D. R., & Schindler, P. S. (1999). Business research methods, McGraw-Hill Publishing Company Limited, New Delhi, India.

6- Desai, G. R. (1981). *A critical analysis of the contribution of education and extension guidance to economic performance of cotton farmers of Karnataka statues,*Unpublished dissertation, University of Agricultural Sciences, Bangalore.

7- Gay, L.R. (1992). Educational research: *Competencies for analysis and application,*Macmillan Publishing Company, Englewood Cliffs.

8- Karbasioun, M., Mulder, M., & Biemans, H. (2008). Changes and problems of agricultural development in Iran. *World Journal of Agricultural Sciences, 4* (6), 759-769.

9- Moulik, T.K., & Rao, C.C.S. (1965). *Self-rating scales for farmers, Measurements in Agricultural Extension,* Indian Agricultural Research Institute, New Delhi, India.

10- Naseri, S., & Bevrani, H. (2013). The role of farm women in agriculture and it's allied felids: A case study of Iran. *International Journal of Agriculture Innovations and Research, 2*(6), 2319-1473.

11- Naseri, S., & Bevrani, H. (2015). Study on participation of Iranian rural women in different agricultural activities based on their characteristics. *Advances in Environmental Biology, 9*(4), 138-142.
12- Ommani, A. R., & Chiizari, M. (2006). An investigation of effective factors on the perception of rural youth regarding teamwork in youth club. 22nd Annual Conference Proceedings of Association International Agricultural Extension Education. Clearwater Beach, Florida.
13- Rahmani, Kh., Ebrahim Habibi, M., & Heidari, H. (2007). Investigation on the health status of inhabitants in Nehbandan district in south of Khorasan province. Institute of Nutrition Sciences and Food Technology, Tehran.
14- Sarada, B. (2007). *Women Educational Developing Countries, Barriers, Benefits and Policies*, World Bank Book, London.
15- Seema, B. (1986). *Role of farm women in the decision making process of farming community in Trivandrum district,* Unpublished Thesis, Kerala Agricultural Sciences University, India.
16- Trivedi, G. (1963). *Measurement analysis of socio-economic status of rural families,*Unpublished dissertation, Indian Agricultural Research Institute, New Delhi, India.

19

# Development and Validation of Sustainable Aquaculture Indicators: Case of Alborz Dam Basin, Mazandaran, Iran

*Ehsan Gholifar [1], Enayat Abbasi [2*], Gholamreza Pezeshki Rad [3], Hassan Salehi [4] and Abdolmotalleb Rezaei [5]*

Abstract

The present study was conducted to examine the development and validation of sustainable indicators related to aquaculture as a social-ecological system. By reviewing the literature, it was found about 148 indicators related to sustainable aquaculture in economic (market-oriented and responsive aquaculture, capability of risk assessment and crisis management, and financial management in aquaculture), social (participation in food security and healthy food, the role of producer and non-governmental organizations, improving the image of aquaculture, and social responsibility), environment (minimizing the impacts of aquaculture at the national level, minimizing the impacts on the environmental conditions, and biodiversity), and institutional (local development and participation in decision-making) components. The extracted components were used in the form of a questionnaire with a five-point Likert type scale according to six criteria. Questionnaires were distributed among aquaculture experts in Alborz Dam Basin in Mazandaran Province, where 32 questionnaires were returned. Research results include ranking indicators based on a total of six criteria mentioned in methodology, and ranking components, as well as their associated indicators related to sustainable aquaculture using AHP. The findings further suggested that economic and environmental aspects also were placed as the first and second ranks among 10 sustainable aquaculture indicators. Accordingly, it is highly recommended that the agencies related to the country's aquaculture development paid attention to these identified dimensions and used them in their planning.

*Keywords:*
***Alborz Dam, basin, development and validation, Iran, sustainable aquaculture indicators***

---

[1] *Ph.D. Candidate, Agricultural Extension and Education Department, Tarbiat Modares University, Iran*
[2] *Assistant Professor, Agricultural Extension and Education Department, Tarbiat Modares University, Iran*
[3] *Professor, Agricultural Extension and Education Department, Tarbiat Modares University, Iran*
[4] *Associate Professor, Iran Fisheries Organization, Tehran, Iran*
[5] *Assistant Professor, Agricultural Extension and Education Department, University of Tehran, Iran*
* *Corresponding author's email: enayat.abbasi@modares.ac.ir*

## INTRODUCTION

Aquaculture has been considered as one of the fastest growths in the field of food production (FAO, 1997). Given the disproportionate exploitation of marine fish resources in recent decades, there has been an increasingly explicit focus on aquaculture to ensure poverty reduction, food security and income generation (FAO, 1999). Developing responsive or sustainable aquaculture can, therefore, contribute significantly to food security and nutrition in the society, as well as a variety of incomes and livelihoods, mainly through the ways that can help to reduce vulnerability and further facilitating water and nutrients management as efficient as possible. Less-privileged groups such as women and workers, especially those having no land for farming, in poor communities can benefit dramatically from aquaculture development in terms of livelihood opportunities related to the processing and service sectors (Bunting, 2013).

Aquaculture has been faced with three main challenges, including sustainable economic growth, environmental considerations and equitable distribution of benefits. An effective response to these challenges requires a coherent interaction, accelerating private investments along with a facilitation role relating to public sectors. Through building partnerships and providing access to financial resources and other resources needed, developing countries can deal successfully with these challenges, revolving primarily around two pivotal aspects, e.g., appropriate management and production and dissemination of sustainability knowledge (World Bank, 2007).

Aquaculture development is considered as a viable source for providing a still increasing population in need for affordable food in Iran. Aquaculture generally is subsumed under agriculture sector as being associated considerably with agriculture where there is water for farming a peaceful coexistence can be established between them. Recent experiences have shown that aquaculture can expedite food security, especially in developing countries. Several past decades have witnessed aquaculture is quickly moving to achieve a growing and dynamic industry (Jafarian, 2006). In Iran, the rapid development of aquaculture breeding (Table 1) clearly supports the rapid growth relating to aquaculture production observed over the last decade. For instance, the amount of aquaculture production reached in 2001 about 73 thousand and 645 tons. In 2011, 10 years later, this level of output reached 285 thousand and 351 tons. Table 1 indicates the development trend of aquaculture and reduction in fishing in the Mazandaran Province as one of pioneer provinces in the field of aquaculture (Statistical Yearbook of the Iranian Fisheries Organization, 2010).

Despite the rapid increase in the development of aquaculture in Iran, given the limited access to natural resources this rapid growth requires a move toward a more sustainable development. Compared to other production systems aquaculture is under more pressure to be sustainable, principally because of using natural resources, such as natural sweet water resources, wetlands, coastal areas, and hunting marine fish to feed farmed fishes. Along with the rapid growth of food production in the world, modern aquaculture has been growing as well. This growth is combined with a shift from focusing solely on production technology development to an emphasis on the economic and environmental aspects of sustainability in aquaculture activities (Stickney, 2002).

The sustainability of aquaculture activities depends on a combination of two groups of internal and external factors. Internal factors include water quality, breeding techniques, operational and location facilities, source of fish eggs, species properties, and natural or artificial access to food resources, while external factors encompass national policies, natural hazards, weather and water change, pollution, market, introduction of foreign species, social and cultural conditions, fish production sites and legal supervision and control. All of these factors influence fish farming operations, marketing products, and distribution of economic and social benefits. The repugnant impacts of internal factors related to the environment and sustainability aspects of aquaculture production can be resolved through the planning and on-farm appropriate management. Most external or off-farm factors require appropriate government policies and management

Table 1
*Statistics Associated to Aquaculture and Fishing Activities in the Period of Ten Years in Mazandaran*

**The fish catches in the Mazandaran Province in the years 2001-2011 (tons)**

| Description | 2001 | 2002 | 2003 | 2004 | 2005 | 2006 | 2007 | 2008 | 2009 | 2010 | 2011 |
|---|---|---|---|---|---|---|---|---|---|---|---|
| Bony fishes | 4837 | 5280 | 7983.5 | 6046.5 | 8316 | 11025 | 10244 | 9987.1 | 8500 | 7360 | 7251 |
| Sturgeon | 350 | 272 | 173 | 170 | 152 | 129 | 87 | 61.2 | 54 | 32 | 32 |
| Kilka | 14785 | 10200 | 8025 | 10260 | 13859 | 13538 | 10301 | 12260 | 20741 | 21216 | 15856 |
| Total | 19972 | 15752 | 16181.5 | 16476.5 | 22327 | 24692 | 20632 | 22308.3 | 29295 | 28608 | 23139 |

**Level of water resources withdrawal in the Mazandaran province in the years 2001-2011 (in tons)**

| Years | 2001 | 2002 | 2003 | 2004 | 2005 | 2006 | 2007 | 2008 | 2009 | 2010 | 2011 |
|---|---|---|---|---|---|---|---|---|---|---|---|
| Mazandaran | 134 | 80 | 240 | 121 | 120 | 165 | 180 | 120 | 206 | 185 | 124 |

**Level of growing cold-water fish in the Mazandaran province in the years 2001-2011 (in tons)**

| Years | 2001 | 2002 | 2003 | 2004 | 2005 | 2006 | 2007 | 2008 | 2009 | 2010 | 2011 |
|---|---|---|---|---|---|---|---|---|---|---|---|
| Mazandaran | 1195 | 1713 | 3187 | 4074 | 4662 | 6864 | 8097 | 9169 | 10514 | 12456 | 13294 |

**Level of growing warm-water fish (carp fish and caviar) in the Mazandaran province in the years 2001-2011 (in tons)**

| Years | 2001 | 2002 | 2003 | 2004 | 2005 | 2006 | 2007 | 2008 | 2009 | 2010 | 2011 |
|---|---|---|---|---|---|---|---|---|---|---|---|
| Mazandaran | 2766 | 20844 | 22233 | 22825 | 24648 | 22959 | 30610 | 29140 | 35950 | 38391 | 41690 |

of government interventions to tackle related problems based on source or an assigned area, planning in the exploitation of land and water, technical aids, information services, supervision and legal controls, and coordination with other industries. The effect of combining internal and external factors can cause aquaculture to be considered crucial when relating to all its system resources (Mokhtari Abkenari et al., 2004).

In this study, to integrate these two groups of internal and external factors and extract indicators, we draw upon social-ecological systems theory introduced by (Ostrom, 2009). This theory includes following sub-systems: the source system (for example, coastal fishing), the source units (for example, marine fish), users (such as fisherman), and government and management systems (including organizations and rules of the coastal fishing). These sub-systems are rather distinct from each another, but in social-ecological systems these sub-systems have to interact with each other to produce the desired result, in which, in turn, these sub-systems and their related components are under the effect of smaller or larger social-ecological systems. To maintain the sustainability of the social-ecological systems, we need the scientific knowledge directed toward studies that consider these systems as a whole and not a separate one, mainly in the form of ecological and social sciences as well as integrated units. In the present study, we also consider a combination of these two groups of variables and their interaction to develop and validate sustainable aquaculture indicators in the Alborz Dam Basin, Mazandaran Province.

## MATERIALS AND METHODS

To achieve development and validation of sustainable indicators related to aquaculture as a social-ecological system, we reviewed the literature (Costa-Pierce & page, 2010, FAO, 2011, 2013, Nunes et al., 2011, Ting et al., 2015) and found about 148 indicators in relation to sustainable aquaculture in economic (market-oriented and responsive aquaculture, capability of risk assessment and crisis management, and financial management in aquaculture), social (participation in food security and healthy food, the role of producer and non-governmental organizations, improving the image of aquaculture, and social responsibility), the environment (minimizing the impacts of aquaculture at the national

Table 2
*Indicators of Sustainable Aquaculture and Their Sub-indicators*

| Indicators | The average of total of criteria | SD | CV |
|---|---|---|---|
| **Market-oriented and responsive aquaculture** | | | |
| 1. The ratio of marketing costs to the total value of sales. | 21.43 | 3.90 | 1.201 |
| 2. The percentage of high value-added products. | 20.37 | 3.18 | 0.997 |
| 3. The difference in price due to product quality. | 19.90 | 3.56 | 1.293 |
| 4. The existence of a clear marketing plan. | 19.34 | 3.43 | 1.342 |
| 5. Include the tastes of consumers in selecting a product type. | 18.50 | 3.53 | 1.390 |
| 6. The existence of a product tracking system. | 19.96 | 3.59 | 1.499 |
| 7. The existence of a quality certificate plan. | 19.03 | 3.09 | 1.352 |
| 8. Innovation capability in products annually. | 20.81 | 3.12 | 1.453 |
| 9. The existence of a unique label to aquaculture. | 21.65 | 2.78 | 1.385 |
| **Capability of risk assessment and crisis management** | | | |
| 10. Market's geographic diversity (number and share of each market from total sales). | 3.43 | 1.10 | 0.322 |
| 11. The number of products (species, size, class, value-added). | 3.84 | 1.32 | 0.344 |
| 12. The cost of insurance to total sales. | 3.68 | 1.28 | 0.348 |
| 13. The proportion of the cost of receiving professional advice to total sales. | 3.25 | 1.19 | 0.366 |
| 14. The number of domestic feed suppliers (also, the percentage of imports). | 3.75 | 1.39 | 0.370 |
| 15. The number of domestic hatcheries (percentage of imported baby). | 3.78 | 1.40 | 0.372 |
| 16. The time scale of aquaculture farm lease. | 3.78 | 1.43 | 0.378 |
| 17. Type of ownership in aquaculture farm. | 3.62 | 1.38 | 0.382 |
| 18. Integration of aquaculture with complementary activities (ecotourism, restaurant, recreational fishing). | 3.25 | 1.27 | 0.390 |
| 19. Supply and product sales, directly or in the contract. | 3.31 | 1.22 | 0.392 |
| 20. The share of each consumer in total sales. | 3.43 | 1.36 | 0.398 |
| 21. The number of renewals in aquaculture farm lease, annually. | 3.59 | 1.43 | 0.399 |
| 22. Use standard ISO 14000 (or other standards). | 3.25 | 1.50 | 0.462 |
| 23. Membership in the cooperatives or organizations for the sale of products. | 3.15 | 1.48 | 0.470 |
| 24. The existence of a law on the disposal of biological wastes. | 3.00 | 1.45 | 0.486 |
| 25. The existence of biosecurity system. | 2.75 | 1.36 | 0.497 |
| 26. The existence of farm health management plans (vaccination plan). | 3.25 | 1.54 | 0.515 |
| **Financial management in aquaculture** | | | |
| 27. The financial costs per kg of fish produced (and percentage of total cost in kg). | 4.15 | 1.11 | 0.267 |
| 28. Debt rate (loan) divided by assets (total farm debts divide by total farm assets). | 3.68 | 0.99 | 0.270 |
| 29. Aquaculture unit production costs (total fixed and variable costs divide by kg of fish produced, and finally divide by executive costs). | 4.18 | 1.17 | 0.281 |
| 30. The energy costs per kg of fish produced (and percentage of total cost in kg). | 3.75 | 1.07 | 0.287 |
| 31. The feed cost per kg of fish produced (and percentage of total cost in kg). | 4.12 | 1.21 | 0.293 |
| 32. The current rate (total current farm assets divide by total current farm liabilities). | 3.96 | 1.17 | 0.297 |
| 33. The rate of farm assets return. | 3.84 | 1.16 | 0.303 |
| 34. Baby fish cost per kilogram of fish produced (and percentage of total cost in kg). | 3.96 | 1.20 | 0.304 |
| 35. Percentage of the net sales (gross profit to revenue * 100). | 3.96 | 1.23 | 0.310 |
| 36. The labor costs per kg of fish produced (and percentage of total cost in kg). | 4.00 | 1.29 | 0.323 |
| 37. Transportation costs per kg of fish produced (and percentage of total cost in kg). | 3.78 | 1.38 | 0.366 |
| 38. The debt rate (loan) divided by capital (total farm debts divide by total farm capitals). | 3.43 | 1.34 | 0.391 |
| 39. Financial support from the government to start aquaculture (land, subsidies, loan). | 3.34 | 1.33 | 0.399 |
| 40. The existence of savings for emergency and unprecedented issues. | 3.18 | 1.30 | 0.410 |
| 41. Costs of environmental monitoring divide by per kg of fish produced (as a percentage of the total cost divided by kg). | 3.12 | 1.31 | 0.420 |
| 42. The existence of training programs for staff and workers related to environmental aspects of their activities. | 3.12 | 1.36 | 0.436 |
| 43. The existence of a plan for managing crisis situations. | 2.90 | 1.27 | 0.441 |
| 44. The existence of training programs for staff related to financial aspects of their activities. | 2.93 | 1.36 | 0.466 |
| 45. Receiving incentives, directly or indirectly, for environmental protection activities. | 2.78 | 1.38 | 0.498 |
| 46. Financial investment for environmental protection divide by per kg of fish produced (as a percentage of the total cost divided by kg). | 2.93 | 1.54 | 0.526 |

| Indicators | The average of total of criteria | SD | CV |
|---|---|---|---|
| **Participation in food security and healthy food** | | | |
| 47. The price of fish in contrast to the national minimum wage. | 3.78 | 1.06 | 0.282 |
| 48. Annual aquaculture production. | 4.18 | 1.20 | 0.287 |
| 49. The rate of production for the local markets (Self-consumption). | 3.84 | 1.39 | 0.362 |
| 50. Percentage of innovation in products, annually. | 3.15 | 1.34 | 0.427 |
| **The role of producer and non-governmental organizations and improving the image of aquaculture** | | | |
| 51. The minimum wage of staff and workers against the national minimum wage. | 3.43 | 1.13 | 0.330 |
| 52. Percentage of trained staff and workers with specialized certification. | 3.50 | 1.27 | 0.362 |
| 53. Membership in farmers' organizations and cooperatives (in number). | 3.31 | 1.20 | 0.363 |
| 54. The existence of ecological labels and product specifications. | 3.06 | 1.41 | 0.461 |
| 55. The existence of a strong social base among the other aquaculture growers. | 2.84 | 1.46 | 0.514 |
| **Social responsibility** | | | |
| 56. The number of work hours by the aquaculture workers. | 3.93 | 1.24 | 0.316 |
| 57. The number of work-related accidents. | 3.31 | 1.14 | 0.346 |
| 58. Percentage of high-quality fish produced. | 3.40 | 1.18 | 0.349 |
| 59. Percentage of women workers | 3.37 | 1.28 | 0.382 |
| 60. The number of identified pathology. | 3.03 | 1.20 | 0.397 |
| 61. The percentage of membership in workers' unions. | 3.28 | 1.32 | 0.403 |
| 62. The rate of visiting relevant agencies and gaining information from them. | 2.78 | 1.18 | 0.425 |
| 63. The rate of unsaleable fishes produced. | 3.12 | 1.38 | 0.443 |
| **Minimizing the impacts of aquaculture at the national level** | | | |
| 64. The feed conversion ratio (kg feed / kg fish). | 4.15 | 1.13 | 0.274 |
| 65. The existence of the hatchery with the masses of indigenous babies. | 3.25 | 0.91 | 0.281 |
| 66. Algal growth and blooms (number of cells per milliliter). | 3.68 | 1.11 | 0.304 |
| 67. Depth (m). | 3.84 | 1.22 | 0.317 |
| 68. The percentage of space used for aquaculture. | 3.62 | 1.15 | 0.319 |
| 69. The number of species introduced in aquaculture. | 3.90 | 1.35 | 0.346 |
| 70. Water volume occupied by per kg of product (kg / m$^3$). | 3.46 | 1.24 | 0.359 |
| 71. Exchange with the open sea (coastline part): distance in meters. | 3.50 | 1.31 | 0.376 |
| 72. Increasing fishing activity around the farm cages. | 2.96 | 1.14 | 0.388 |
| 73. Oxygen saturation (percent). | 3.46 | 1.34 | 0.388 |
| 74. Consumption of marine fish (tons per year). | 3.50 | 1.36 | 0.390 |
| 75. Loss of nursery and spawning habitat. | 3.25 | 1.29 | 0.398 |
| 76. Vegetable consumption (tons per year). | 3.40 | 1.38 | 0.407 |
| 77. Turbidity / clarity (sushi disc). | 3.56 | 1.45 | 0.409 |
| 78. The microbiological indexes (coliforms). | 3.40 | 1.49 | 0.440 |
| 79. Rehabilitation index and the spawning mass (biomass). | 3.37 | 1.49 | 0.444 |
| 80. Carrying and storage capacity of the ecosystem. | 2.93 | 1.41 | 0.481 |
| 81. Supervision on the quality of fish larvae produced in hatcheries. | 3 | 1.50 | 0.500 |
| 82. Modified catching of target species in the area (monitoring of fishing activities). | 2.96 | 1.15 | 0.525 |
| **Minimizing the impacts of aquaculture at the national level** | | | |
| 83. Total Phosphorus (kg). | 4.02 | 0.98 | 0.244 |
| 84. Accumulation of heavy metals (one millionth gram). | 3.96 | 1.14 | 0.290 |
| 85. Oxidation-reduction potential (PH). | 3.81 | 1.14 | 0.301 |
| 86. Antiphrastic kg per ton of fish produced (kg). | 3.78 | 1.15 | 0.305 |
| 87. The use of non-indigenous species. | 3.71 | 1.22 | 0.329 |
| 88. Use GMO species. | 3.31 | 1.09 | 0.329 |
| 89. Turbidity (sushi disc). | 3.78 | 1.26 | 0.334 |
| 90. Use locally-based production sources. | 3.71 | 1.25 | 0.336 |

| Indicators | The average of total of criteria | SD | CV |
|---|---|---|---|
| 91. Antibiotic kg per ton of fish produced (kg). | 3.84 | 1.29 | 0.337 |
| 92. Chlorophyll (mg / m³). | 3.65 | 1.23 | 0.338 |
| 93. Use food nutrients along with chemical antioxidants. | 3.50 | 1.21 | 0.348 |
| 94. The level of nutrients lost to total (percentage). | 3.31 | 1.17 | 0.355 |
| 95. Disinfectants kg per ton of fish produced (kg). | 3.71 | 1.39 | 0.376 |
| 96. Total particulate organic matter (mg / M³). | 3.62 | 1.38 | 0.382 |
| 97. Total organic carbon (organic) (TOC, mg / m³). | 3.56 | 1.38 | 0.390 |
| 98. Total dissolved organic matter (mg / M³). | 3.62 | 1.45 | 0.401 |
| 99. Stool sedimentation rate (grams per day). | 3.18 | 1.33 | 0.418 |
| 100. Use antifouling biocides in aquaculture. | 3.40 | 1.43 | 0.421 |
| 101. The level of nutrient balances (kg). | 3.21 | 1.40 | 0.438 |
| 102. Use certified organic fish feed. | 3.15 | 1.41 | 0.449 |
| 103. The level of farm's disseminated pathogens in other farms. | 2.96 | 1.40 | 0.472 |
| **Local development** | | | |
| 104. The number of reports related to the environmental crises over the last five | 3.50 | 1.13 | 0.324 |
| years. | 3.25 | 1.19 | 0.366 |
| 105. Percentage of the seasonal workers in aquaculture in comparison to seasonal | 3.62 | 1.33 | 0.369 |
| workers with supplement activities. | 3.06 | 1.13 | 0.370 |
| 106. The number of staff and workers (both direct and indirect). | 3.62 | 1.36 | 0.375 |
| 107. Receiving subsidies for ecological services. | 3.50 | 1.31 | 0.376 |
| 108. Percentage of full-time and permanent workers. | 3.15 | 1.32 | 0.419 |
| 109. Percentage of seasonal workers. | 2.81 | 1.40 | 0.498 |
| 110. The rate of participation with local policy-making organizations and institutions (Dehyari, Gubernatorial). | | | |
| **Participation in decision making by aquaculture growers** | | | |
| 112. Percentage of the workers and technicians familiar with the rules. | 3.43 | 1.24 | 0.362 |
| 113. The number of participating in advisory sessions. | 3.25 | 1.16 | 0.357 |
| 114. The number of constructed measures in partnership relationships. | 2.75 | 1.19 | 0.433 |
| 115. The number of disputes resolved at the local level. | 2.78 | 1.12 | 0.405 |
| 116. Effective participation in local decision-making processes. | 2.90 | 1.27 | 0.441 |

level, minimizing the impacts on the environmental conditions and biodiversity), and institutional (local development and participation in decision-making) components. The extracted components were used in the form of a questionnaire with a five-point Likert type scale according to six criteria: (a) measurement capability: is the indicator quantifiable and measurable? is there data to measure it?; (b) scientific relevance: to what extent indicator is scientifically valid?; (c) policy relevance: is whether indicator related to the country's strategic and important issues for management and decision making or not?; (d) sensitive to changes: to what extent is sensitive to changes in its encompassing environment and react to it?; (e) simplicity: is indicator understandable for people without specialized knowledge?; and (f) economic justification: is collection and calculation of data in a time and cost framework possible? Questionnaires were distributed among aquaculture experts which were selected purposively, where 32 questionnaires were returned. Respondents were experts in National Fisheries Organization (n=15), National Institute of Research and Science for Fisheries (n=8), Faculty Members of Fisheries in University of Tehran (n=1), Sari's Fisheries Organization (n=4), Academic Center for Education, Culture and Research (n=2), and Faculty members at the Sari University (n=2). Next, the opinions of respondents were received, where most important indicators ranked using the Coefficient of Variation (CV) and the mean. The Analytic Hierarchy Process

(AHP) was applied to determine the weight and importance of each of the indicators in their corresponding components. The multi-criteria decision-making methods primarily draw upon the several criteria instead of using a measure of optimality. Available models in this area are categorized in two models,, that is, multi-objective and multi-criteria. Multi-objective models intend to design objectives; however, multi-attribute models are used to select the top priority option. The decision space for the first method is continuous while the later benefit from the discrete space used for the decision-making.

AHP method is considered as one of the most common methods for Multi-Criteria Decision Making or MCDM, used primarily to solve unstructured problems in different areas of the study. This method is used mainly in two settings: (a) finding the weight or relative importance of the indicators that have been used in a given study, and (b) rankings options in relation to several indicators. In this study, we draw upon the first application of AHP method, where paired comparisons between indicators are established. After obtaining the relative weight of each of the indicators, the Inconsistency Rate (IR) is calculated. To this end, first, we should calculate the vector of the weighted sum in which the paired comparison matrix will be multiplied by the vector of relative weight, resulting in a matrix so-called the vector of the weighted sum. Finally, the consistency vector will be obtained through dividing the elements of the vector of the weighted sum to relative weights.

## RESULTS

In this section, findings are presented in two sub-sections (a) ranking indicators based on a total of six criteria outlined in methodology, and (b) ranking components and their associated indicators relating to sustainable aquaculture using AHP.

### Ranking indicators based on a total of six criteria

The average of the total of six criteria (0-5 for each criteria), (measurement capability; scientific relevance; policy relevance; sensitive to changes; simplicity; and economic justification) in relation to each variable was calculated and presented in Table 2.

### Ranking components and their associated indicators related to sustainable aquaculture using AHP

In order to study sustainable aquaculture indicators, we first should culminate in paired comparisons from utilizing questionnaires completed by 32 experts. After norming the pairwise comparison matrices, the indicators were ranked, mainly through obtaining the arithmetic mean of each row of the matrix and multiplying the relative weights in the obtained mean. The next step is to calculate the largest eigenvalues of pairwise comparisons matrix in order to come to indicators and the IR. For the paired comparisons in the present study, IR was close to 1.0, showing that the obtained relative weights are the valid and reliable collective opinions. Figures 1, 2 , and 3shows the relative weight of dimensions and their associated indicators related to sustainable aquaculture, respectively

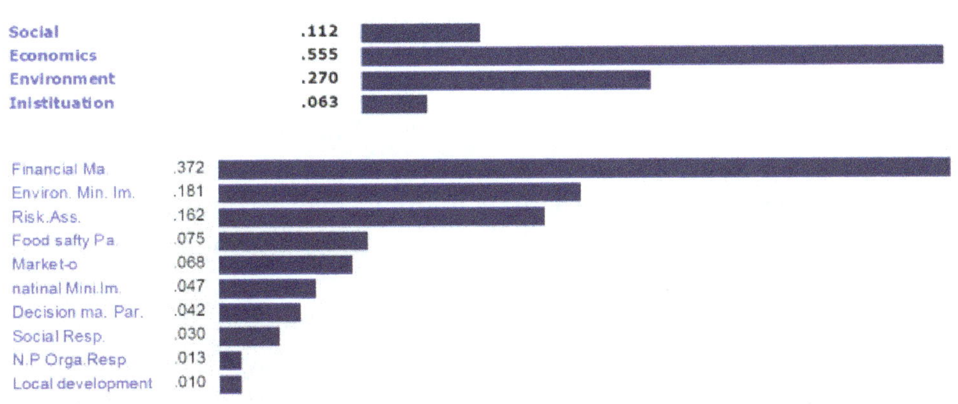


*Figure 2.* The relative weight of each indicator of sustainable aquaculture

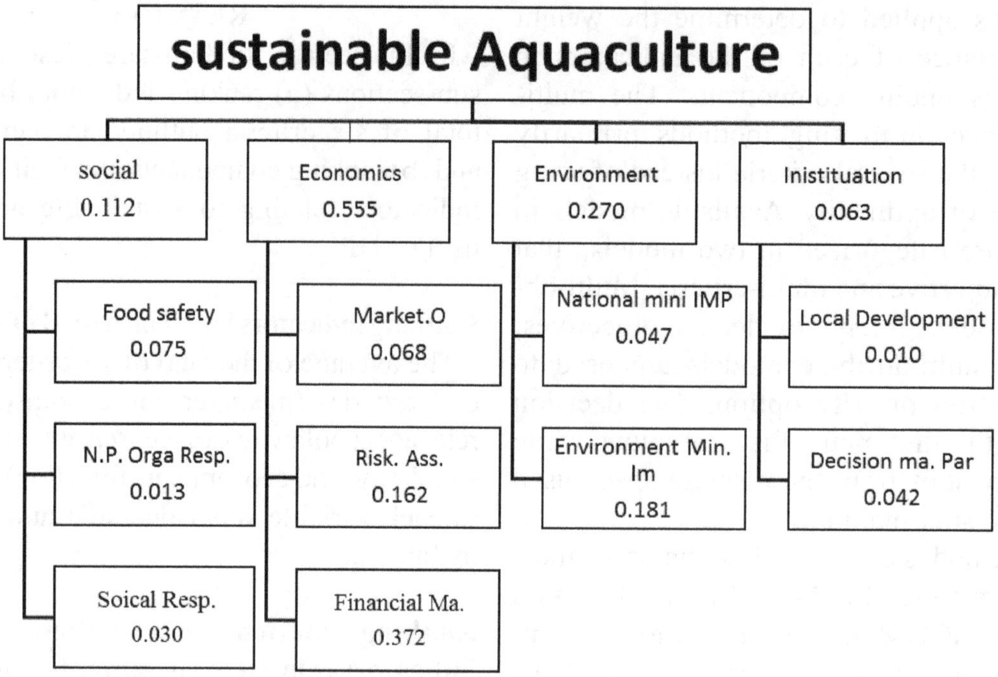


*Figure 2.* Hierarchal map of each indicator of sustainable aquaculture

## CONCLUSION AND RECOMMENDATIONS

The loss of potential water resources, forests, and fisheries has been a critical problem around the world. Understanding the processes that lead to the improvement or exacerbation of natural resources are frustratingly limited, mainly because of the various languages and concepts used in scientific disciplines to describe complex socio-ecological systems. There is no ability to access multidisciplinary knowledge without having a common framework to organize the findings (Ostrom, 2009). In this way, the principles of "systematic approach to managing the sustainability of aquaculture" has been developed by (Bunting, 2013). Components or features that interact in an aquaculture system include moral, cultural, social-political, economic, climatic, agricultural- environmental, water and weather, production, resource utilization, mass balance, and biological.

Generally, the stages of using a system-based approach include the definition of cultural practices under investigation and the determination of the scope and boundaries of the system based on the physical area occupied, as well as the comprehensive environmental impact assessments or zero assessment to 100 products, including input, by products, and waste disposal. The next steps in this approach include studying the relationships between the components within the system, evaluating the context, describing the system, checking performance, controlling system behaviour, describing the practical implications, and the effects of the system output. The subsequent combination of these components should lead to a better understanding of the context of the overall system, making it possible to sort different components of the system and provide more details about the choices made, as well as the required predictions related to the outcomes and consequences of the system (Bunting, 2013). In the light of such a systemic view, the present study aimed to develop and validate indicators of sustainable aquaculture in the Alborz Dam Basin, Mazandaran Province. One hundred and sixteen sub-indicators (four components and 10 indicators) were identified and ranked. The ranked priorities were: economic, environmental, social, and institutional components. What is certain is that since most of aquaculture growers in the study area were holding small-scale aquaculture, it is necessary to give more attention to economic aspects and provide financial facilities for them to use modern methods related to aquaculture in accordance with the environmental considerations, a finding supported by experts' opinions.

The findings further indicated that economic and environmental aspects also were placed in the first and second ranks among 10 sustainable aquaculture indicators. Therefore, it is highly recommended that attention to these identified dimensions and planning according to their corresponding indicators should be the focus of agencies related to developing the country's aquaculture development agenda, where developing a comprehensive plan for sustainable aquaculture development based on these identified priorities can accelerate the growth of sustainable aquaculture in the country to provide healthy products for the consumer along with taking into account all environmental effects it matters.

## ACKNOWLEDGEMENT

Authors would like to thank aquaculture experts in Alborz Dam Basin in Mazandaran Province for their cooperation in this study and their time completing the questionnaires.

## REFERENCES

1- Bunting, S.W. (2013). Principles of sustainable aquaculture.London and New York, Rout ledge press.

2- Costa-Pierce, B.A., & Page, G.G. (2010). Sustainability science in aquaculture In: Costa-Pierce, B.A. (ed.),Ocean Farming and Sustainable Aquaculture Science and Technology. Encyclopedia of Sustainability Science and Technology. Springer Science.

3- FAO (1997). Aquaculture Development. Technical guideline for responsible fisheries. Rome, No. 5

4- FAO (1999). Indicators for sustainable development of marine capture fisheries. Technical Guidelines for Responsible Fisheries, Rome, No.8. FAO (2011). Indicators for the sustainable development of finfish Mediterranean aquaculture: highlights from the InDAM Project, Rome. Available at http://agris.fao.org/agris-search/search.do? recordID=XF2006446683.

5- FAO (2013). Indicators for sustainable development of aquaculture and guidelines for their use in the Mediterranean background information regional workshops on the identification of reference points for economic, environmental, social and governance indicators on aquaculture. Izmir 9-19 December 2013.

6- Jafarian, H. (2006). Aquaculture development using by Probiotics in Iran. Fisheries Journal. 2 (4), 36-48.

7- Mokhtari Abkenari, A., Chizari, M. & Salehi, H. (2004). Investigation of Iranian aqua culturist attitude towards sustainable aquaculture.

8- Agricultural Education Extension Science Journal, 2 (2), 14-20.

9- Nunes, J.P., Ferreira, J.G., Bricker, S.B., O'Loan, B., Dabrowski, T., Dallaghan, B., Hawkins, A.J.S., O'Connor, B., & Carroll, T.O. (2011).

10- Towards an ecosystem approach to aquaculture: Assessment of sustainable shellfish cultivation at different scales of space, time and complexity.Aquaculture, 315 (13), 369–383.

11- Ostrom, E. A. (2009). General framework for analyzing sustainability of social-ecological systems. Science, 325 (5939), 419-422.

12- Statistical Yearbook of the Iranian Fisheries Organization. Iranian Fisheries Organization (2010). Planning and Development management. Tehran.

13- Stickney, R. R., & McVey, J. P. (2002). Responsible marine aquaculture, New York, CABI Publishing.

14- Ting, K,H., Kun-Lung, L., Hao-Tang, J., Teng-Jeng, H., Chi-Ming, W., & Wen-Hong, L. (2015).

15- Application of a Sustainable Fisheries Development Indicator System for Taiwan's Aquaculture Industry. Aquaculture, 437(15), 398–407.

16- World Bank (2007). Changing the face of the waters: the promise and challenge of sustainable aquaculture. Agriculture and rural development. Washington, DC: World Bank.

# 20

# Enhancing Cocoa Beans Quality through Improved Cultural Practices in Major Cocoa Producing States of Nigeria

*Adigun James Kehinde [1], Ashimolowo Olubunmi R [1], Adeogun Stephen Oluseun [2], Mesele Samuel Ayodele [3] and Fabusoro Eniola [1]*

Abstract

In West Africa, low adoption of agricultural innovations by farmers has been partly due to the complexity of the innovation structure or new knowledge beyond the farmers' socio-economic adaptive capacities. Different attempts have been made to improve cocoa beans quality in Nigeria among which was the recommendation of 'improved cultural practices' or 'best known practices' (ICMP). This had recorded little success because of the complexity and enormity of the ICMP. The present study, therefore, made an attempt to resolve this, in the major cocoa growing states of Nigeria, through farmers-researcher participatory approach with the aid of interview guide. Collected data were analyzed using descriptive statistics and the Principal Component Analysis procedures in SPSS to identify the minimum essential practices out of the overall ICMP for sustainable cocoa production. The result showed that cocoa farmers practiced 47 activities recognized as ICMP which were grouped as nursery, on farm, harvesting or post-harvest activities. The highest overall percentage use of any of these activities was 3 %, while the least being smaller than 1 %. It was found that 13 out of 47 activities contributed 71.45 % to the overall ICMP, whereas the current cumulative percent use of these 13 activities was 28 %. The individual percent use of each of the 13 activities was about 3 %. The highest contributors among the 13 activities were coded ba1 and ba11 while the least was bb4. Efforts to intensify the use of these 13 activities rather than the 47 activities should be encouraged among farmers for sustainable and high quality cocoa production.

***Keywords:***
***best practice, cocoa, cocoa bean quality, improved cultural practice, Nigeria***

[1] *Department of Agricultural Extension and Rural Development, Federal University of Agriculture Abeokuta Nigeria*
[2] *Department of Agricultural Administration, Federal University of Agriculture Abeokuta Nigeria*
[3] *Department of Soil Science and Land Management, Federal University of Agriculture Abeokuta Nigeria*
* *Corresponding author's email: ayodelemesele@hotmail.com*

## INTRODUCTION

Annually, about 3 million tonnes of cocoa beans are used for the production of a wide range of chocolate and cocoa flavored foods globally (Badrie et al., 2015, Hajra & Yang 2015). In West Africa, the average cocoa farm has been estimated to vary from 3 to 4 hectare plot, operated by a family that lives on the farm or nearby (Akanni and Dada, 2012). Estimates place the number of West African cocoa farms at 1.5 to 2 million (Yusuf, 2014). Nigeria comes fourth after Ivory Coast, Ghana and Indonesia being the world leading cocoa producer and exporter (Egwuatu, 2015). Cocoa contributes about 20.6 % to Nigeria foreign exchange earnings (Government of Nigeria, 2015). Many factors, however, constrained the quantity and quality of cocoa beans production making it difficult for farming families to realize the true potential of cocoa farming (Aneani et al., 2012).

These factors could be generally grouped under three major categories which are genetic, edaphic and management factors. The genetic factors have to do with the varieties of cocoa seeds or seedlings and this varied widely in Nigeria ranging from very poor and low yielding varieties to exotic or high yielding inbred lines. The availability of these high or improved varieties to the smallholder farmers still remains a source of concern in Nigeria. While much improvement or success has been recorded in the area of genetics (McMahon et al., 2009), the edaphic or management aspect of cocoa production still remain under developed.

The edaphic factor has to do with the soil fertility and fertility constraints as well as the climatic conditions such as the rainfall amount, relative humidity, temperature and the amount of sunshine hours (Shamshuddin et al., 2011). Basically the management factors include agronomic cultural practices such as the time of planting, site selection, type of soil amendments (Onwuka et al., 2007) and its methods of application (Padi & Owusu, 1998). All other factors constant, these practices could help improve the quantity and quality of cocoa beans and reduce the infestation of pests and disease (McMahon et al., 2009).

The International Cocoa Organization (ICCO, 2008) has however recognized this and had given out a long lists of activities tagged as best known practices or improved cultural management practices which were supposed to be adopted by farmers so as to achieve sustainable cocoa production, covering the aspects of economic, social and environmental sustainability. These improved cultural management practices have been disseminated among the smallholder farmers through agricultural extension agents, farmer organizations and national research institutes but the success of this in terms of its adoption has been considerably low, hence the low ranking of the Nigerian Cocoa beans in the world markets.

ICCO (2008) further stipulates that no effort should be spared to implement recommended best known practices or ICMP in order to ensure high quality cocoa. From choosing the structure and texture of the soil, preparing and establishing the farm, selecting the planting material, managing the farm, harvesting, through to processing and quality control, were all noted as vital factors and should all be implemented by cocoa producers. Though the ICMP has a sound scientific background but its wide spread adoption has been constrained by farmers' socio-economic conditions particularly in the Cocoa producing states of Nigeria (Oluyole, 2009; Aneani et al., 2011).

In a bit to identify why this recommendation had recorded little success. This research made considerable efforts to identify 47 different improved cultural management practices under the ICCO recommendations to the cocoa growers. In cocoa growing states in Nigeria, there are farmer organizations sometimes called father's schools, where farmers are trained and assisted in sustainable cocoa production. These 47 improved cultural management practices are mandated to be adopted by the farmers on their farmers. While many of those trained farmers tried to adopt this, the quality of their quantity and quality of their cocoa beans are yet to receive a significant improvement. The huge list of 47 improved cultural management practices could somehow create drudgery and discouragement to the farmers (Oluyole, 2009). This study therefore seeks to resolve this through factor and principal components analysis by ascertaining those practices out of the 47 recommended practices that could give similar contributions as the 47 recommended practices to cocoa beans quality. Similarly, the research questions were formulated to know the percentage of use of the various ICMP by farmers, and what combinations of improved cultural management practices would give optimum cocoa beans quality which farmer could adopt when having difficulties in adopting the overall ICMP.

## MATERIALS AND METHODS

### Description of study location

The study was carried out in South West Nigeria. South West Nigeria comprises six states, namely, Ekiti, Lagos, Ondo, Ogun, Osun, and Oyo States. It is bound by the Atlantic Ocean in South, Kwara State in the North, and the Republic of Benin in the West. The Southwest region, which has a land area of about 114,271 square-kilometer (about 12% of total land mass of Nigeria) (Government of Nigeria, 2015), lies between latitude 4o 20' and 9o 23' North of the equator and longitude 2o 25' and 6o 31' East.

The target population of this study consisted of farmers from Cocoa Grower Association of Nigeria (COGAN) in Ondo and Osun States. Multi-stage sampling technique was employed to select respondents for this study. The first stage is the purposive selection of Ondo and Osun States being the highest cocoa producing states in South West Nigeria. The second stage involves the purposive selection of the two highest cocoa producing Local Government Areas in each selected state, that is, Idanre and Ondo-East Local Government Areas from Ondo State and Ife-south and Ife-east Local Government Areas from Osun State. In Ondo State, 16 Local Governments produce cocoa and fourteen Local Governments from Osun State (NCARD, 2010; NCDC, 2010a, 2010b) out of which 10% of each is selected. The third stage involves the purposive selection of three cocoa producing communities where the members of Cocoa Grower Association of Nigeria domicile were selected from the Local Governments Area of each state. The fourth stage involved the random selection of 50% cocoa farmers registered with cocoa grower association in the selected communities.

The sample size for this study was, therefore, two hundred and eighty-three respondents.

The primary data for this study were collected from the respondents using interview guide where four facilitators from Sustainable Tree Crop Production unit of IITA were employed to get the target respondents from the selected local government areas. Well-trained enumerators were also employed to help aid in data collection. Relevant information was also sourced from the literatures, past research works, internet, and libraries. The variables were measured with a 3-point rating scale of regularly (2), occasionally (1), and never (0).

List of ICMPs and their codes

Nursery operations or activities (Ba):

Selection of suitable nursery site close to plantation (Ba1)

Preparation of ground surface-level off and install drains (Ba2)

Construction of nursery shed (Ba3).

Preparation of parameter fence (Ba4)

Filling the polythene bags (15cmx30cm) with potting mix or top soil (Ba5)

Line bags in nursery (row of 8- 40cm air gap) to maintain air space (Ba6)

Selection of improved seed variety (Ba7)

Clean seeds and dip in fungicide i.e. thiram (Ba8)

Collection of germ seeds at three to five days and plant into pots (Ba9)

Removal of twisted, deformed seedling (Ba10)

Regular browsing on the nursery beds (Ba11)

Hardening-off plants ten to fourteen days before transplanting (Ba12)

ICMPs of farm activities (Bb)

Pruning of cocoa trees during raining season (Bb1)

Weeding the cocoa farm at least twice a year (Bb2)

Regular removal of dead (mummify pods) (Bb3)

Regular removal of mistletoes (Bb4)

Regular removal of stagnant water from the farm (Bb5)

Removal of cocoa pod husk from the farm (Bb6)

Removal of coupons from the trees (Bb7)

Planting of improved/disease resistant variety to replace the older trees (Bb8)

Fertilizer application (three bags/ acre) (Bb9)

Regular removal of mosses (Bb10)

Regular removal of epiphytes (Bb11)

Spot application of recommended fungicides/ insecticide (Bb12)

Application of plant extract (Bb13)

Burying of removed diseased pods outside cocoa farm (Bb14)

Spraying before pre-harvest interval (Bb15)

Regeneration of coupon where necessary (Bb16)

Maintenance of planting space of 3.1mx3.1m (Bb17)

Sourcing for improved seeds from research institute and ADPs (Bb18)

Harvesting activities (Bc);

Pods are harvested at two weeks interval (Bc1)

Sharp harvesting hook/sickle is use to harvest yellow pods (Bc2)

The stalk between the tree and the pod is carefully cut (Bc3)

The bark of the trees are not cut or wounded with knife/cutlass (Bc4)

Pods are not pulling off by hand during harvest (Bc5)

ICMPs of on-farm processing/post-harvest activities; (Bd)

Harvested pods are stored for a few days (Bd1)

Wounded pods are not store more than one day (Bd2)

Off-farm pods breaking (Bd3)

Use of club or wooden mallet to break the pods (Bd4)

Beans are fermented for five to six (5-6) days (Bd5)

Use of banana leaves to cover heap or fermentation box (Bd6)

Turning of the beans once daily from day three (Bd7)

Dry the fermented beans for seven days (Bd8)

Use of a raise platform or flat cement concrete to dry beans (Bd9)

Thorough sorting during drying (Bd10)

Bagging dried beans with jute bag (Bd11)

Store the dried beans in a cool dry place free from smoke (Bd12)

Principal Component Analysis procedure (CPA) was employed to analyze the data collected.

Model specification

$P_1 = a_{11}X_1 + a_{12}X_2 + \ldots\ldots\ldots\ldots a_{1k}X_k$

$P_2 = a_{21}X_1 + a_{22}X_2 + \ldots\ldots\ldots\ldots a_{2k}X_k$

$P_3 = a_{31}X_1 + a_{32}X_2 + \ldots\ldots\ldots\ldots a_{3k}X_k$

$P_4 = a_{41}X_1 + a_{42}X_2 + \ldots\ldots\ldots\ldots a_{4k}X_k$

P = Principal component of ICMPs activities

$X_j$ = variables (j=1, 2, 3……k).

## RESULTS AND DISCUSSION

### Contributions of some nursery activities to ICMPs

The principal component analysis of the nursery practices (see Table 1) showed that activities such as Ba1, Ba11 or Ba9 contributed 50.85% with the percentage use of 10.0%, 10.0%, 5.0% respectively to the improved cultural management practices (ICMPs). Ba8 or Ba9 activity contributed 16.07%, while Ba9 or Ba8 in the third component contributed 13.26% with the percentage use of 5.0% each.

Farmers engaged in different nursery activities under the ICMP. The results demonstrated that farmers had greater use of Ba10 among other nursery activities. The percentage contribution of the Ba10 activity to the overall ICMP was, however, very low. It was found that practices such as Ba1, Ba8 and Ba9, which cumulatively contributed 80.14 to the ICMPs, had low level of adoption (10, 5 and 5.0%, respectively). It could, therefore, be argued that the combination of Ba1, Ba8 and Ba9, which gave greater return to ICMPs, should be promoted among farmers rather than the general activities which individually or cumulatively contributed less to the ICMPs. This fidning concords with the report of Vos and Ritchie (2003) who identified some crop growing methods that could be adopted for greater yield. An increase in the percentage use of these components (Ba1, Ba8 & Ba9) could help raise the productivity level and the quality of cocoa beans, whereas reducing the

Table 1
*Percentage Contributions of Some of the Nursery Activities to the ICMP*

| Nursery Practices | % Contribution to ICMP | Cumulative % | % Use |
|---|---|---|---|
| Component 1 (Ba1 or Ba11 or Ba3) | 50.85 | 50.85 | 10, 10 or 9 |
| Component 2 (Ba8 or Ba9) | 16.07 | 66.92 | 5, 5 |
| Component 3 (Ba9 or Ba8) | 13.26 | 80.18 | 5, 5 |

drudgery associated with many activities among the cocoa farmers.

**Contributions of some of the on-farm activities to the ICMP**

From the results presented in Figure 1, on-farm activities such as Bb8, 7, 17, 4, and 5 had the greatest use of 8 % each among the cocoa farmers. Further analysis showed that five activities such as Bb 8, 7, 14, 13, 17, 4 and 11 out of the 18 on-farm activities had the highest contributions to the ICMPs. A comparative analysis of the % contributions and % use (see Table 2) pin-points that activities such as Bb 8, 7, 17 and 4 had significant contributions to the ICMP and are well used among the farmers while activities such as Bb 14, 13 and 11, even though they had made great contributions to ICMPs and were less practiced. Table 2 also showed that 7 out of the 18 on-farm activities (Figure 1) contributed 71.12 % to the ICMPs. Increased awareness and use of these activities could, therefore, raise the adoption of ICMPs, reduced farmer's drudgery due to practicing so many activities, and eventually improve the quality of cocoa beans produced. This result is in conformity with the submission of Adigun (2013) who opined that farmers need to be informed that proper practices of cultural management will reduce chemical application, increase yield, and enhance quality.

**Contributions of harvesting activities to ICMP**

Figure 2 shows the percentage distribution of use of the five harvest activities among farmers. The percentage use of the various activities was 20 % each. This indicated that there was an even distribution in terms of use of the activities among farmers. The principal components analysis of these activities (see Table 3) revealed that Bc 4 had 43.9 % contributions to the ICMPs. Bc4, 1 and 3 activities cumulatively contributed 83.09% to ICMPs. The % use of these activities was 20 %. A slight increase in the % use of Bc4 while maintaining the good use of Bc1 and 3 would, therefore, be needed to increase the cocoa quality of the study area.

**Contributions and use of some of the post-**

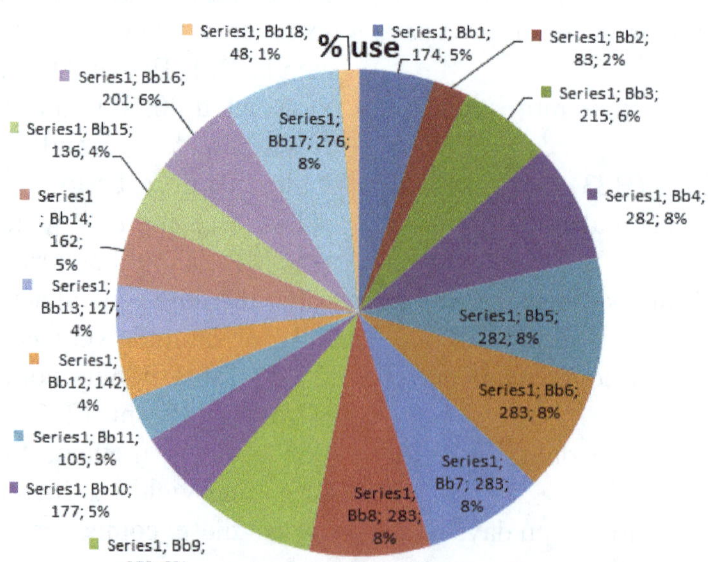


*Figure 1.* Percentage Use of the Various on Farm Activities by the Farmers

Table 2
*Percentage Contributions of Some of the on Farm Activities to the ICMP*

| On-farm activities | % Contribution to ICMP | Cumulative % | % Use |
|---|---|---|---|
| Bb 8 | 17.68 | 17.68 | 8 |
| Bb 7 | 13.88 | 31.56 | 8 |
| Bb 14 | 10.63 | 42.19 | 5 |
| Bb 13 | 9.74 | 51.93 | 4 |
| Bb 17 | 6.85 | 58.78 | 8 |
| Bb 4 | 6.50 | 65.28 | 8 |
| Bb 11 | 5.84 | 71.12 | 3 |

**harvest activities to ICMPs**

Among the various post-harvest activities used by the farmers in the study area, Bd 12 and bd 3 had the highest and the lowest use of 11 % and 1 %, respectively (see Figure 3). Assessment of the 12 post-harvest activities showed that 6 out of the 12 cumulatively gave 71.41 % contributions to the ICMPs (Table 4). These activities are Bd12, 5, 10, 9, 4 and 6. This implied that the intensive use of the 6 activities rather than the 12 could help achieve the objectives of the ICMPs that is, to minimize the problem of pest and disease, declined yield, and low or poor quality produce. Similarly, the percentage use of the 6 activities more or less followed the same trend as their percentage contributions to ICMPs, denoting that the farmers use more of those activities that gave them greater return to the ICMPs than those that had fewer contributions. Such Findings are inconsistent with those of Olujide and Adeogun (2006) who stated that few farmers appear to be adopting farm hygiene and management techniques to minimize the problem of pest and disease, declined yield, and low or poor quality produce. This is, however, a desirable trend that should be encouraged beyond the study area for higher and widespread cocoa quality production in the country.

**Contributions of some selected activities to**

Table 3
*Percentage Contributions of Harvesting Activities to ICMP*

| Harvesting activities | % Contribution to ICMP | Cumulative % | % Use |
|---|---|---|---|
| Bc 4 | 43.91 | 43.91 | 20 |
| Bc 1 | 21.10 | 65.01 | 20 |
| Bc 3 | 18.07 | 83.09 | 20 |

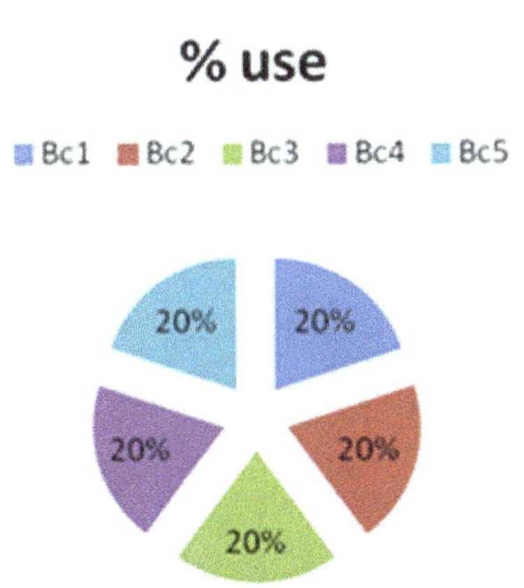


*Figure 2.* Percentage Use of the Various on Farm Activities by the Farmers

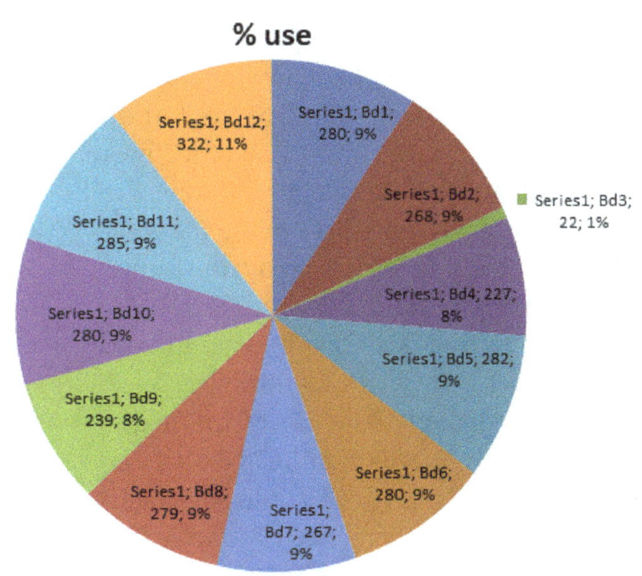


*Figure 3.* Percent Use of Post-Harvest Activities

Table 4
*Contributions and Use of Some of the Post-Harvest Activities to ICMP*

| Post-Harvest activities | % Contribution to ICMP | Cumulative % | % Use |
|---|---|---|---|
| Bd 12 | 18.55 | 18.55 | 11 |
| Bd 5 | 13.97 | 32.52 | 9 |
| Bd 10 | 11.91 | 44.44 | 9 |
| Bd 9 | 10.09 | 54.53 | 8 |
| Bd 4 | 9.13 | 63.65 | 8 |
| Bd 6 | 7.76 | 71.41 | 9 |

Table 5
*Percentage of Contributions of Some Selected Activities to ICMP and their % Use*

| Post-Harvest activities | % Contribution to ICMP | Cumulative % | % Use |
|---|---|---|---|
| Ba 1 or 11 | 15.72 | 15.72 | 2 |
| Bc 4 | 10.65 | 26.36 | 3 |
| Bb 13 | 8.17 | 34.53 | 1 |
| Bb 8 or 7 | 6.47 | 41.00 | 3 |
| Bb 6 or 17 | 5.49 | 46.50 | 3 |
| Bb 14 | 4.57 | 51.07 | 2 |
| Ba 8 | 4.57 | 51.07 | <1 |
| Bd 11 | 3.96 | 58.67 | 3 |
| Bd 1 | 3.94 | 61.62 | 3 |
| Bb 11 | 2.77 | 64.39 | 1 |
| Bd 8 | 2.56 | 66.94 | 3 |
| Bb 15 | 2.30 | 69.24 | 1 |
| Bb 4 | 2.21 | 71.45 | 3 |
| N0=13 | | | Sum=28 |

**ICMPs and their % use.**

Figure 4 showed that cocoa farmers of the study area practiced 47 activities recognized as ICMP be it nursery, on-farm, harvesting or post-harvest activities. The overall percentage use of any of these activities was 3 % while the least being <1 %. The principal component analysis of these activities to determine those activities that played the major role in achieving the objectives of the ICMP revealed that 13 out of 47 activities achieved 71.45 % of these objectives (Table 5). The highest contributors among the 13 activities were Ba1 and Ba11 while the least was Bb4. As at the time of this study, the % use of each of the 13 % activities ranged from < 1 – 3 %. Table 5 also indicated that the cumulative % use of these 13 activities identified was 28 %. Due to the relatively low use of the major activities, their net benefits from ICMP could as well be low. This study has however demonstrated that though the farmers practiced 47 activities recognized as ICMP, their net return was low because 13 out of these activities could actually achieved the objectives of the ICMP. This study is a response to the opinion of Tony (2004) that government of cocoa producing countries, the cocoa industry, international donors and other

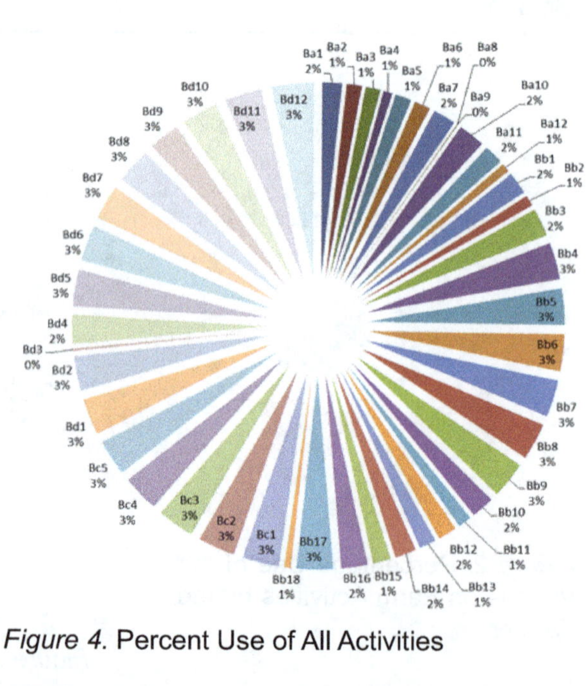


*Figure 4.* Percent Use of All Activities

stakeholder to invest in research into better, non-chemical management system to control pest and diseases. Farmers could therefore raise their productivity and their cocoa quality, by concentrating on those 13 activities identified in this study rather than the 47 activities currently in practice.

## CONCLUSION

This paper recognized that cocoa farmers in West Africa particularly Nigeria continually produced poor quality cocoa beans with very low premium price despite the introduction and some level of adoption of the improved cultural management practices (ICMP). The study, therefore, made attempts to analyze the percentage farmers' use of the ICMP and then quantifies the contributions of each of the activities to the overall ICMP. Juxtaposing these two levels of analyses, it was found that among the recommended 47 ICMPs in cocoa farming system, only 13 activities played the major role, whereas their current farmers' cumulative use is 28 %. Efforts to intensify the use of these 13 activities rather than the 47 activities should be encouraged among farmers for sustainable and high quality cocoa beans production.

## REFERENCES

1- Adigun, JK. (2013). *Analysis of cocoa farm cultural management techniques among cocoa farmers in South-West Nigeria.* A Seminar paper presented to the Department of Agricultural Extension and Rural Development, Federal University of Agriculture Abeokuta, Nigeria. Pp 97.

2- Akanni, K. A., & Dada, A. O. (2012). Analysis of Labour-Use Patterns among Small-Holder Cocoa Farmers in South Western Nigeria. *Journal of Agricultural Science and Technology, B, 2*(1B). Aneani, F., Anchirinah, V.M., Owusu-Ansah, F., & Asamoah, M. (2011). An analysis of the extent and determinants of crop diversification by cocoa (Theobroma cacao) farmers in Ghana. *African Journal of Agricultural Research 6*(18), 4277-4287.

3- Aneani, F., Anchirinah, V. M., Owusu-Ansah, F., & Asamoah, M. (2012). Adoption of some cocoa production technologies by cocoa farmers in Ghana. *Sustainable Agricultural Research, 1*(1), 99-103.

4- Badrie, N., Bekele, F., Sikora, E., & Sikora, M. (2015). Cocoa agronomy, quality, nutritional, and health aspects. *Critical Reviews in Food Science and Nutrition, 55*(5),620-659.

5- Bateman, R. (2008). Pesticide Use in Cocoa: A Guide for Training, Administrative and Research Staff (No. 632.95/B329). ICCO.

6- Egwuatu, C. (2015) Repositioning Cocoa As A Major Cash Crop In Nigeria. Online article, 05 March from http://www.nigerianorientnews.com/?p=4521.

7- Government of Nigeria (2015) National Bureau of statistics. Online article. Retrieved 20 May from http://www.nigerianstat.gov.ng/

8- Hajra, N. G., & Yang, C. W. M. (2015). Diversification of the Tea Products Global Scenario. *Journal of Tea Science Research, 5*(1), 23-26.

9- McMahon, P., Iswanto, A., Agung W.S., Sulistyowati, E., & Wahab, A. (2009). On-farm selection for quality and resistance to pest/diseases of cocoa in Sulawesi: Performance of selections against cocoa pod borer, Conopomorpha cramerella. *International Journal of Pest Management,55*(4), 325–337.

10- Olujide, M.G., & Adeogun, S.O. (2006). Assessment of cocoa growers farm management practices in Ondo State, Nigeria. *Spanish Journal of Agricultural Research. 4*(2), 173-179.

11- Oluyole, K. (2009). The influence of technological changes on labour Availability: A case study of cocoa farming households in Ogun state, Nigeria. *African Journal of Food Agricultural Nutrition and Development, 9*(7), 3-9.

12- Onwuka, Emmanuel Chike, & Agatha Eguavoen (2007). Globalization and Economic Development: The Nigerian Experience. *Journal of Social Sciences, 14*(1), 45-51.

13- Padi, B., & Owusu, G. K. (1998). Towards an integrated pest management for sustainable cocoa production in Ghana. Cocoa Research Institute of Ghana. *Ghana Journal of Agricultural Science, 18*, 33-42

14- Shamshuddin, J., Anda, L.M, Fauziah, C.I., & Syed omar, S.R. (2011). Growth of Cocoa planted on highly weathered soil as affected by application of basalt and/or compost. *Communication in Soil Science and Plant Analysis, 42*, 2751–2766.

15- Tony, L. (2004). *Opportunities and threats to cocoa consumption.* In: Cocoa future, a source book of some important issues facing cocoa industry (Flood J., Murphy R., eds). CABIFEDERACAFE, USDA, Chinchina (Columbia), pp. 25-32.

16- Vos, J.G.M., & Ritchie, B.J. (2003). *Discovery learning about cocoa. An inspirational guide for training facilitators.* CABI Biosciences UK. pp. 4-7.

17- Yusuf, S. A. (2014). Role of Agriculture in Economic Growth & Development: Nigeria Perspective. *Spanish Journal of Agricultural Research, 12*(1), 17-27.

# 21

# Assessment of Small Scale Farmers' Skills Regarding Integrated Pest Management (IPM) in District Sargodha-Pakistan

*Ejaz Ashraf [1]*, Abu Bakar Muhammad Raza [1], Samiullah [2] and Muhammad Younis [3]*

Abstract

A survey study was conducted to assess the knowledge /awareness level in IPM technology among farmers. Four villages were randomly selected from Sargodha district for data collection. Thirteen farmers from each village were selected randomly and sample size was 52 respondents. More than 92% of respondents have no advisory services either from public or private sector. The findings imply that respondents need knowledge for all levels of competence in IPM technology. They need to get high-level of competence for application of this technology in the field. In addition, they have little exposure to long-term training opportunities due to low education level for applications of this technology. More than 77% of respondents think that government agricultural policies and no access to information sources regarding integrated pest management at grass-root level are main constraints. The findings from correlation and regression analyses indicate that age and knowledge/awareness level are negatively correlated. It may be concluded that elder respondents have less adaptability to new ideas and techniques as compared to young respondents. However, training and information, education, and experience play a significant role in enhancing the knowledge/awareness level of respondents in IPM technology.

***Keywords:***
***Pest Management, Training, Information, Knowledge, Awareness***

[1] *Assistant Professor, Agriculture. Extension Education. University College of Agriculture, University of Sargodha, Punjab-Pakistan.*
[2] *Lecturer in Statistics, University College of Agriculture, University of Sargodha, Punjab-Pakistan.*
[3] *University College of Agriculture, University of Sargodha, Punjab-Pakistan*
* *Corresponding author's email: ejazashraf60@hotmail.com*

## INTRODUCTION

Technologies are always being used as essential tool for development and so does Integrated Pest Management (IPM) in modern agriculture. This technology could be used as alternative for curtailing the use of pesticides in crop management. Morse *et al.*, (2000) described that IPM is constantly referred to as one of the main elements in the development of sustainable agriculture. They further described in their study that goals and philosophies of sustainable agriculture and those of IPM have significant similarities. The crisis of uncontrolled use of pesticides required urgent solution, and IPM is regarded as an alternative pest management and solution to the present problems of our growers. The variety of crop protection methods are being used such as crop rotation, biological, and limited use of chemicals.

The IPM technology is driven by many principles, which require careful application of effective pest management. Integrated pest management (IPM) is regarded as the future plant protection model for developing countries and ideal pest management approach (Altieri, 1993). It is expected that IPM has potential to sustain and improve yields, to lower the dependence on pesticides, to decrease the cost of production, and to reduce negative impacts on the environment and human population (Erbaugh *et al.*, 2001). Extensive research is going on in different parts of the world to examine the benefits of this technology, and to check the awareness and knowledge level of farmers for adaptation of IPM technology.

There is a dire need to implement this technology at grass root level for pesticides free crop yields. However, the question is how the goal of implementation could be achieved. The knowledge and awareness about the use of this technology among end users is significant for its application. Moreover, training and information may also play an important role in enhancing the knowledge level of growers in IPM technology.

Knausenberger *et al.*, (2001) pointed out that access to information on IPM practices and implementation can result in IPM becoming a sustainable crop protection method in Africa. Kyamanywa (2001) noted that in Uganda, extension agents recommend the use of chemicals and overlook cultural tactics due to limited information on its application. The role of extension workers is more important than any other factor in adoption of IPM technology at grass root level. Providing knowledge and information to growers are the core responsibilities of extension workers. The extension system could play a significant role in proper adoption of IPM approaches and for sustainable agricultural. Norris *et al.,* (2003) indicate that the economic reliability of IPM practices is particularly crucial, and to a great extent, determines adoption. IPM practices must reduce the risk of crop loss or encounter rejection by farmers.

The modern research tells that pest management is crucial for any farming system. Plants and animals must be protected from damage caused by insects, weeds, nematodes and other pathogens. Due to relaxed policies in pesticide use regulations in developing countries, farmers are continuously relying on chemical without considering their negative effects on human and animal health and also for environment. It is important to create awareness and boost up knowledge level of growers for using IPM technology as a solution to the prevailing problems in agriculture sector. IPM is regarded as a practical alternative in pest management for overcoming issues that arise from increased pesticide use (Afreh-Nuamah, 2001).

## MATERIALS AND METHODS

A survey study was conducted to assess the knowledge /awareness level in IPM technology of small scale farmers. Erbaugh *et al.*, (2001) conducted a similar study to evaluate farmers' knowledge and awareness regarding IPM in a collaborative research support project in Uganda. The present study was conducted in Sargodha district of Punjab province of Pakistan. The two stage random sampling procedure was used to collect data from respondents. Four villages were randomly selected from Sargodha district and then13 farmers from each village were selected randomly. Therefore, the sample size was 52 respondents. A survey questionnaire was used as an instrument for this study. The instrument was pre-tested on 10 farmers and then finalized by a panel of experts after making number of corrections. In addition, instrument was also tested for reliability by calculating the Cronbach's alpha using the data obtained from pilot study and found reliable. The instrument consisted of four sections demographic profiles of respondents, constraints in using IPM, training and information, and knowledge/awareness level of respondents.

## Objectives of the Study

The study was conducted to achieve the following objectives.

1- To describe the demographic profiles of the respondents

2- To identify the knowledge/awareness level of respondents in specific concepts of IPM technology.

3- To identify the opportunities of training and information in IPM technology for respondents.

4- To describe constraints those limit the applications of IPM technology.

5- To identify the relationship among the variables such as training/ information for IPM technology, demographic profiles, and knowledge/awareness level of respondents in specific concepts of IPM technology.

## Data Analysis

The collected data was coded and entered into computer for further analysis. The Statistical Package for Social Sciences (SPSS) 15.0 was used for analysis. The descriptive statistics such as mean, standard deviation, frequency distribution and percentages were computed for general description of the data. The multiple regression analysis was also used to predict how much of variance in dependent variable of knowledge / awareness level accounted for by independent variables of training/information and demographic profiles of the respondents. Multiple regression procedures also make it possible to observe the relationships between each of independent variable and dependent variable while controlling for other variables in the model (Urdan, 2001). Inter-correlations also computed to show strength of mutual relationships between the variables.

## RESULTS, DISCUSSIONS, AND CONLUSIONS

The demographic profiles were age, education, years of experience in farming, and advisory services. Average age of the respondents was

Table 1: Means, standard deviations, and ranks of knowledge/awareness level in specific Concepts of IPM technology as perceived by the respondents

| Knowledge / Awareness level in IPM Technology | N | Mean* | SD | Rank |
|---|---|---|---|---|
| Harmful effects of Pesticides use | 52 | 3.58 | 0.99 | 1.0 |
| Prevention is better than cure | 52 | 3.27 | 1.43 | 2.0 |
| Important Insect/Pests Knowledge | 52 | 2.77 | 0.73 | 3.5 |
| Pruning and Thinning can minimize pest population | 52 | 2.77 | 1.00 | 3.5 |
| Resistant varieties against different Insect/Pests | 52 | 2.56 | 0.61 | 5.0 |
| Timely sowing can reduce Insect/Pests population | 52 | 2.52 | 0.80 | 6.0 |
| Damage Symptoms of Insect/Pests | 52 | 2.48 | 0.80 | 7.0 |
| Insect/Pest/Control | 52 | 2.38 | 0.77 | 8.0 |
| Susceptible Stages for Insect/Pests attack on plant | 52 | 2.17 | 0.90 | 9.0 |
| Use of light traps Or Pheromone traps | 52 | 2.12 | 1.04 | 10.0 |
| Inter-cropping and crop rotation is essential for pest control | 52 | 2.04 | 0.93 | 11.0 |
| Techniques of Control Insect/Pests | 52 | 1.96 | 0.79 | 12.0 |
| Other methods to control attack of Insect/Pests | 52 | 1.75 | 0.84 | 13.0 |
| Pesticides as a last option to control attack of Insect/Pests | 52 | 1.67 | 0.78 | 14.0 |
| Alternative use of pest control measures are effective | 52 | 1.63 | 0.82 | 15.0 |
| Use of Fertilizers to control Insect/Pests population | 52 | 1.62 | 0.57 | 16.5 |
| Options available to control Insect/Pests other than pesticides | 52 | 1.62 | 0.72 | 16.5 |
| Selective use of Pesticides to control attack | 52 | 1.58 | 0.80 | 18.0 |
| Inter-cultural Practices to control attack of Insect/Pests | 52 | 1.37 | 0.56 | 19.0 |

*Mean: 1= None, 2=Low, 3=Moderate, 4=High, 5=Very high

46 years. More than 67% of the respondents were of the age from 40 to 59 years. Of the 52 respondents, almost 35% were educated between grade eight (Middle) to grade ten (Metric) and only 10% were above metric. These results indicate that respondents belong to mature age group with low education level. Approximately 46% of respondents have farming experience from 21 to 30 years. Almost 19% of respondents have 31 to 40 years farming experience Unfortunately more than 92% of respondents have no advisory services either from public or private sector.

The results indicated that respondents have moderate knowledge/awareness level in two concepts of IPM technology such as "Harmful effects of Pesticides use (3.58) and Prevention is better than cure (3.27)". The respondents have low level of knowledge and awareness in "Important Insect/Pests (2.77) and Pruning and Thinning can minimize pest population with mean of (2.77)". Of the 19 specific concepts of IPM technology, the respondents were considered having "No" knowledge/awareness in 8 specific concepts of IPM technology as shown in the following table.

These findings imply that respondents have none to moderate knowledge and awareness in all specific concepts of for IPM technology which does not provide enough competence for applications of this technology in the field. The respondents need to get high-level of competence for applications of this technology. Ashraf (2007) reported similar findings in his doctoral study while assessing in-service educational needs of agricultural officers for adaptation of remote sensing technology for precision agriculture in the province of Balochistan-Pakistan.

The respondents were asked to rate the accessibility of training and information opportunities on a 5-point Likert scale ranging from none to very high (1-5). The means ranged from 1.21 (none) for "ETL/EIL knowledge" to 3.13 (moderate) for "other farmers as source of information" shown in the table given below.

The respondents described that they have "none" to "moderate" access to the training/information opportunities for IPM and its applications in agriculture. These findings indicate that respondents have limited access to training/information opportunities. The results further confirm that extension system in Pakistan does not provide enough training and information to clients for applications of IPM technology in the field. One of the reasons is, the extension officers themselves have limited options of training for this technology. It also implies that respondents have limited interaction with other professionals/researchers for applications of IPM technology. Consequently, they have little exposure to the long-term training opportunities due to low educational level.

The respondents also pointed out the constraints in applications of IPM technology. Few of them feel that inputs and finance are major constraints However, 77% respondents think that government agricultural policies and no access to information sources regarding integrated pest management

Table 2: Means, standard deviations, and ranks for training and information opportunities for IPM technology

| Sources of Training/Information for IPM | N | Mean* | SD | Rank |
|---|---|---|---|---|
| Other farmers | 52 | 3.13 | 0.79 | 1 |
| Electronic Media | 52 | 2.50 | 0.64 | 2 |
| Print Media | 52 | 1.88 | 0.78 | 3 |
| Interactions with researchers | 52 | 1.65 | 0.81 | 4 |
| Extension worker | 52 | 1.50 | 0.61 | 5 |
| Training Sessions | 52 | 1.48 | 0.77 | 6 |
| Bulletins | 52 | 1.46 | 0.58 | 7 |
| Seminars | 52 | 1.44 | 0.70 | 8 |
| Publications | 52 | 1.36 | 0.60 | 9 |
| Short Courses | 52 | 1.28 | 0.54 | 10 |
| ETL/EIL Knowledge | 52 | 1.21 | 0.41 | 11 |

*Mean: 1=None, 2=Low, 3=Moderate, 4=High, 5=Very high

Table 3: Regression analysis for Knowledge/awareness (DV)

| Variable | Coefficients | S.E. | t | Sig. |
|---|---|---|---|---|
| Training and Information | 1.044 | 0.190 | 5.482* | <0.001 |
| Experience | 0.804 | 0.142 | 5.663* | <0.001 |
| Education | 3.680 | 0.795 | 4.628* | <0.001 |
| Age | -0.614 | 0.162 | -3.779* | <0.001 |
| Constant | 18.570 | 6.025 | 3.081* | 0.003 |

$R^2$ =.783; Adj. $R^2$ =.765; df= 4, 47; F= 42.505 *Significant at .05 level

at grass-root level are main constraints.

A regression model was used with independent variables of training/information, experience, education, and age of respondents. The model was statistically significant $p < 0.0001$ and explained 78.34% of the variance for knowledge/awareness level. The results indicated that independent variables such as training and information, experience, education, and age played a significant role in knowledge /awareness level among respondents in specific concepts of IPM technology. In addition all demographic variables were statistically significant.

The findings from correlation and regression analyses also indicate that age and knowledge/awareness level are negatively correlated. It may be concluded that elder respondents have less adaptability to new ideas and techniques as compared to young respondents. However, training and information, experience, education, and age play major role in enhancing knowledge/awareness level of respondents in IPM technology. Hussain *et al.,* (2011) concluded similar findings in their study that in Pakistan education plays an effective role in the adoption of IPM technology. They further recommended that government may take actions to upgrade the education as well as training programs in IPM for cotton producers.

## RECOMMENDATIONS

The following are few recommendations for implementation and dissemination of this technology at grass-root level in Pakistan.

1- Continuous training programs are required for propagation of this technology among small scale farmers.

2- Refresher courses are the keys for success

3- Doing IPM practices onsite on farmers land.

4- Permanent coordination among farmers and researchers is required for the solution of the problems.

5- Educational institutions of higher learning need to include courses and offer training programs for Extension staff for up-gradation of their skills in IPM technology.

6- The Government should be responsible for implementing agricultural policies on priority bases.

## REFERENCES

1- Afreh-Nuamah, K. (2001). The status of IPM in Ghana. Retreived December 14, 2010 from http://www.ag.vt.edu/ail/proceedings/ghana.htm

2- Altieri, M.A. (1993). Crop protection strategies for subsistence farmers, Boulder, Colorado: Westview Press.

3- Ashraf, E. (2007). In-service educational needs of agricultural officers for adaptation of remote sensing technology for precision agriculture in the province of Balochistan, Pakistan, Unpublished doctoral dissertation, Mississippi State University at Mississippi State.

4- Erbaugh, J.M. (2001). Activities of the IPM CRSP in Sub-Saharan Africa. Retreived December 14, 2010, from http://www.ag.vt.edu/ail/proceedings/usaid.htm

5- Hussain, M., Zia, S., & Saboor, A. (2011). The adoption of integrated pest management (IPM) technologies by cotton growers in the Punjab. Journal of Soil Environ. 30(1): 74-77.

6- Knausenberger, W.I., Schefers, G.A., Cochrane, J., & Gebrekidan, B.(2001). Africa IPM-Link: An initiative to facilitate IPM information networking in Africa. Retreived December 14, 2010 from http://www.ag.vt.edu/ail/proceedings/usaid.htm

7- Kyamanywa, S. (2001). Current status of integrated pest management in Uganda. Retrieved December 14, 2010 from http: //www.ag.vt.edu/ail/proceedings/usaid.htm

8- Morse, S., McNamara, N., Acholo, M. & Okwoli, B. (2000). Visions of sustainability: stakeholders, change and indicators. Hampshire, England: Publishing Company.
9- Norris, R.F., Caswell-Chen, E.P.; Kogan, M. (2003). Concepts in integerated pest management. New Jersey: Pearson Education, Inc.
10- Urdan, T. C. (2001) Statistics in plain English. New Jersey: Lawrence Erlbaum Association, Publishers.

# 22

# Analysis of Technical Efficiency of Smallholder Cocoa Farmers in Cross River State, Nigeria

*Agom, Damian Ila[1], Susan Ben Ohen[1], Kingsley Okoi Itam[1] and Nyambi N. Inyang[2]*

Abstract

The technical efficiency involved in cocoa production in Cross River State was estimated using the stochastic frontier production function analysis. The effects of some selected socio-economic characteristics of the farmers on the efficiency indices were also estimated. The study relied upon primary data generated from interviewing cocoa farmers using a set of structured questionnaire. A multi-staged random sampling technique was adopted in selecting two hundred (200) cocoa farmers from Ikom Agricultural Zone in the state. The data on the socio-economic characteristics of the farmers were analyzed using descriptive statistics, while the stochastic production function, using the Maximum Likelihood Estimating (MLE) techniques was used in estimating the farmer's technical efficiency and their determinants. Result of the analysis showed that farmers were experiencing decreasing but positive returns to scale in the use of the farm resources. The efficiency level ranged between 0.20 and 0.93 with a mean of 0.69. The result of the generalized Likelihood Ratio (LR) tests confirmed that the cocoa farmers in the area were technically inefficient. The major contributing factors to efficiency were age of farmers, farm size, level of education, sex of farmer and age of the farms. The study observed that there is enough room to improve efficiency with the farmers' current resource base and available technology and concluded that policies that would directly affect these identified variables should be pursued.

*Keywords:*
*Cocoa production, Technical efficiency, Stochastic frontier, Likelihood ratio, Maximum likelihood ratio*

[1] *Deptartment of Agric Economics, University of Calabar, Calabar, Nigeria.*
[2] *Commercial Agric Development Project (CADP), Cross River State, Nigeria.*
* *Corresponding author's email: agomd@yahoo.com*

## INTRODUCTION

Agriculture in developing countries is characterized by low productivity leading to low farm incomes. In Nigeria, cocoa production is characterized by several problems that lead to low productivity. This has resulted to a fall in percentage share of cocoa output. As Amos (2007) notes, two reasons are said to be responsible for the fall in percentage share of cocoa output. First is the negligence of the agricultural sector by the past administration due to the discovery of the petroleum resources that now accounts for the bulk of foreign exchange earnings. Second is the endemic problem in the cocoa industry. Therefore, increasing productivity will increase the percentage share of cocoa production. Analysis reveals that increasing agricultural production has probably been the simple most important factor in determining the speed and extent of poverty reduction. Most of these evidences are derivable from the Green Revolution in Asia (Adeniran, 2007). In China rapid productivity gains achieved largely through technological advances of the Green Revolution directly increased producers income and labourers' wages by lowering the price of food and by generating new livelihood opportunities as success in agriculture provides the basis for economic diversification. The importance of productivity is that it gives a measure for efficiency.

In Nigeria, there have been studies on farm level efficiency in tree crop production and very few have focused on cocoa production. Among these are studies by Giroh *et al.,* (2008) who carried out analysis of the technical inefficiency of gum Arabic based cropping patterns among farmers in the gum Arabic belt of Nigeria and that of Amos (2007) whose study is analysis of productivity and technical efficiency of small holder cocoa farmers in Nigeria. The authors employed the stochastic frontier production function analysis in their studies. However we do not have such studies in Cross River State which is a major cocoa producing area in Nigeria. Recent studies carried out on cocoa production in Cross River State examined the Socio-economic variables and cocoa production (Oluyole and Sanusi, 2009). Fertilizer use and cocoa production and Investment in cocoa production in Nigeria: .A Cost and Return analysis of three cocoa production management system in Cross River State cocoa belt by Nkang *et al.,* (2009). None of these studies examined the technical efficiency of the cocoa farmers.

### Objectives:

This study was carried out to provide estimates of levels of technical efficiency of cocoa farmers in Cross River State using farmers in Ikom Agricultural zone, where there is a high concentration of cocoa farmers in the state. The study was interested in whether the cocoa farmers were fully technically efficient, the current level of efficiency and factors that influenced efficiency. The study therefore was to estimate the level of and determinants of technical efficiency among cocoa farmers in Cross River State.

## MATERIALS AND METHODES

### Study area

This study was conducted in the two major Cocoa Producing Local Government Areas in Cross River State; Etung and Ikom. Cross River State is located in the Niger Delta region of Nigeria. It is bounded in the North by Benue State, in the South by Atlantic Ocean, in the East by Cameroon Republic and on the West by Akwa Ibom State, Abia and Ebonyi States. The state lies within latitude 40° 4, South and 60°30, North and between longitude 8° and 9° 00" East of the equator. It has three distinct ecological zones, the mangrove forest to the south, the tropical rainforest in the middle and the guinea savanna to the north. The annual mean rainfall ranges between 1500mm and 2000 mm.

### Sampling procedure and sample size

The multistage random sampling technique was adopted for this study. The first stage involved a purposive selection of two (2) Local Government Areas in the Ikom Agricultural Zone- Ikom and Etung. This is because Ikom and Etung are the major cocoa producing areas in Cross River State. The second stage involved the random selection of Five (5) villages from each of the selected Local Government Areas, giving a total of Ten (10) villages. The villages selected in Ikom were Akparabong, Ayukasa, Okondi, Alok and Ajassor, while those selected in Etung were Effraya, Agbokim, Bendeghe Ekim, Abijang and Abia. A simple random selection of twenty (20) farmers from each of the selected villages was carried out making up a total of one hundred (100) farmers from each of the two Local Government Areas which gave us 200 cocoa farmers for the study. Information was obtained on socio-economic characteristics of the farmers, output, labour, farm size and prices of variables using a set of structured questionnaire.

**Model specification**

Descriptive statistics including mean, standard deviation and variances were used to analyse the socio-economic characteristics of farmers while the stochastic production function was used to analyze the level of technical efficiency. The production technology of the cocoa farmers was assumed to be specified by the Cobb-Douglas frontier production function (Tadesse, and Krishnamorthy, 1997; Amos, 2007).

The specified cocoa production function was given as follows;

$$\text{In } Y = \text{In } \beta_o + \beta_1 \text{In}X_{1i} + \beta_2 \text{In}X_{2i} + \beta_3 \text{In}X_{3i}\ \beta_4 \text{In}X_{4i} + V_i - U_i \quad (1)$$

Where;

$Y$ = Quantity of cocoa produced (kg)

$X_1$ = Farm size (hectares)

$X_2$ = Quantity of fertilizer (kg)

$X_3$ = Quantity of fungicide (litres)

$X_4$ = Labour (man days)

$B_o$ = Y - intercept

$B_1$ to $\beta_4$ are coefficients to be estimated and $V_i$ and $U_i$ are error terms. It is expected that $\beta_1$, $\beta_2$, $\beta_3$, and $\beta_4$ will have positive signs.

**Determinants of technical efficiency**

The influence of some socio-economic factors on the computed technical efficiency was determined by incorporating the socio-economic factors directly in the frontier model, because they have influence on efficiency (Kalirajan, 1981). The technical efficiency model was specified as:

$$Ui = ą_o + ą_1Z1í + ą_2Z2í + ą_3Z3í + ą_4Z4í + ą_5Z5í + ą_6Z6í + ą_7Z7í \quad (2)$$

Where

$U_i$ = Technical efficiency

$Z_1$ = Sex of farmer (dummy)

$Z_2$ = Marital status (dummy)

$Z_3$ = Age of farmer (years)

$Z_4$ = Education (years spent in school)

$Z_5$ = Family size

$Z_6$ = Age of farm (years)

$Z_7$ = Farm size (ha)

$ą_o$ = y - intercept

$ą_1$ to $ą_7$ are coefficients that were estimated. It was expected that $ą_3$ would have a negative sign, while $ą_2$, $ą_4$, $ą_5$, $ą_6$ and $ą_7$ would have positive signs. The sign of ą1 was indeterminate.

In determining the level of technical efficiency of the cocoa farmers and analyzing the determinants of technical efficiency among the cocoa farmers, a generalized likelihood ratio (LR) test was used to test the hypothesis of full technical efficiency effects defined as

$$LR = -2 \text{ In } (\log H_1 - \log H_2) \quad (3)$$

Where, H1 is the log – likelihood function of the average function. $H_2$ is the log- likelihood function of the frontier function. The value has a mixed chi-square distribution with degrees of freedom equal to the number of parameters plus one. A computer programme frontier version 4.1 by Coelli (1994) was used in the computation, while the testing of the parameters was done at 1 and 5 percent levels of significance.

## RESULTS AND DISCUSSION

**Socio-economic characteristics of the sampled cocoa farmers**

The distribution of sampled cocoa farmers according to their sex, age, marital status, educational level, family size and farming experience is presented in Table 1. The results indicates that majority (88.5%) of the farmers were males while few (11.5%) were females. The cultural setting of the area allows the males to have easy access to land especially, where majority of them are the heads of their respective households.

The table also indicates that majority of the farmers (44.5%) were within the 47 to 57 years age bracket. This was closely followed by the farmers with age 36-46 years (32%). Farmers that were in the minority constituted 1.5% and these farmers were above the 68 years age bracket. This shows that about 86% of the farmers were in their most economically active age bracket (25-57) years. However, there was a widespread of farmers among all the age groups, implying that cocoa farming was embraced by all the age groups. The results also showed that most of the farmers (68.5%) were married while 31.5% were single (Table 1).

Furthermore, the distribution of the cocoa framers according to their educational level (number of years spent in school), shows that majority (76.5%) of the farmers had attained one level of formal education or the other. The mean level of educational attainment (years spent in school) of the farmers in the area was about 6 years, with 11% of the farmers having tertiary education. This is an indication that some graduates were involved in cocoa farming in the study area. This is a good pointer to improved productivity as the level of education is a tool with which an individual could be efficient at whatever endeavour being undertaken by the individual (Oluyole and Usman, 2006). As regards family size, a high proportion (86.5%) of the farmers had family sizes of 5 persons and above, while 13.5% had less than 5 persons in their household. The mean family size of the cocoa farmers was 7 persons. Effiong (2005) reported that a relatively large household size enhances the availability of family labour which reduces constraints on labour cost in agricultural production.

The table also shows that a very high proportion (99%) of the farmers had between 5 and above thirty (30) years of experience in cocoa farming. The mean farming experience was about 15 years. Farmers sometimes count more on their experience than educational attainment in order to increase their productivity (Nwaru, 2004). The result implies that a good number of the farmers are experienced farmers and therefore are expected to have higher technical efficiencies.

Table 1: Distribution of socio-economic characteristics of sampled cocoa farmers

| Variables | Frequency | Percentage |
|---|---|---|
| Sex: | | |
| Male | 177 | 88.5 |
| Female | 23 | 11.5 |
| Total | 200 | 100 |
| Age of farmer (years): | | |
| 25 – 35 | 19 | 9.5 |
| 36 – 46 | 64 | 32.0 |
| 47 – 57 | 89 | 44.5 |
| 58 – 68 | 25 | 12.5 |
| >68 | 3 | 1.5 |
| Total | 200 | 100 |
| Means | 47.70 (8.91) | |
| Marital status: | | |
| Single | 63 | 31.5 |
| Married | 137 | 68.5 |
| Total | 200 | 100 |
| Mean | 2.02(1.61) | |
| Educational level (school years): | | |
| 0 | 47 | 23.5 |
| 6 | 81 | 40.5 |
| 8 | 14 | 7.0 |
| 12 | 36 | 18.0 |
| 14 | 14 | 7.0 |
| 16 | 8 | 40 |
| Total | 200 | 100 |
| Mean | 6.45 (4.49) | |
| Family size: | | |
| <5 | 27 | 13.5 |
| 5 – 7 | 76 | 38 |
| 8 – 10 | 54 | 27 |
| 11 – 13 | 24 | 12 |
| 14 – 16 | 9 | 4.5 |
| 17 – 19 | 7 | 3.5 |
| >19 | 3 | 1.5 |
| Total | 200 | 100 |
| Mean | 7. 11 (4.66) | |
| Farming experience (years) | | |
| <5 | 2 | 1.0 |
| 5 – 10 | 74 | 3.7 |
| 11 – 15 | 54 | 2.7 |
| 16 – 20 | 33 | 16.5 |
| 21 – 25 | 19 | 9.5 |
| 26 - 30 | 6 | 3.0 |
| >30 | 12 | 6.0 |
| Total | 200 | 100 |
| Mean | 15.22 (7.70) | |

Source: Field Survey 2010
Note: Values in parentheses are standard deviations.

### Mean output and other production variables in cocoa production in Cross River State

The statistics of the production variables obtained from cocoa farmers in the study area are summarized in table 2. The mean output of cocoa farmers in the area was 2428.10kg/annum/farmer. This is relatively high compared to figures of less than two tonnes recorded elsewhere. This may be related to the age of the farms as most farms in Cross River State are within the productive age of 11 to 40 years (Table 3) compared to farms in other parts of Nigeria. For labour, the

Table 2: Summary statistics of output and other variables for sampled Cocoa farmers

| Variables | Minimum | Maximum | Mean | Standard Deviation |
|---|---|---|---|---|
| Output (kg) | 760.00 | 76200.00 | 2428.10 | 2343.53 |
| Farm size (ha) | 1.00 | 40.00 | 6.90 | 6.35 |
| Fertilizer (kg) | 0.00 | 200.00 | 18.50 | 37.09 |
| Fungicide (litres) | 0.00 | 2500.00 | 1142.80 | 644.37 |
| Labour (man days) | 18.00 | 93.00 | 51.49 | 14.65 |
| Age of farmer (years) | 25.00 | 71.00 | 47.70 | 8.91 |
| Family size | 1.00 | 26.00 | 7.00 | 4.66 |
| Farming experience (years) | 4.00 | 50.00 | 15.22 | 7.70 |

Source: Derived from Field Survey Data 2010

mean man- days used by the farmer during the production season were 51.50 man days. Farm sizes in Nigeria have been described as small, medium or large scale, if they fall into categories of less than 5ha, between 5ha and 10ha, or more than 10ha, respectively (Upton,1972). Most of the farms in Nigeria are of small to medium scale categories. The average farm size among the cocoa farmers in the study area is 6.90 hectares scattered in different locations in the locality, hence the small holdings. It was observed that majority of the cocoa farmers in Cross River State did not use fertilizer in cocoa production. The mean fertilizer used by the farmers was 18.50, which is very low. The result is in line with Oluyole and Sanusi, (2009) which reported that 98.13% of cocoa farmers in Cross River State did not use fertilizer in cocoa production.

An average of 1142.80 litres of fungicide was applied by the cocoa farmers (Table 2). The high rate of application may be due to the less resistance of the cocoa variety to infection.

**Classification of cocoa farmers according to age groups and their mean output in the study area**

The sampled farms were grouped according to their age groups and mean output and majority (92%) of the farms are within the productive age of 11 to 40years age group (Table 3). The mean output initially increased within this age as the cocoa trees get fully established, and thereafter output declines as shown in the table.

**Maximum likelihood estimates of stochastic production frontier function for cocoa farmers in Cross River State**

The coefficient of the Maximum Likelihood Estimates (MLE) of the parameters of the stochastic frontier models of cocoa farmers is shown in Table 4. The variance parameters of the stochastic frontier production function are represented by sigma squared ($\delta^2$) and gamma (y). From the table, the sigma squared is 0.171 and significantly different from zero at one percent level. This indicated a good fit and correctness of the distributional form assumed for the composite error term. Gamma (y) indicates that the systematic influences that are unexplained by the production function are the dominant sources of random error. The gamma estimate which is 0.327 and significant at five percent level shows the amount of variation resulting from the technical efficiencies of cocoa farmers

Table 3: Age group of farmers and mean output

| Age group | Frequency | Percentage | Mean output | Standard deviation |
|---|---|---|---|---|
| 5 - 10 | 5 | 2.5 | 6305.2 | 6305.2 |
| 11- 20 | 81 | 40.5 | 8064.1 | 8064.1 |
| 21-30 | 81 | 40.5 | 8797.0 | 8797.0 |
| 31- 40 | 22 | 11.0 | 10448.0 | 10448.0 |
| 41- 50 | 8 | 4.0 | 7465.1 | 7465.1 |
| >50 | 3 | 1.5 | 6573.0 | 6573.0 |
| | **200** | **100** | **(2428.1)** | **(2428.1)** |

Source: Field Survey Data 2010

Note: The values in parenthesis are the mean output and standard deviation of the farms respectively and do not represent the total mean output and standard deviation of the groups.

Table 4: Maximum likelihood estimates of the stochastic production function for cocoa production

| Variable | Parameters | Coefficients | Standard errors | Standard errors |
|---|---|---|---|---|
| Constant | $\beta_0$ | 7.450 | 0.141 | 52.837*** |
| Farm size | $\beta_1$ | 0.155 | 0.016 | 9.688*** |
| Quantity of fertilizer | $\beta_2$ | 0.013 | 0.001 | 1.30 |
| Quantity of fungicide | $\beta_3$ | 0.002 | 0.004 | 0.50 |
| Labour | $\beta_4$ | 0.009 | 0.003 | 3.00*** |
| Diagnostic statistics | | | | |
| Gmma (Y) | Y | 0.327 | 0.144 | 2.71** |
| Sigma square | $\delta_2$ | 0.171 | 0.037 | 4.622*** |
| Log likelihood function | | -88.08 | | |
| Likelihood ratio (LR) | $\lambda$ | 71.42 | | |

Source: Computed from Field Survey Data 2010 using Frontier 4.1
Note: *** P < 0/01, ** P < 0/05

in the study area. This means that more than 32% of the variation in farmer's output is due to difference in technical efficiency.

The result further shows that the signs of all the estimated coefficients of the stochastic production frontier are positive which is consistent with a priori expectation. This implies that there is a positive relationship between the level of output of cocoa and farm size, the quantity of fertilizer, fungicide and labour used. This is expected as the level of production depends largely on the quantities of these inputs used on the farm. This can only be up to a level that is considered optimal after which farmers will be operating at sub optimal level. However, the coefficients of the slope, farm size and labour were significant at one percent indicating that farm size and labour are important determinants of cocoa output.

## Determinants of technical efficiency in cocoa production

The analysis of the efficiency model shows that the signs of the estimated coefficients in the efficiency model have important implications on the technical efficiency of cocoa farmers in the study area. The coefficient of sex (z←←1) had a positive sign indicating that the male farmers obtain higher levels of technical efficiency than their female counterparts in the area. Cocoa farming is dominated by males in the area. This is so because cocoa farming is a tedious job and requires more strength which females may not be able to provide (Oluyole and Sanusi, 2009).

The coefficient of marital status ($Z_2$) is indeterminate. However, it is negative as shown in Table 5.

The coefficient of age ($z_3$) is also positive. This does not agree with a prior expectation. As farmers age, there is a tendency that productivity will continue to fall owing to their declining strength. However, this result could be attributed to the fact that most of the farmers in the study area started farming at early age. Hence, the older they are the more experienced and efficient they would be, since farmers' experience increases with the number of years spent in farming, it implies that the longer the time spent in

Table 5: Determinants of Technical Efficiency

| Variable | Parameter | Coefficient | Standard error | t-ratios |
|---|---|---|---|---|
| Constant | $a_0$ | -1.278 | 0.699 | -1.828 |
| Sex of farmer | $a_1$ | 0.009 | 0.415 | 0.022 |
| Marital status | $a_2$ | -0.209 | 0.135 | -1.548 |
| Age of farmer | $a_3$ | 0.011 | 0.136 | 0.801 |
| Educational level (sch yrs) | $a_4$ | 0.054 | 0.204 | 0.265 |
| Family size | $a_5$ | -0476 | 0.236 | -2.016** |
| Age of farm | $a_6$ | -0.012 | 0.723 | -0.017 |
| Farm size | $a_7$ | 0.081 | 0.010 | 8.10*** |

Source: Computed from Field Survey Data 2010 using Frontier 4.1
Note: *** P < 0/01, ** P < 0/05

Table 6: Elasticity of production and returns to scale

| Variables | Elasticity |
|---|---|
| Farm size | 0.155 |
| Quantity of fertilizer | 0.086 |
| Quantity of fungicide | 0.420 |
| Labour | 0.094 |
| RTS | 0.179 |

Source: Field Survey, 2010

farming the more experience they are.

Furthermore, the coefficient of educational level ($z_4$) was positive but not significant. This implies that the level of technical efficiency of the farmer increases with the level of education but not significantly. However, the result agrees with a prior that technical efficiency should increase with increases in years of schooling of the farmers since education and adoption of innovation were expected to be positively correlated.

The coefficient of family size ($z_5$) has a negative sign and is significant at the five percent level. This implies that increase in family size will lead to decrease in technical efficiency. This result does not agree with a prior expectation. However, given that the cocoa farmers utilized more of hired labour than family labour and less labour (weeding and pruning) is required once the cocoa has been established, it could be acceptable. The farmers pay wages that are more than the value of their marginal production and hence would be inefficient as a result of allocative inefficiency (Idiong, 2006).

The negative coefficient of age of farm ($Z_6$) implies that efficiency decreases as the farms get older. Amos, (2007) had a similar result. Lastly the coefficient of farm size ($Z_7$) was positive and significant at the one percent level. This implies that technical efficiency increases with the size of farm. This result agrees with those of Giroh *et al.,* (2008) and Amos (2007). Large farm sizes if properly managed should have higher efficiency. However, there is a threshold where returns to scale decreases with increase in farm size.

**Elasticity of production and returns to scale**

Typical of the power function (Cobb-Douglas), the estimated coefficients for the specified function can be explained as the elasticities of the explanatory variables.

The values of the variables indicates that a 10 percent increase in the size of farm, fertilizer, fungicide and labour will lead to a 1.5, 0.9, 4.2 and 0.9 percent increase respectively in

Table 7: Test of hypotheses that cocoa farmers in Cross River State are fully technically efficient ($\gamma$= 0)

| Efficiency | Likelihood function ($\lambda$) | Log likelihood ratio (LR) | Critical $X^2$ 0.05 | Conclusion |
|---|---|---|---|---|
| Technical | -88.08 | 10.712 | 10.371 | Reject |

Source: Derived from Table 4. Critical $X^2$ was obtained from Kodde and Palm (1986).

Table 8: Frequency distribution of technical efficiency estimates

| Efficiency level | Frequency | Percentage |
|---|---|---|
| 0.20 – 0.30 | 2 | 1 |
| 0.31 – 0.40 | 3 | 1.5 |
| 0.41 – 0.50 | 20 | 10 |
| 0.51 – 0.60 | 17 | 8.5 |
| 0.61 – 0.70 | 38 | 19 |
| 0.71 - 0.80 | 91 | 45.5 |
| 0.81 – 0.90 | 26 | 13 |
| >90 | 3 | 1.5 |
| Total | 200 | 100 |
| Mean | 0.69 | |
| Minimum | 0.20 | |
| Maximum | 0.93 | |

Source: Derived from output of the computer programme, Frontier 4.1 by Coelli (1994)

output of cocoa (Table 6). The value of the returns to scale (RTS) was 1.79 indicating that the farmers were operating in the region of decreasing but positive returns to scale (stage II of the production function). Therefore, increasing the units of inputs will not be the best option to the farmers as it will add less to total cocoa output (Table 7).

**Technical efficiency estimates of the sampled cocoa farmers in Cross River Sate**

The predicted farm specific technical efficiencies (TE) ranged between 0.20 and 0.93 (Table 8). The mean efficiency of the cocoa farmers was 0.69. The 69% mean efficiency indicates that in the short run, there is a scope of increasing cocoa production by about 31% by adopting the technologies and techniques practiced by the best cocoa farmers in the study area (Table 8).

The efficiency distribution also indicates that majority of the cocoa farmers (79%) were having efficiency of between 61% and 90% while a few of them (21%) were less than 60% efficient in their production process. The high levels of efficiency may be due to the long years of farming experience of the farmers.

## CONCLUSION

This study reveals that, cocoa farmers in Cross River State are not fully technically efficient in their resource use. The policy variables that were identified as having significant effects (positive and negative) on the efficiency levels of the cocoa farmers are farmers age, family size, farm sizes, educational level, and age of the farm. It is believed that farmers' technical efficiency in resource use could increase since cocoa farmers in the area were not fully technically efficient; hence, there is room for improvement in the level of this efficiency if the important policy variables are addressed. Majority of farmers were in the productive age bracket and this was directly related to technical efficiency of cocoa farmers in the area. It is therefore important that a policy that would make cocoa farming attractive to persons within this age bracket. The 69% mean efficiency indicates that in the short run, there is a possibility of increasing cocoa production by about 31% by adopting the technologies and techniques practiced by the best cocoa farmers in the study area.

## REFERENCES

1- Adeniran, K. O (2007). Perspective on the role of Agriculture in meeting the Millennium Development Goal. World Bank Report.

2- Amos, T. T. (2007). An Analysis of Productivity and Technical Efficiency of Samallholder Cocoa Farmers in Nigeria. Journal of Social Science. 15 (2): 127-133.

3- Coelli, T. J. (1994). A Guide to Frontier 4.1: A Computer Programme for Stochastic Frontier Production. Department of Economics. University of New England, Amirdale.

4- Effiong, E. O. (2005). Efficiency of Production in Selected Livestock Enterprises in Akwa Ibom State, Nigeria. Unpublished Ph.D Dissertation, Department of Agricultural Economics, Michael Okpara University of Agriculture, Umudike, Nigeria.

5- Giroh, D. Y., Valla, W., Mohammed, A., & Peter, O. (2008). Analysis of the Technical Inefficiency of Gum Arabic Based Cropping Patterns among Farmers in the Gum Arabic Belt of Nigeria. Journal of Agriculture and Social Science. 4:125-128.

6- Idiong, I. C. (2006). Evaluation of Technical Allocative and Economic Efficiencies in Rice Production Systems in Cross River State, Nigeria. Unpublished Ph.D Thesis, Department of Agricultural Economics. Michael Okpara University of Agriculture, Umudike, Nigeria.

7- Kalirajan, K. (1981). The Economic Efficiency of Farmers Growing High Yielding, Irrigated Rice in India. American Journal of Agricultural Economics. 63(3): 566-569.

8- Kodde, F.C., & Palm, D.C. (1986). Wald Criteria for Jointly Testing Equality and Inequality Restrictions. Econometrical. 54: 1243- 1248.

9- Nkang, N. M., E. A. Ajah, S. O. Abang and E. O. Edet (2009). Investment in Cocoa Production in Nigeria: A Cost and Return Analysis of Three Cocoa Production Management Systems in Cross River State Cocoa Belt. African Journal of Food Agricultural Nutrition and Development. 2(2): 35-40.

10- Nwaru, J. C. (2004). Rural Credit Market and Arable Crop Production in Imo State of Nigeria. Unpublished Ph. D Dissertation, Michael Okpara University of Agriculture, Umudike, Nigeria.

11- Oluyole, K. A. and J. M. Usman (2006). "Assessment of Economic Activities of Cocoa Licensed Buying Agents (LBAs) in Odeda Local Government Area of Ogun State Nigeria". Akoka Journal of Technology and Science Education. 3(1): 130-140.
12- Oluyole. K. A. and R. A. Sanusi (2009). Socio-economic Variables and Cocoa Production in Cross River State Nigeria. Hum Ecol. 25 (1): 5-8.
13- Tadesse, B. and S. Krishnamurthy (1997). Technical Efficiency in Paddy Farms of Tamil Nadu: An Analysis Based on Farm Size and Ecological zone. Agricultural Economics. 16: 185 – 192.
14- Upton, M. (1972). Farm Management in Nigeria. Occasional Report Department of Agricultural Economics, University of Ibadan Nigeria.

# 23

# Effectiveness of Extension Services in Enhancing Outgrowers' Credit System: A Case of Smallholder Sugarcane Farmers in Kisumu County, Kenya

*Abura Odilla Gilbert [1], Barchok Kipngeno Hillary[1] and Onyango Christopher Asher[2]*

***Keywords:***
***Public extension service, Private extension service, Outgrowers' Credit system, Effectiveness, Western Kenya***

**Abstract**

The purpose of this study was to investigate the role of extension services in enhancing effectiveness of outgrowers' credit system in Kisumu County, Kenya. The study specifically sought to determine whether public and private extension services play a significant role in enhancing effectiveness of out-growers' credit system among smallholder sugarcane farmers. A total of 110 small scale farmers were randomly selected for the study. A closed ended questionnaire was used to collect data from farmers. Both descriptive and inferential statistics were used for data analysis. The findings indicated that both public and private extension services were insignificant in enhancing effectiveness of outgrowers' credit system. Further, the findings indicated that there was no significant difference between public and private sector in provision of extension services. The findings suggest that for outgrowers' credit system to be effective in terms of creation of awareness about credit, accessibility, timely supply of credit, supervision of credit and provision of extension advice on credit utilization, both public and private extension services should be intensified and coordinated to avoid duplication. The results also suggest that sugarcane factory extension division should be strengthened just like in the coffee and tea sub-sectors.

[1] *Chuka University College, P. O. Box 109, Chuka, Kenya.*
[2] *Egerton University, P. O. Box 536, Egerton, Kenya.*
* *Corresponding author's email: gilbura@yahoo.com*

## INTRODUCTION

The importance of agriculture to the African economies is stressed due to the fact that agriculture remains the principal occupation of the majority of people, constitutes the largest production sector, and produces an average of 32% of GDP, major sources of raw materials for industries and a significant purchaser of the countries manufacturers and services (Agbamu, 2005). In Kenya for example, the economy is heavily dependent upon agricultural sector and as the World Bank (2007) report indicates, the country's future will considerably depend on productivity of smallholder farms. Agriculture is by far the single largest economic sector in Kenya and accounts for about 30% of GDP, over 60% of the exports, 75% of the total labour force and provides 80% of industrial raw materials (Economic Survey, 2007; Kenya Sugar Research Foundation, 2007, Government of Kenya, 2005). Since independence, smallholder agriculture gained ground from mere provision of subsistence and minimal marketed surplus to account for over three quarters of agricultural production and 85% of agricultural employment (GoK, 2005, World Bank, 2007). Sugarcane farming is one such subsector that contributes to the national economy. According to Guda *et al.,* (2001) smallholder farmers accounts for 89% of the total area under sugarcane farming in Kenya. This provides an investment opportunity. However, this is only possible if the problems affecting it are addressed. Some of the main problems include shortage of sugarcane due to lack of systemic and synchronised sugarcane development, poor crop husbandry practices, poor cane varieties and qualities in some factory zones, poor harvesting methods, poor management in some factories affecting factory efficiency and output and inadequate agricultural production credit among others (Guda *et al.,* 2001). However the most notable problem is complaints due to delayed payment dues to the farmers after cane delivery. These problems have elicited diverse reactions from the farmers. The most severe reactions are cases of burning the cane crop by the farmers in their own farms, so as to turn to other lucrative enterprises like maize or bean seed production (Agribusiness Development Support Project Annual Report, 2001).

Agricultural extension is considered to be an important service in increasing agricultural productivity and attaining sustainable development (Kibet, *et al.,* 2005). Its role is to help people identify and address their needs and problems. There is a general consensus that extension services if successfully applied, should result in outcomes which include observable changes in attitudes and adoption of new technologies, and improved quality of life based on indicators such as health, education and housing. It has been recognized that agricultural extension accelerates development in the presence of other factors such as markets, agricultural technology, availability of supplies, production incentives and transport (Kibet, *et al.,* 2005). Koyenican (2008) equates help in extension to empowering all members of the farm households to ensure holistic development. This is because agricultural extension brings about changes, through education and communication in farmers attitude, knowledge and skills.

The performance of the public agricultural extension service in Kenya has been a very controversial subject (Gautam and Anderson, 1999). The system has been perceived as top-down, uniform (one-size-fits-all) and inflexible and considered a major contributor of the poor performing agricultural sector (Government of Kenya, 2005). Thus there has been a desire to reform extension in to a system that is cost effective, responsive to farmer's needs, broad based in service delivery, participatory, accountable and sustainable. As a result of ineptness in the public extension system, private agricultural extension system has emerged comprising of private companies, non-governmental organizations (NGO's), community based organizations (CBO's) and faith based organizations (Nambiro *et al.,* 2005 and Rees *et al.,* 2000).

Agricultural extension as a public sector institution has an obligation to serve the needs of all agricultural producers, either directly or indirectly (Anderson, 2007). This is because public sector extension is a public good. The Kenya government has tried a number of extension models and styles, including the progressive (farmer approach) model, integrated agricultural rural development approach, farm management, training and visit, farming systems approaches and farmer field schools. All these approaches have emerged with varying level of success for different groups. However, the effectiveness of extension services in enhancing effectiveness of outgrowers' credit system among sugarcane farmers in Kisumu County of Kenya has not been examined. Thus the present study was set with the premise that both outgrowers' credit and extension service are instruments for promoting agricultural development and that an efficient and effective extension service is important in enhancing effectiveness of outgrowers credit system. Credit to farmers is an important instrument in improving productivity. Indeed as Wangia (2001) noted, it is a prerequisite to the adoption of improved agricultural technologies for the smallholder farmers. Nevertheless, for credit system to help the smallholder farmers it should be tied to improved technologies, remunerative prices for the farmers' output and good extension network (Ogunsumi, 2004).This paper presents results on the role of extension services in enhancing effectiveness of outgrowers' credit system among smallholder cane farmers in Kenya.

**Conceptual Framework**

This study formulated a conceptual model that encompassed major variables and their possible patterns of influence on each other and eventually on effectiveness of outgrowers' credit system. The effect of the extension services namely public and private services are mediated by level of education, farm size and farmers period of residence. What this structural model indicates therefore is that the moderator variables influence adoption of sugarcane technologies disseminated by extension agents whether from public or private sector. In view of this model, the theory underpinning this study is that, adoption is complex and multifaceted process. While the main activity of extension centers on increasing production, this study concentrated on implementation of such activities. These are; creation of awareness about credit, accessibility to credit, timely supply of credit, supervision of credit and provision of extension advice on credit utilization (Table 1).

## MATERIALS AND METHODES

Ex-post facto survey design was adopted for this study. Kisumu county in Kenya was purposely selected because of its uniqueness in that the county boasts of three major sugar factories.

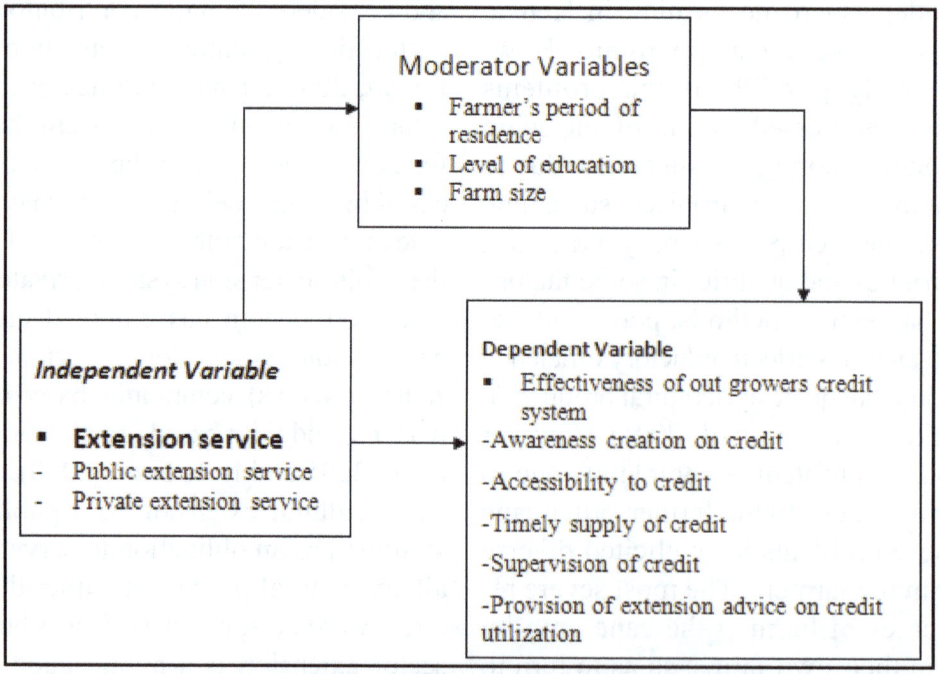


Figure 1: The Conceptual Framework on the Role of Extension in Enhancing Effectiveness of Outgrowers' Credit System

Table 1: Creation of Awareness on credit facility

| Activity | | SD | D | U | A | SA | Total |
|---|---|---|---|---|---|---|---|
| **Creation of Awareness** | **f (n=23)** | **2.0** | **17.0** | **-** | **2.0** | **2.0** | **23.0** |
| | **%** | **8.7** | **73.9** | **-** | **8.7** | **8.7** | **100.0** |

Key (SD= Strongly Disagree, D= Disagree, U= Undecided, A= Agree, SA= Strongly Agree)

Table 2: Land Preparation

| Activity | | SD | D | U | A | SA | Total |
|---|---|---|---|---|---|---|---|
| **Land Preparation** | **f (n=23)** | **2.0** | **14.0** | **-** | **6.0** | **1.0** | **23.0** |
| | **%** | **8.7** | **60.2** | **-** | **26.0** | **4.3** | **100.0** |

Key (SD=Strongly Disagree, D=Disagree, U=Undecided, A=Agree, SA= Strongly Agree)

These are; Miwani, Muhoroni and Chemelil which were established in the years 1923, 1966 and 1968 respectively. The County has favourable moderate climatic conditions, with temperatures averaging 27° C and receives bimodal rainfall ranging from (560 -1630) mm per annum. Kisumu County comprises of the main topographical land formations namely, the Nandi hills, the Nyando plateau and Kano plains which are sandwiched between two hills. The Kano plains comprise predominantly black cotton clay soils derived from igneous rocks. The County's altitude range from 1000-1860 M above sea level. The target population was the sugarcane farmers in Kisumu county. A total of 110 smallholder cane farmers were randomly selected for the study but only 108 farmers questionnaire were useful for analysis. A closed ended questionnaire was used to collect data by personal interviews. The information gathered was analysed using both descriptive and inferential statistics.

## RESULTS AND DISCUSSIONS

### Public extension Services

Tables 1, 2 and 3 below shows data on different extension activities done by public sector extension service for cane farmers. Table 1 show result on the effect of public extension services in relation to creation of awareness. Creation of awareness was one of the activities used to measure public extension services. To elicit information on creation of awareness, the farmers were asked to respond to statement designed to elicit negative responses on performance resulting to creation of awareness. A 5-point likert scale was constructed to record these responses. Table 1 show results on public extension services in

Table 3: Appropriate Input Use

| Activity | | SD | D | U | A | SA | Total |
|---|---|---|---|---|---|---|---|
| **Appropriate Input Use** | **f (n=23)** | **2.0** | **7.0** | **6** | **6.0** | **2.0** | **23.0** |
| | **%** | **8.7** | **30.4** | **26.0** | **26.0** | **8.7** | **100.0** |

Key (SD=Strongly Disagree, D=Disagree, U=Undecided, A=Agree, SA=Strongly Agree)

Table 4: Creation of awareness on credit facility

| Activity | | SD | D | U | A | SA | Total |
|---|---|---|---|---|---|---|---|
| **Creation of Awareness** | **f** | **16.0** | **51.0** | **-** | **9.0** | **4.0** | **80.0** |
| | **%** | **20.0** | **63.8** | **-** | **11.3** | **5.0** | **100.0** |

Key (SD=Strongly Disagree, D=Disagree, U=Undecided, A=Agree, SA=Strongly Agree)

Table 5: Land Preparation

| Activity | | SD | D | U | A | SA | Total |
|---|---|---|---|---|---|---|---|
| Land Preparation | f (n=80) | 9.0 | 12.0 | 1.0 | 35.0 | 23.0 | 80.0 |
| | % | 11.25 | 15.0 | 1.25 | 43.8 | 28.8 | 100.0 |

Key (SD=Strongly Disagree, D=Disagree, U=Undecided, A=Agree, SA=Strongly Agree)

Table 6: Use of Appropriate Inputs

| Activity | | SD | D | U | A | SA | Total |
|---|---|---|---|---|---|---|---|
| Timely Use of Appropriate Inputs | f (n=80) | 3.0 | 17.0 | - | 44.0 | 16.0 | 80.0 |
| | % | 3.75 | 21.3 | - | 55.0 | 20.3 | 100.0 |

Key (SD=Strongly Disagree, D=Disagree, U=Undecided, A=Agree, SA=Strongly Agree)

relation to creation of awareness.

The results in Table 1 indicated that majority of farmers (82.6%) receive information on credit facility from public extension providers.

Advice on land preparation was one of the activities used to measure public extension services. To elicit information on land preparation, the farmers were asked to respond to statement designed to elicit positive knowledge of performance resulting to land preparation. A 5-point likert scale was constructed to record these responses. Table 2 show results on public extension services in relation to land preparation.

The result in table 2 showed that land preparation as an activity has not been satisfactorily addressed by public service extension officers as reflected by the high percentage (68.9%) of farmers who disagreed with the positive statement. This suggests that perhaps the declining cane production in Kisumu County is due to poor and inadequate land preparation, culminating from inadequate machinery to prepare land for cane growing.

Advice on appropriate use of inputs was one of the activities used to measure public extension services. To elicit information on the use of appropriate input, the farmers were asked to respond to statement designed to elicit positive knowledge of performance resulting to use of appropriate input. A 5-point likert scale was constructed to record these responses. Table 3 show results on public extension services in relation to appropriate use of input.

The results in table 3 showed that 39.1% of the farmers have not received services on appropriate use input from public sector extension services. This suggests that farmers do not know whether the inputs they use are appropriate or not.

**Private extension service**

Private sector extension may play a predominant extension role for particular inputs, particular enterprises / commodities and for particular farmer's in particular geographical areas. This enables farmers to benefit from increased incomes and economic security. Tables 4, 5, and 6 shows data on different extension activities done by private sector extension service for cane farmers.

Table 4 show result on the effect of private extension services in relation to creation of awareness. Creation of awareness was one of the activities used to measure private extension services. To elicit information on creation of awareness, the farmers were asked to respond to statement designed to elicit negative knowledge of performance resulting to creation of awareness. A 5-point likert scale was constructed to record these responses. Table 4 show results on private extension services in relation to creation of awareness.

The results in table 4 showed that majority of farmers (83.75%) received information on credit facility.

Advice on land preparation was one of the activities used to measure private extension services. To elicit information on land preparation, the

Table 7: Comparison between role of public and private extension services in Kisumu County

| | | Public Extension Service (N=23) | | | | Private Extension Service (N=80) | | | |
|---|---|---|---|---|---|---|---|---|---|
| | | LP | UAI | CA | Average | LP | UAI | CA | Average |
| SA | f | 1.0 | 2.0 | 2.0 | 2.0 | 23.0 | 16.0 | 4.0 | 13.2 |
| | % | 4.3 | 8.7 | 8.7 | 8.7 | 28.8 | 20.3 | 5.0 | 16.5 |
| A | f | 6.0 | 6.0 | 2.0 | 5.0 | 35.0 | 44.0 | 9.0 | 28.2 |
| | % | 26.0 | 26.0 | 8.7 | 21.5 | 43.8 | 55.0 | 11.3 | 35.3 |
| U | f | - | 6.0 | - | 2.0 | 1.0 | - | - | 00.8 |
| | % | - | 26.0 | - | 8.7 | 1.25 | - | - | 01.0 |
| D | f | 14.0 | 7.0 | 17.0 | 12.5 | 12.0 | 17.0 | 51.0 | 27.4 |
| | % | 60.2 | 30.4 | 73.9 | 53.8 | 15.0 | 21.3 | 63.8 | 34.3 |
| SD | f | 2.0 | 2.0 | 2.0 | 1.8 | 9.0 | 3.0 | 16.0 | 10.4 |
| | % | 8.7 | 8.7 | 8.7 | 7.74 | 11.25 | 3.75 | 20.0 | 13.0 |
| Total | f | 23.0 | 23.0 | 23.0 | 23.0 | 80.0 | 80.0 | 80.0 | 80.0 |

Key (SD=Strongly Disagree, D=Disagree, U=Undecided, A= Agree, SA=Strongly Agree)
LP- Land Preparation
UAI- Use of Appropriate Inputs
CA- Creation of Awareness

farmers were asked to respond to statement designed to elicit positive knowledge of performance resulting to land preparation. A 5-point likert scale was constructed to record these responses. Table 5 show results on public extension services in relation to land preparation.

The results in 5 showed that majority of the farmers (72.5%) received extension services on land preparation from private sector. Perhaps this is due to the diverse nature of private sector extension where land preparation machines are readily provided to try and promote return on investment as well as enabling the farmers to increase their income through increased cane production.

Advice on appropriate use of inputs was one of the activities used to measure private extension services. To elicit information on the use of appropriate input, the farmers were asked to respond to statement designed to elicit positive knowledge of performance resulting to use of appropriate input. A 5-point likert scale was constructed to record these responses. Table 6 show results on public extension services in relation to appropriate use of input.

The results in table 6 showed that majority of the farmers (75%) receive advice on use of appropriate input from private sector.

### A comparison of public and private extension services

Table 7 shows a comparison of the role public and private extension play with respect to various cane farming activities. The relevant information was elicited by asking the farmers to state from whom they receive extension services from, followed by their responses to statements designed to elicit positive knowledge of performance to various cane farming activities. A 5-point likert scale was used to record these responses.

The results in table 7 indicate that, public extension has a lesser role in enhancing effectiveness of outgrowers' credit system as compared to private extension services. This was because the majority of the public extension recipients (61.5%) either disagreed or strongly disagreed compared to 47% private extension recipients in terms of advising farmers on various farm activities.

Table 8: Chi-square test for Effectiveness of Outgrowers Credit System by Type of Extension

| | Value | df | Asymp. Significance 2 sided |
|---|---|---|---|
| Pearson Chi-square | 14.952 | 18 | 0.667 |
| No. valid cases | 103 | | |

Significance set at ($\alpha = 0.05$)

**Hypotheses testing**

The null hypothesis tested stated that there is no significant difference between the public and private extension services in terms of enhancing effectiveness of out-grower's credit system in Kisumu County. Both the public and private sector extension services were measured with respect to advice given to farmers on land preparation, use of appropriate inputs and creation of awareness. The information on public and private extension services with their effects on effectiveness of out-grower's credit was elicited by use of farmers' questionnaire. Testing of this hypothesis was carried out by use of chi-square test and the results are presented in Table 8.

Results in Table 8 indicate that, there was no significant difference between public and private sector extension services. This was because the Pearson chi-square value (14.952) was not significant at $\alpha = 0.05$ ($p>0.05$). The null hypothesis was thus accepted. This result suggests that public and private sector extension services are inadequate in terms of quantity despite the fact that cane farmers require it to realize a positive change.

## DISCUSSION

Akroyd and Smith (2007) noted that lack of agricultural services has negatively impacted on food production. Consequently, in many parts of less developed countries, agricultural extension services often bypass or do not reach the rural farmers (FAO, 1997). In most countries extension services provided by the government are supplemented by private sectors. Milu and Jayne (2006) acknowledged that, in developing countries, the private sector extension is extremely diverse. Depending on the particular economic and political situation, the private sector may consist of individual farmers/ farm enterprises of all sizes, agricultural input industries, agro-services enterprises, processing industries, marketing farms and multinational firms. It may also include a wide range of agricultural production and marketing co-operatives, farmers associations and private and voluntary organizations. Despite their differences, all these organizations share a common market orientation. They all try to make profit by selling goods and services. As a result all these private sector organizations have a strong incentive to deliver goods and services (including agricultural extension) efficiently and effectively so as to enhance their ability to survive. Firms that supply agricultural inputs such as seed, chemical fertilizer, pesticides may provide farmers with a wide range of technical and managerial information (through various outreach mechanisms) both to assure that their products are used correctly and also increase agricultural production and income to the farmers. These also motivate customers to buy more products in future (Milu and Jayne 2006). Examples of these private extension agencies are the Muhoroni Sugarcane Outgrowers Company and Chemelil Outgrowers Company, which are currently operating and supporting farmers.

The findings of this study indicated that majority of farmers (82.6%) receive information on credit facility from either from public or private extension providers. The results agree with the findings by Khasiani (1992) who indicated that agricultural technologies might not be adopted if the farmers are not aware of its existence. He continued that lack of awareness acts as a hindrance to the effective participation in agricultural activities. Similarly, Madhur (2000) argued that, impact would be limited if extension is unable to appreciably increase the level of farmers awareness. Further, the results also supports the findings by Mbata (1991) who acknowledged that through extension services the small-scale farmers should be made to understand that credit supervision is for his / her own interest and that, through supervision, credit would be better managed and used for the intended purposes which in turn will increase his productivity and raise their capital base.

The findings of the study also showed that majority of the farmers received extension services on land preparation from private sector. Perhaps this is due to the diverse nature of private sector extension where land preparation machines are readily provided to try and promote return on investment as well as enabling the farmers to increase their income through increased cane production. However, among the farmers who received advice from public extension officers it was noted that the services were not satisfactory. This suggests that perhaps the declining cane production in Kisumu County is due to poor and inadequate land preparation, culminating from inadequate machinery to prepare land for cane growing.

The results further showed that majority of the farmers receive advice on use of appropriate inputs from private sector. This suggests that probably a few farmers may be benefiting as the private sector normally targets potential farmers to maximize the profit from their products. Absence of dependable information to farmers on inputs, on credit and marketing would erode the credibility of extension, hence the rate of adoption by, farmers would be low (FAO, 1994, Khasiani, 1992). However there should be a positive correlation between farmers' link with information sources and adoption (World Bank, 1992 & Chitere, 1995).

This result suggests that public and private sector extension services are inadequate in terms of quantity, that is, in terms of extension agent contact with the farmer despite the fact that cane farmers require it to realize a positive change. Perhaps inadequate extension from government is due to the retrenchment of many staff in the Ministry of Agriculture in the study area (Owuor, 2002).

## CONCLUSIONS

Based on the findings of the study a number of conclusions were drawn:

-Public extension service has a role in enhancing effectiveness of the outgrowers' credit system among smallholder cane farmers in Kisumu County. However, the sector needs to enhance the following: provision of advice with respect to, land preparation and use of appropriate inputs.

- That except for inadequate quantity of extension, private extension service plays a greater role in enhancing effectiveness of outgrowers' credit system.

- That in terms of enhancing effectiveness of outgrowers' credits system there was no significant difference between public and private sector extension services.

Recommendations

From the findings of the study, the following recommendations were suggested.

- Intensification of both public and private extension services.
- Strengthening factory extension division.
- Increasing the number of extension personnel.
- Establishing the contribution of extension among other factors in cane production.

## REFERENCES

1- Agbamu, I. U. (2005). 'Problems and Prospects of Agricultural Extension Services in Developing Countries.' In: Adedoyin, S. F. (ed) op cit Pp 159-169.

2- Agribusiness Development Support Project, (2001). Annual Report. Kisumu: Lagrotech Limited.

3- Akroyd, S., & Smith, L. (2007). The Decline in Public Spending to Agriculture – Does it Matter? Briefing Note, No. 2, Oxford Policy Management Institute, Oxford.

4- Anderson, J. R. (2007). Agricultural Advisory Services. Background Paper for the World DevelOpment Report 2008. http://siteresources.worldbank.org/INTWDR2008/Resources/2795087-1191427986785/Anderson AdvisoryServices.pdf.

5- Chitere, P. A. (1995). Extension Education and Farmers Performance in Improved Crop Farming in Kakamega District (Kenya). Agricultural Administration. 18: 39-57.

6- Economic Survey (2007). Central Bureau of Statistics. Ministry of Planning and National Development. Government of Kenya. 2007.

7- Food and Agriculture Organization. (1997). Effectiveness of Agricultural Extension Services in reaching Rural Women in Africa, Volume 2. Italy, Rome: FAO.

8- Gautam, M. & Anderson, J. R. (1999). Reconsidering the Evidence on Returns to T&V Extension in Kenya. Policy Research Working Paper 1098, the World Bank, Washington D. C.

9- Guda, E., Otieno, L.O., Ko'bonyo, P., Okumu, B., Ohito, D., Odera, J., Ogallo, O.S., Rasugu, O., & Odudo, J. (2001). Business and Investment Insight: (Abstract). Maroko Investments Advisory Services Publications.

10- Government of Kenya. (2005). Review of the National Agricultural Extension Policy (NEAP) and its Implementation. Volume II – Main Report and Annexes. Ministry of Agriculture and Ministry of

Livestock and Fisheries Development. Nairobi.
11- KESREF (2007). Kenya Sugar Research Foundation. Strategic Plan 2009-2014.
12- Khasiani, C. (1992). Towards Legitimisation of African Women Indigenous Knowledge in Natural Resource Management. Award News Kenya: Issue Number 4. December, 1993.
13- Kibet, J. K., Omunyinyi, M. E., & Muchiri, J. (2005). Elements of Agricultural Extension Policy in Kenya. Challenges and Opportunities. African Crop Science Conference Proceedings. 7: 1491 - 1494.
14- Koyenikan, M. J. (2008). Issues for Agricultural Extension Policy in Nigeria. International Journal of Agricultural Extension. 12:51-61.
15- Madhur, G. (2000). Agricultural Extension: The Kenya Experience, an Impact Evaluation. Washington D. C: The World Bank.
16- Mbata, J. N. (1991). Agricultural Credit Scheme in Nigeria, A Comparative Study of the Supervised and Non- Supervised Agricultural Credit Scheme as a Tool for Agricultural Development in Rivers State Nigeria. Discovery and Innovation. (Abstract).
17- Milu, M., & Jayne, T.S. (2006). Agricultural Extension in Kenya: Practice and PolicyLessons. Tegemeo Institute of Agricultural Policy and Development, Egerton University.
18- Nambiro, E., Omiti, J., & Mugunieri, L., (2005). Decentralization and Access to Agricultural Extension Services in Kenya. SAGA Working Paper.
19- Ogunsumi, L.O. (2004). Analysis of Sustained Use of Agricultural Technologies on Farmers' Productivity in Southwest, Nigeria. Ph.D. Dissertation, Department of Agricultural Economics and Extension, Federal University of Technology, Akure, Nigeria.
20- Owuor, G. (2002). The Effect of Financial Self-help Groups Credit on Agricultural Production, A Case of Ukwala Division in Siaya District. (Unpublished MSc Thesis), Njoro: Egerton University.
21- Rees, D. M., Wekundah, F., Ndungu, J., Odondi, A.O., Oyure, D., Andima, M., Kamau, J., Ndubi, F., Musembi, Mwaura, L., & Joldersma, R. (2000). Agricultural Knowledge and Information in Kenya-Implications for technology dissemination and development. ODI Agricultural Research and Extension Network Paper (Abstract).
22- Wangia, C. (2001). Micro- Finance Experience in Kenya. In: Anandajayasekeram, Dixon, Kashuliza, Ng'anjo, Tawonezvi, Torkelsson, Wanzira (Eds). Micro – Finance Experience of FARMESA Member Countries in East and Southern Africa. Farmesa, Harare, Zimbabwe.
23- World Bank. (1992). Trends in Agricultural Diversification: Regional Perspectives. Technical Paper No. 180, Washington D.C: World Bank.
24- World Bank (2007). World Development Report 2008: Agriculture for Development, World Bank Washington D.C.

# Production Efficiency of Farmers under National Fadama II Project in Oyo State, Nigeria

*O.L Balogun[1], A. Adeoye[1], S.A Yusuf[1], R.J Akinlade[1] and A. Carim-Sanni [2]*

Abstract

The study examines production efficiency of farmers under National Fadama-II Project in Oyo State Nigeria. Primary data were collected from two hundred and sixty-four farmers using multistage sampling technique. The analytical framework used for the study include: descriptive, infrastructure index, gross margin and stochastic frontier production function. Average infrastructural index in the area was 0.42. The gross margin for IDV was ₦445, 968.30 while for IUV for under-developed in Fadama villages is ₦357, 805.00. Gross margin was higher for Fadama II farmers than non Fadama-II farmers in IDV. The mean technical efficiency were 0.69 and 0.59 for Fadama and non-Fadama farmers respectively. The result showed that technical inefficiency of female Fadama-II farmers reduced by 0.19% while that of non-Fadama II farmers by 1.23%. Similarly, extension contact, marital status and infrastructural status reduced technical inefficiency of Fadama-II farmers by 2.8%, 0.3% and 2.6% respectively. Presence of infrastructure of Fadama-II project has imparted on efficiency of resource use among the beneficiary. There is therefore need to improve on Community Driven Development programme like Fadama-II and on coming Fadama-III project or any developmental project, so as to further impart more technical and economic knowledge to farmers.

***Keywords:***
***Rural infrastructure, Infrastructural Index, Gross margin, Agricultural production, Technical inefficiency***

[1] *Department of Agricultural Economics, University of Ibadan, Nigeria..*
[2] *National Institute of Social and Economic Research Ibadan, Nigeria.*
* *Corresponding author's email: blarrybunmi@yahoo.com*

## INTRODUCTION

Nigeria is a food-deficit country that on many occasions has been dependent on food imports (Adeoye, 2010). Its agricultural sector has ceased to be an important contributor to foreign exchange earnings; even its contribution to employment has declined. According to Ojo and Akanji (1996), the growth index of agricultural production for crops has shown a decline from 7.4 percent in 1986 to 3.4 percent in 1995. Most studies show that aggregate food production in Nigeria has been growing at about 2.5 percent in recent years, but the annual rate of population growth has been at 3.5 percent (Ajibefun and Abdulkadir, 1999). This situation has not always been so and in fact; there is a great national optimism that the current predicament would be reversed and Nigeria return to full status of a major food basket in the region.

Successive governments in Nigeria have studied the decline in agriculture and have attempted various efforts to re-activate the sector. They include National Agricultural Land development Authority (1991-1999); Presidential Initiatives on cocoa, cassava, rice, livestock, fisheries and vegetables (1999 - 2007) and Poverty Alleviation Programs. Most such efforts have been frustrated by inadequate attention to the critical social and economic root causes of the decline. For instance part of the plight of the traditional Nigerian farmers, which constitute the bulwark of its agricultural landscape, has been the loss of income resulting from wastage and spoilage due to poor infrastructure base. This has been disincentive to the traditional farmers and therefore has been discouraged from extending his acreage. Furthermore, what was a poor infrastructure based deterred the transfer/transportation of farm produce to the urban centers where better prices could be attained. Limited potentials of traditional preservation and processing methods are formidable constraints to increase output.

Consequently development economists have not focused on infrastructure as much as they have on directly productive activities such as agriculture and industry. Also in the course of conducting research on food policies and agriculture in general, it has increasingly been recognized that development of infrastructure, particularly rural infrastructure, bears enormous implications for policy outcome (IFPRI, 2008). Underlying the trend of poor performance in the agricultural sector, is the problem that the farming systems are upland subsistence agriculture that depend mainly on vagaries of weather while the potentials for irrigation using underground and surface water remain underdeveloped. In Nigeria, various agricultural programs and policies have been instituted in the past, and were meant to improve sustainable productivity and farmers' income, consequently the quality of lives of the rural households. One of such projects is the National Fadama (Fadama is a Hausa word for low-lying flood plains, usually with easily accessible shallow groundwater) Development Project II. However, despite the beneficial goals of the project in phases, some communities are yet to participate and benefit from the services offered in the study area. This is because they lack the required basic infrastructures, reducing their production efficiencies and capacities to meet market demands.

In 2001 a New Agricultural Policy and the Integrated Rural Development Policy were initiated to ensure national food security, attain self-sufficiency in basic food production, enhance employment opportunities and achieve high growth rate for the economy. In order to fast track the gains of the 2001 New Agricultural Policy, there came the Presidential Initiatives in Agriculture (PIA) in 2004 and the National Special Food Security Program (NSFSP) and FADAMA II in 2005. The PIA gave priority to four different crop-based expansions of production and utilisation programs (e.g. cassava, rice, tree crops and vegetable oil) and livestock and fisheries programme with a view to curtail the huge foreign exchange expended in their importation and their importance in the revival of industries based on their raw materials. However, FADAMA II are targeted at the resource poor rural farmers and aimed at raising their agricultural productivity and production to eliminate their poverty and through them attain food security. The inadequacy in the provision of rural infrastructure, lack of maintenance culture coupled with inconsistency in policies regarding infrastructural development is expected to have a negative impact on agriculture, which is the major occupation in the rural areas and main sustenance of development in Nigeria. In this regards, it becomes pertinent to know how effectively Fadama II project has solved this problem of rural infrastructural development. This study therefore, attempts to examine production efficiency of farmers under National Fadama II Project in Oyo State Nigeria.

## Theoretical Framework and Literature Review

Infrastructural investments in transport (roads, railways and civil aviation), power, irrigation, watersheds, hydroelectric works, scientific research and training, markets and warehousing, communications and informatics, education, health and family welfare play a strategic but indirect role in the development process. Unlike sectoral development, say agriculture or industry, infrastructure does not directly increase output, but makes a significant contribution towards growth by increasing the factor productivity of land, labour and capital in the production process.

Theoretically, economists proceed from the premise that the creation of infrastructure by generating external economies leads to widespread benefits. For example; Figure 1, shows how traditional theory conceptualizes the effect of infrastructural development on production for a competitive market economy. In a situation of inadequately developed infrastructure, firms are confronted with higher marginal cost ($MC_1$) at every level of production, and, given the market price of their output, produce at $Q_1$. with an improvement in infrastructure, the marginal cost curve shifts downward to the right ($MC_2$), resulting in a total cost savings of area abcd for the earlier level of output, $Q_1$, and an increase in output from $Q_1$ to $Q_2$. The cost reduction occurs through the interaction of infrastructure with directly productive inputs of firms/farms thereby increasing efficiency of production. This may, however, come in a variety of ways, such as reduction in transfer costs, improved diffusion of technology, new combinations of inputs and outputs, better input prices, increased specialization and commercialization, and improved entrepreneurial capacity, all realized through infrastructural investment. The cost reduction is the outcome of an interaction between directly productive inputs of other firms.

$MC_1$= Marginal Cost with infrastructure deficiencies

$MC_2$=Marginal Cost with adequate infrastructure

According to Idachaba and Olayide (1980), rural infrastructures constituted the substance of rural welfare, which is the improvement of the socio-economic life of a community. Idachaba and Olayide (1980) observed that a realistic national development programme should be able to cater for a majority of the nation's populace, which according to him, is formed in the rural areas in less developed countries. However, World Bank (2002) asserted that the provision of social amenities in the rural areas could help in the achievement of an increased rural production and income. Also, Ekong (2000) explains that the spread of needed infrastructure and introduction of appropriate technology in rural areas would markedly improved rural agriculture and industrial output. There is a consensus among authors in terms of general objectives of infrastructural development in rural areas that is the improvement of the standard of living of the rural poor and their integration into the life of the nation.

Peng (2002) who pointed out that road construction could reduce the expenditure of agricultural production while Fang and Zhang (2004) revealed that the potential of agricultural production can be release through rural infrastructure investment.

A recent study by Zongang Li and Xiaomin Liu (2009) on the effect of rural infrastructural development on agricultural production technical efficiency using data from Second Agricultural Census of China indicated that telecommunication, road, good water supply, conducting vocational/technical education and electricity were all positively associated with agricultural production technical efficiency expect telecommunication. These studies demonstrate that investment in infrastructures is essential to increase farmers' access to input and output markets, to stimulate the rural non-farm economy and vitalize rural towns, to increase consumer demand in rural areas, and to facilitate the integration of less-favoured rural areas into national and international economies.

## MATERIALS AND METHODES

**The Study Area**: The study was carried out in Oyo State one of the states selected for Fadama II project in the south western geo political zone, Nigeria. It is bounded in the west by Benin Republic, in the south by Ogun State, in the east by Osun State and in the north by Kwara State. According to the 2006 Census, Oyo State population stood at 5,591,589. Oyo State has thirty-three Local Government Areas (LGAs) in which only 10 participated in Second National Fadama project.

Agriculture is the major source of income for the greatest number of people of the State. Apart from the primary roles of providing food and shelter, employment, industrial raw materials, it remains an important source of internally generated revenue in the State. The state has distinct wet and dry seasons, which characterize its humid tropical climate, with the dry season extending from November to March. Annual rainfall varies from about 500 mm in the northern belt to 1,100 mm in the forest belt. The climate favours the growth of food crops like yam, cassava, millet, maize, fruits, rice and plantains. Cash crops such as cocoa, citrus, tobacco and timber also abound in the state.

**Source of data and sampling procedure:** Primary data were collected for the purpose of this study using structured questionnaire. Some of these include: socio economic and demographic characteristics, Infrastructure proxy variable (such as distance of getting to various infrastructure such as road, market facilities, processing equipment and the access to sanitation etc.) and total production inputs and output quantities and their respective prices of Fadama and non-Fadama crop farmers. A multi-stage stratified random sampling procedure was adopted for the study. The stratification sampling procedure helped in avoiding selection bias that could arise from comparison between participating and non-participating Fadama II project LGAs. The sampling frame was stratified into two strata: Beneficiaries' local government areas and Non -beneficiaries' local government areas (LGAs) that have some social economic and biophysical characteristics comparable to the beneficiaries' LGAs. The first stage of selection involved random selection of two LGAs out of ten that participated in Fadama II project and two LGAs from the remaining twenty-three local government areas that are non participants. In the next stage, 17 villages were randomly selected from each of these LGAs. The last stage involved selection of four farmers from each village. In all, a total of 160 farmers/respondents were chosen in each stratum (given total of 320 farmers/respondents for Fadama II and non-Fadama farmers). A total of 320 respondents were interviewed while two hundred and sixty four questionnaires were retrieved for analysis.

**Analytical tools:** The analytical techniques in the data analysis include: descriptive statistics, infrastructure index, gross margin and stochastic frontier production function.

**Descriptive statistics:** Descriptive statistics (mean, frequency table, percentages).

Composite measure of infrastructure development (Infrastructure Index): The infrastructural index used for this study is based on the sampled village level data adopted from Fakayode *et al.*, (2008) and comparable to method developed by Sen (1990). A total cost of access (TC) was computed by summing the individual cost of access (TCi) to the some six basic infrastructure elements in the study area. These six are those provided by Fadama II project. These infrastructure elements/facilities include market, motorable road, potable borehole, box Culvert, VIP toilet and processing unit.

A total cost of infrastructure availability (TC) was computed by summing the average cost (ACi) of getting a particular infrastructural facility in the 68 villages. ACi was however obtained as an average individual transportation cost was (IDci) of the respondents in each of the 68 villages. For instance, a village may be located 2 kilometers from processing unit center and yet access to the center may be difficult than for a village located 5 kilometers away, if the latter has a better transport system, which is normally reflected in the transportation cost.

An Average Total Cost (ATC) of getting to each of the six infrastructure elements across the villages was obtained by dividing the total cost (TC) by the total number of village (N). ACi was finally weighted with ATC to obtain the weight Wi for each infrastructure and across all the villages. The infrastructure index (INF) was finally obtained by finding the average of the Wis of the six infrastructural facilities for each of the 68 villages.

Algebraically:

$$ACi = \frac{\sum_{i=1}^{n} IDi}{n} \quad \ldots\ldots (1)$$

$$TC = \sum_{i=1}^{N} ACi \quad \ldots\ldots (2)$$

$$ATC = \frac{TC}{N} \quad \ldots\ldots (3)$$

$$Wi = \frac{ACi}{ATC} \quad \ldots\ldots (4)$$

$$INF = \sum_{1}^{6} (Wi.TCi \sum_{1}^{6} Wi \quad \ldots\ldots (5)$$

Where:

$ID_{ci}$ = Individual transportation cost of getting to each Infrastructure by the respondents in each village

$AC_i$ = Average cost of transportation in each village.

TCi = Total cost of transportation to a particular infrastructure i across villages.

ATC=Average total cost of transportation across villages.

Wi = Weight of Average transportation cost in each village.

INF = Infrastructural Index

N = Total number of villages.

M = Total number of infrastructure facilities.

n = Number of respondents in each village.

The infrastructural Index (INF) indicates the degree of under-development, thus, the higher the value of the INF, the less developed the village considered. Further approach to grouping the villages into developed and underdeveloped areas was to sum the infrastructural index for all the 68 villages and the average obtained. The villages with value above the average were said to be under-developed and those below average were said to be developed.

Gross Margin Analysis: The gross margin of an enterprise is the difference between the total value of production and the variable cost. In this study, the gross margin/farmer in the developed and underdeveloped areas for both Fadama and non-Fadama farmers were estimated and compared to determine the profitability of their enterprises.

Gross Margin can be expressed mathematically as;

$$GM = TR - TVC \quad \ldots\ldots (6)$$

Where:

GM=Gross Margin/farmer (₦)

TR= Total Revenue (₦)

TVC= Total Variable Cost (₦)

TVC includes the cost of: Land area (ha), Labour (man-days), Chemical, Seeds and fertilizer, Land clearing etc

TR includes the cost of all sales in the production

**T-test analysis:** T-test analysis was used for the testing of hypothesis that rural infrastructure has a significant effect on agricultural production between the beneficiaries and non-beneficiaries or otherwise.

$$t = \frac{\bar{x} - \mu}{\frac{\sigma}{\sqrt{n}}}$$

Where $\bar{x}$ is the sample mean $x_1, x_2 \ldots x_n$ taken from a normal distribution of $\mu$ and $\sigma^2$. $\sigma^2$ is an estimate of $\sigma$ n is sample size. $\mu$ is the mean while $\sigma$ is the estimated variance.

### Test of Difference between Means

Test of difference between means was employed to determine whether the difference in the profit made by Fadama II beneficiary and non –beneficiary farmers in the developed and underdeveloped areas was significantly different from zero. The null hypothesis stated as; there is no significant difference in the average profit of Fadama II beneficiary and non–beneficiary farmers in the developed and underdeveloped areas is given by;

$$H_O: XP \neq XNP$$

Where:

$$XP = \frac{\sum \Pi P}{nP} \quad \text{and} \quad XNP = \frac{\sum \Pi NP}{nNP}$$

The relationship for the test of difference between means is given by:

$$Z_{score} = \frac{XP - XNP}{S(XP - XNP)}$$

Where the standard error; S(XP – XNP) is given by:

$$\frac{\sqrt{S2P}}{NP} + \frac{S2NP}{NNP}$$

XP = Average profit of Fadama II participated farmers

XNP = Average profit Non-fadama participated farmers

SP and SNP = Standard deviations of XP and XNP

**Stochastic Frontier Production Model:** The stochastic frontier production function was used to determine the effect of rural infrastructure on the crops farmer's efficiency of beneficiaries and non-beneficiaries communities. The stochastic frontier developed by Battese and Coelli (1995) was used to estimate the MLE equation. The Stochastic Frontier model production function is defined by:

$Y1 = F(X1, \beta) \exp(Vi - Ui)$......................(7)

where i = 1, 2,-------------N, where $V_i$ is a random error.

Yi = output or dependent variable (grain equivalent)

Xi = Vector of inputs or independent variable

β = the regression coefficient

Ui = the stochastic error term

Therefore, Cobb-douglas production function in the stochastic frontier form, the model can be expressed thus:

$Yi = \beta 1\ X2^{\beta 2}\ X3^{\beta 3}\ X4^{\beta 4}\ X5^{\beta 5}\ X6^{\beta 6}\ X7^{\beta 7}\ e^{ui}$...... (8)

Where:

Y = Output (grain equivalent)

$X_1$= Farm size (ha)

$X_2$= Labour (man day)

$X_3$= Seeds (kg)

$X_4$= fertilizer (kg)

$X_1$ to $X_4$ are productive variables

U = is the stochastic disturbance

However, if this mode is Log-linearised, the new equation is obtained as thus:

$LnY_i = Lnb_o + b_i LnX_i ..... + b_n l_n X_n + (V_i - U_i)$...(9)

Where bo = constant term

$(V_i - U_i)$ Stochastic error term

$V_i$ = symmetric error

$U_i$ = inefficiency

$b_i - b_n$ = production coefficient of the variables

**Inefficiency model**

The inefficiency model can be stated as:

$|Ui| = \theta_o + \alpha i \sum Mi + di$

Where

$|Ui|$ = inefficiency of ith farmer

$M_1$ to $M_5$ are technical variables

$M_1$ = sex of the farmer (Male=1, 0 otherwise)

$M_2$ = age (years)

$M_3$ = Education level of the farmer (years)

$M_4$ = assess to extension facilities (weekly)

$M_5$ = Participation in Fadama II project

$M_6$ = infrastructural index

$d_i$ = error term

## RESULTS

Table 1 presents the distribution of respondents by marital status. The bulk of the respondents (87.1%) are married regardless of the category of respondents (81.1% for Fadama and 98.8% for non-Fadama farmers). The implication of this is that, there is likely to be more family labour available for farm work. However, majority of respondent farmers (Fadama and non-Fadama) farmers are older than 50 years. This is the active age when farmers can carry out the physical rigor of farm activities. This has implication for agricultural production because farm work requires physical energy and strength.

Education status shows that the largest percentage of the respondents (83.3%) had primary education and more. Education has an important implication particularly for the adoption of new technology and practice (Akinbile and Ndaghu,

Table 1: Socio-economic characteristics of the respondents

| Variable | Fadama | Non -Fadama | All |
|---|---|---|---|
| **Marital status** | | | |
| Single | 1.7 | 1.1 | 1.5 |
| Married | 81.1 | 98.8 | 87.1 |
| Widowed | 12.1 | - | 8.0 |
| Divorced | 5.2 | - | 3.4 |
| **Total** | **100.0** | **100.0** | **100.0** |
| **Age** | | | |
| < 30 | 1.7 | 1.1 | 1.5 |
| 30-50 | 62.6 | 50.0 | 58.3 |
| 51-70 | 35.6 | 48.9 | 40.2 |
| **Total** | **100.0** | **100.0** | **100.0** |
| **Educational level** | | | |
| No Formal | 20.7 | 8.9 | 16.7 |
| Primary | 39.1 | 65.6 | 48.1 |
| Secondary | 29.9 | 24.4 | 28.0 |
| Tertiary | 10.3 | 1.1 | 7.2 |
| **Total** | **100.0** | **100.0** | **100.0** |
| **Household size** | | | |
| 1-5 | 16.1 | 2.2 | 11.4 |
| 6-10 | 73.0 | 84.4 | 76.9 |
| 11-15 | 8.6 | 13.3 | 10.2 |
| > 15 | 2.3 | - | 1.5 |
| **Total** | **100.0** | **100.0** | **100.0** |
| **Gender** | | | |
| Male | 69.0 | 88.9 | 75.8 |
| Female | 31.0 | 11.1 | 24.2 |
| **Total** | **100.0** | **100.0** | **100.0** |
| **Membership of organization** | | | |
| Members | 66.7 | 46.7 | 59.8 |
| Non members | 33.3 | 53.3 | 40.2 |
| **Total** | **100.0** | **100.0** | **100.0** |
| **Farm size(ha)** | | | |
| < 1.00 | 8.0 | 7.8 | 8.0 |
| 1.00-2.00 | 66.7 | 60.0 | 64.4 |
| 2,00- 4.00 | 21.8 | 31.1 | 25.0 |
| > 4.00 | 3.4 | 1.1 | 2.7 |
| **Total** | **100.0** | **100.0** | **100.0** |
| **Farming experience** | | | |
| <10 | 42.5 | 20.0 | 34.8 |
| 11-12 | 36.8 | 38.9 | 37.5 |
| 21-30 | 17.2 | 36.7 | 23.6 |
| > 30 | 3.4 | 4.4 | 3.8 |
| **Total** | **100.0** | **100.0** | **100.0** |
| **Employment status** | | | |
| Full time | 55.20 | 80.0 | 63.6 |
| Part time | 44.80 | 20.0 | 24.4 |
| **Total** | **100.0** | **100.0** | **100.0** |

Table 2: Average amount spent on market in the study area

| Status | ≤ ₦ 40 | ₦41- ₦ 60 | ₦ 61-. ₦ 80 | ≥ ₦ 80 | Average |
|---|---|---|---|---|---|
| **Fadama** | ₦ 8.42 | ₦52.00 | ₦ 74.29 | ₦133.5 | ₦44.44 |
| | 0.55km | 27 km | 3.36 km | 24.25km | 22.27km |
| | 0 mins | 0.82mins | 19.23 min | 61.25min | 27.02min |
| **Non-fadama** | ₦20.86 | ₦53.57 | ₦80.00 | ₦100.00 | ₦55.23 |
| | 1.47 km | 3.00km | 3.00km | 4.71km | 2.56km |
| | 0.16 min | 6.67 min | NA | 16.67min | 7.03min |
| **All** | ₦ 12.97 | ₦52.65 | ₦74.69 | ₦126.80 | ₦32.39 |
| | 1.1 km | 2.24 km | 2.0 km | 2.66 km | 1.50km |
| | 3.53 min | 20.58min | 22.01min | 27.60min | 9.05min |
| **Average amount spent on motorable roads** | | | | | |
| **Status** | ≤ ₦ 40 | ₦ 41- ₦ 60 | ₦ 61-. ₦ 80 | ≥ ₦ 80 | Average |
| **Fadama** | ₦ 11.91 | ₦50.00 | ₦72.00 | ₦138.24 | ₦28.58 |
| | 1.11 km | 1.50 km | 2.80 km | 4.94 km | 4.25 km |
| | 2.10 mins | 7.50 mins | 19.00 mins | 12.94 mins | 4.25 mins |
| **Non-fadama** | ₦14.05 | ₦50.00 | NA | ₦185.00 | ₦34.02 |
| | 0.92 km | 2.00 km | NA | 3.10 km | 1.18km |
| | 0.0.45 mins | 7.50 mins | NA | 18.50mins | 2.61 mins |
| **All** | ₦ 12.67 | ₦ 50.00 | ₦ 72.00 | ₦ 155.56 | ₦ 30.38 |
| | 1.04 km | 1.67 km | 2.80 km | 4.26km | 0.45 km |
| | 0.26 mins | 7.50 mins | 19.00 mins | 21.67 mins | 3.63 mins |
| **Average amount spent on water** | | | | | |
| **Status** | ≤ ₦ 40 | ₦ 41- ₦ 60 | ₦ 61-. ₦ 80 | ≥ ₦ 80 | Average |
| **Fadama** | Na | NA | NA | NA | NA |
| | 1.12km | NA | NA | NA | NA |
| | NA | NA | NA | NA | NA |
| **Non-fadama** | Na | | NA | NA | |
| | NA | NA | NA | NA | NA |
| | NA | NA | NA | NA | NA |
| **All** | | | | | |
| | NA | NA | NA | NA | NA |
| | NA | NA | NA | NA | NA |
| | NA | NA | NA | NA | NA |
| **Average amount spent on possessing unit in the study area** | | | | | |
| **Status** | ≤ ₦ 40 | ≤ ₦ 40 | ₦ 61-. ₦ 80 | ≥ ₦ 80 | Average |
| **Fadama** | ₦ 8.42 | ₦ 8.42 | ₦ 74.29 | ₦133.5 | ₦44.44 |
| | 0.55km | 0.55km | 3.36 km | 24.25km | 22.27km |
| | 0 mins | 0 mins | 19.23 min | 61.25min | 27.02min |
| **Non-fadama** | ₦20.86 | ₦20.86 | ₦80.00 | ₦100.00 | ₦55.23 |
| | 1.47 km | 1.47 km | 3.00km | 4.71km | 2.56km |
| | 0.16 min | 0.16 min | NA | 16.67min | 7.03min |
| **All** | ₦ 12.97 | ₦ 12.97 | ₦74.69 | ₦126.80 | ₦32.39 |
| | 1.1 km | 1.1 km | 2.0 km | 2.66 km | 1.50km |
| | 3.53 min | 3.53 min | 22.01min | 27.60min | 9.05min |

Source: Field Survey (March 2009)

Note: NA – NOT AVAILABLE, ₦150.00 =$1.00

2000). In all, most of the households have at least 6 members which is higher than the national average for all respondents (Fadama and non-Fadama). The national average household size is 5 (NBS, 2007). The size of the household is an importance variable especially in a situation where human power is a major source of power for carrying out farming activities. Notwithstanding Fadama respondents shows a relatively higher percentage of women participation in the farming activities than men. This was attributed to their participation in Fadama I project, a project that gave equal chances to both man and woman and with the provision of some incentives such as market expansion and rehabilitation/construction of rural roads that links to the city, which particularly motivate women to agricultural activities. The implications of more women participation in farming activities increases the population in the agricultural production, thereby reduces food prices, by making food available and improves the standard of living (Nkonya *et al.,* 2008). The result further shows that majority of respondents/farmers belonged to organization. Membership of associations is common among Fadama II more than non-Fadama farmers. Belonging to farmers' organization enable respondents/farmers to have access to information, cheaper inputs, extension services, profitable and other intangible benefits that enhance efficiency in production.

The distribution of the respondent's farm size shows that average farm size for the entire groups was 2 hectares and most farmers have farming experience of at least 10 years while majority of respondents/farmers are full time farmers.

Table 2 shows that infrastructure facilities in the study are those related to agriculture available in both Fadama and non-Fadama areas. These include: Market, motorable road, Boreholes, VIP toilet, Box culvert and processing services center.

Fadama farmers spent an average of ₦44.44 and 27.02 minutes respectively to access market infrastructure provided by the project in beneficiary communities while in non-beneficiary communities spent more on the average to access the same facility. The infrastructure facilities in the study are those related to agriculture available in both Fadama and non-Fadama areas. These include: Market, motorable road, Boreholes, VIP toilet, Box culvert and processing services center. The study revealed that Government and Non-Governmental agents provided available infrastructure facilities in non- Fadama areas. Fadama farmers spent an average of ₦44.44 and 27.02 minutes respectively to access market infrastructure provided by the project in beneficiary communities while in non-beneficiary communities spent more on the average to access the same facility. It shows that Fadama farmers spent the least average amount to various infrastructure elements. Thus the distance barrier is reduced, as transport cost is at minimal in Fadama participating LGAs. Thereby, Fadama participating villages had better access to various infrastructural facilities provided and they were found to be significantly better off in a number of areas including agricultural production, household incomes, and health. The findings support Bhatia and Rai (2008), that the measure of access to various infrastructures is the physical distance in kilometers or transport cost between the households and the centers where these services are provided.

Table 3 shows the average length of time individuals wait for motor vehicle. It was observed that average waiting time for Fadama LGAs is lower compare to non-Fadama LGAs at 10.44 minute, compared with Fadama LGAs of 5.70 minutes. Across LGAs it is 6.80 minutes. In

Table 3: Average time taken to wait for motor vehicle transport

| Status | Average waiting time (minute) | Standard deviation (minute) |
|---|---|---|
| Fadama | 5.70 | 4.5462 |
| Non – Fadama | 10.44 | 4.8452 |
| All | 6.80 | 5.0182 |

Source: Field Survey (March 2009)

Table 4: Distribution of villages by degree of infrastructure development

| Range of index number | Number of villages | | | Percentages | | | Ranking level |
|---|---|---|---|---|---|---|---|
| | Fadama | Non-Fadama | All | Fadama | Non-Fadama | All | |
| ≤0.10 | 20 | 3 | 23 | 29.41 | 4.41 | 32.35 | Highly developed |
| 0.11-0.3 | 13 | 8 | 21 | 19.12 | 11.76 | 32.35 | Moderately developed |
| 0.31-0.5 | 9 | 12 | 21 | 13.24 | 17.64 | 30.88 | Moderately under-developed |
| ≥0.51 | 2 | 1 | 3 | 2.94 | 1.47 | 4.41 | Highly under-developed |
| **Total** | 44 | 24 | 68 | 64.71 | 35.28 | 100 | |

Source: Field Survey (March 2009)

Table 5: Gross margin analysis

| | Developed (naira) | | Under-developed (naira) | |
|---|---|---|---|---|
| **Variable inputs** | **Fadama** | **Non-fadama** | **Fadama** | **Non-fadama** |
| Cost of labour | 49481.0 | 41350.10 | 31464.00 | 22000.00 |
| Cost of fertilizer | 17148.00 | 25850.50 | 13660.00 | 34640.30 |
| Cost of planting mats | 7103.70 | 17801.20 | 11056.00 | 19650.00 |
| Cost of land clearing | 26869.00 | 50,000.00 | 30455.00 | 67500.00 |
| Total variable cost | 100,601.70 | 135,001.80 | 86,635.00 | 143,790.30 |
| Total revenue | 546,570.00 | 499,150.00 | 4.44,440.00 | 486,360.00 |
| Gross Margin | 445,968.30 | 364,148.20 | 357,805.00 | 342,569.70 |

Source: Field Survey (March 2009)

order to have a vivid exposition of the degree of under-development, index of infrastructure in Table 4 shows that the index of infrastructure ranges between 0.04 and 0.53 for all the LGAs with an average of 0.17, 0.24 and 0.42 for Fadama, non-Fadama and the entire 68 villages respectively. It further reveals that Fadama villages were more highly infrastructural developed compared with non-Fadama villages.

Cost structure and Gross margin were analyzed and compared to isolate the effect of rural infrastructural development on the profitability of Fadama beneficiaries and non –beneficiaries in developed and underdeveloped areas In Fadama LGAs, cost of labours is higher in developed villages than in the underdeveloped villages and for all other variable input, except for the Cost of land clearing. Total variable cost is however higher in the developed villages than in the underdeveloped villages. Despite the higher total variable cost in the developed villages, gross margin was higher in the developed villages than in the underdeveloped villages. Table 5 shows that the total variable cost was estimated at ₦100,601.70 in the developed villages and ₦86635.00 in the underdeveloped villages while the gross margin was estimated in Fadama community to be ₦ 445, 968.30 in the developed villages and ₦ 357,805.00 in the underdeveloped villages. On the other hand, all variable factors cost is lower in the developed villages except for the cost of labour under non-Fadama LGAs. Thus, total variable cost is however higher in the underdeveloped villages than in the developed villages. This might be as a result of poor road that make market accessibility difficult and at the time increases the cost of procuring inputs. Despite the higher total revenue in both developed and the underdeveloped villages, gross margin/farmer was lower because of the higher total variable cost in both developed and the underdeveloped when comparing villages. Table 5 further shows that the total variable cost was estimated at ₦135, 001.80 in the developed villages and ₦143, 790.30 in the underdeveloped villages while the gross margin was estimated in Non-Fadama community to be ₦364,148.20 in the developed villages and ₦ 342,569.70 in the underdeveloped villages. This result therefore, shows a higher return for Fadama participants in both developed and underdeveloped villages than the non-Fadama, a

Table 6: Test of Difference between means

| Status | Mean gross margin (₦) | Mean Difference (₦) | T -value | Prob. |
|---|---|---|---|---|
| Fadama | 88,163.3 | 66584.8 | 3.216 | 0.000 |
| Developed | 445,968.30 | | | |
| Underdeveloped | 357,805.00 | | | |
| Non-fadama | 21,578.50 | | | |
| Developed | 364,148.20 | | | |
| Underdeveloped | 342,569.70 | | | |

Source: Field Survey (March 2009)

result, which must have been made possible by the presence of infrastructure provided by Fadama II project. Test of difference between the gross margins of Fadama beneficiaries and non–beneficiaries is shown in Table 6. Result shows that there is significant different in the gross margins of Fadama beneficiary and non–beneficiary farmers at 1% and the gross margins for Fadama beneficiaries are higher than that of the non–beneficiary.

The result of efficiency score among the farmers in Table 7 shows that the mean technical efficiency of 0.693, 0.588, and 0.681 for Fadama, non-Fadama and the entire respondents respectively. The two groups of famers are operating some distance from the frontier with fadama technical efficiency. 68% of the Fadama farmers have technical efficiency above 0.84 while more that two-thirds of the Fadama farmers are operating not too far from the frontier. About 22.2% and 43.2% of the farmers have technical efficiency exceeding 0.84 for non-Fadama and the entire respondents respectively.

Table 8 shows the result of technical inefficiency of Fadama farmers. The result showed that farm size significantly leads to technical inefficiency whereas labour leads to reduced technical inefficiency at 1% level of significance. The implication is that increase in investment on labour in farming activities tends to give higher productivity and enhance the technical efficiency. In the case of seed input used, the estimate was also positive and significant, implying that as quantity of seed sown increases, crop yield increases for the Fadama II farmers. However, the coefficient of infrastructure index was negative. This implies that, as the level of infrastructural development in an area increases, technical inefficiency of farmer's increase. This result agrees with (Bhatia *et al.,* 2004 and Wanmali, 1985) who reported that the farther the services from the households, the less they were used. In case of extension contact, the variable was negative and significant at 1%. It indicates that the involvement of extension agent tends to reduce the technical inefficiency for crops production. This agrees with Olayide (1985) that development of infrastructure and extension work is a precondition for the adoption and diffusion of new agricultural technology.

Also, the coefficients of educational level and years of experience were negative and these factors led to decrease in technical inefficiency of farmers. This agrees with the findings of Ojo and Ajibefun (2000) that education and year of experience increase the rate of adoption of im-

Table 7: Summary of technical efficiency of farmers

| Interval | Percentages | | |
|---|---|---|---|
| | Fadama | Non- Fadama | All |
| 0.04 – 0.23 | 5.70 | 4.44 | 3.78 |
| 0.24 – 0.43 | 12.07 | 23.33 | 13.67 |
| 044 – 0.63 | 20.11 | 33.33 | 21.59 |
| 0.64 –0.83 | 22.99 | 16.67 | 17.80 |
| 0.84-1.03 | 68.2 | 22.20 | 43.1 |

Source: Field Survey (March 2009)

Table 8: Estimation of the technical efficiency of farmers

| Variables | | CoefficientCoefficient | |
|---|---|---|---|
| | Fadama | Non- Fadama | All |
| Constant | 6.348 (11.28)*** | 8.930 (47.77) | 7.564 (9.03***) |
| Farm Size | -0.494 (-4.40)*** | 0.321 (1.10) | 0.0136 (0.10) |
| Labour | 0.534 (13.32)***_ | -8.280 (-0.21) | 0.588 (3.18)*** |
| Chemical | 0.750 (0.39) | 0.026 (0.21) | 0.133 (0.48) |
| Seeds | 0.128 (2.56) | 0.139 (1.81*) | 0.166 (2.23)** |
| Equipment | -0.0749 (-0.64) | | 0.159 (0.65) |
| **Inefficiency function** | | | |
| Constant | 1.636 (1.00) | -1.348 (-1.37) | 0.713 (-0.09) |
| Age | -0.833 (-1.22) | 0.565 (2.17)** | -0.779 (0.39) |
| Gender | -0.190 (-2.19)** | -1.226 (-1.81)* | -1.0920 (-1.30) |
| Marital Status | -0.273 (-2.89)*** | | 0.255 (0.24) |
| Educational Level | -.0.395 (-1.06) | 1.747 (1.67)* | 0.346 (0.79) |
| Extension Contact | -2.803 (-2.86)*** | | 3.335 (1.91)* |
| Infrastructure status | -2.644 9 (-2.09)** | -0.259 (0.94) | 3.761 (-1.78)* |
| Fadama Status dummy | 0.378 (0.72) | 0.259 (0.94) | 0.836 (0.99) |
| Years of Experience | -1.268 (-1.54) | 1.137 (3.90)*** | 0.590 (1.28) |
| **Member of an organization** | | | 0.325 (0.45) |
| Sigma-Square | 7.769 (5.56) | 2.865 (7.29)*** | 20.0504 (2.84)*** |
| Gamma | 0.9999 (1530819.9) | 0.999(869317.01)*** | 0.976 (91.18)*** |

Source: Field Survey (March 2009) t- ratios are in parentheses, Log likelihood function = -0.307
*** Significant at 1 percent, ** Significant at 5 percent, * Significant at 10 percent
*** P < 0/01 **P < 0/05 *P < 0/1

proved production. The result also shows that the coefficients age and gender of farmers were negative; implying that older farmers have higher technical inefficiency than younger farmers and farming activities is labour intensive with gender bias.

Infrastructural index are significant at 5% while extension contact and marital status were significant at 1%. For non-Fadama farmers, the result on the impact of socio-economic variables on technical inefficiency reveals that the coefficient of gender was negative at 10%. The implication is that being female decreases inefficiency among the farmers. In other hand, the result further shows that educational level and years of experience were negative and significantly explained inefficiency of the fadama farmers. The shows that additional year of education and experience of the farmers decreased inefficiency in fadama farmers whereas for the non fadama farmers it increased their inefficiency. Years of participation in Fadama farmers is negatively related to technical inefficiency. This implies a decreased level of technical inefficiency

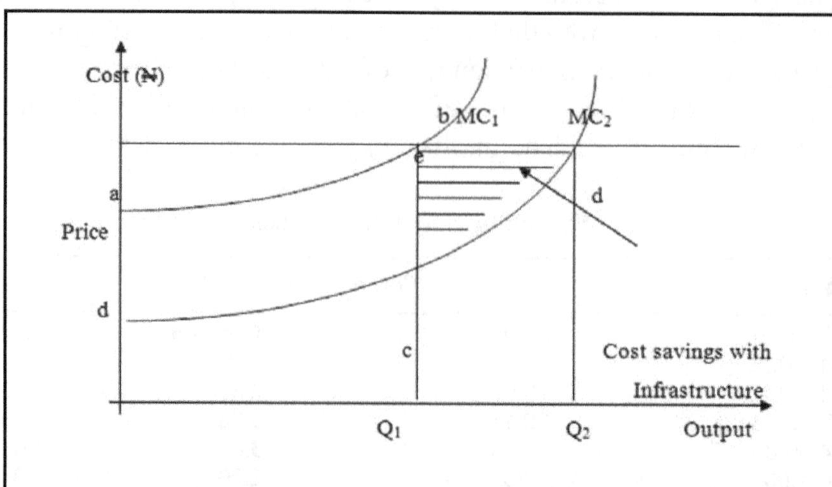


Figure 1: Infrastructure Provision and Efficiency of Production

as the farmers spend more years in the programme. The likely reason for this is that the farmer may develop more skill with time as long as they are in the scheme. This finding supports Muhammad-Lawal, *et al.*, (2009) that more skill will be acquired as long as the farmer remains in the program. Infrastructure index was not significant meaning it did not contribute to inefficiency of the farmers.

## CONCLUSIONS AND RECOMMENDATIONS

The efficiency of production is enhanced by high level of infrastructural development. Though infrastructure is seen as an indirect capital input, it is essential for improving farmers' production efficiency. In order to improve on the level of efficiency of farmers, improved infrastructural development is a sine qua.

Presence of infrastructure has implications for efficiency of resource use. Fadama participating villages had better access to various infrastructural facilities provided and they were found to be significantly better off in a number of areas including agricultural production, household income and also the participation of women in the economy also they obtain higher price for produce and to buy a larger proportion of consumption needs from the market when compare to non-Fadama participating villages. Thus development of infrastructure has a positive effect/impact on the wholesome lives of the people in the areas. Therefore more infrastructural facilities should be provided to aids development most especially in non-Fadama area.

## ACKNOWLEDGEMENT

The authors gratefully acknowledge the funding support of Fadama Support for the Research program.

## REFERENCES

1- Adeoye, A. (2010). Impact of Rural Infrastructural Development on Agricultural Production under Fadama II project in Oyo State. Unpublished M.Sc thesis, University of Ibadan.

2- Akinbile, L. A., and Ndaghu A. A. T. (2000). Poverty Level and Poverty Alleviation Strategies of Farm Families Michika L.G.A of Adamawa State, Nigeria. Journal of Economics and Rural Development, 14 (2), 101-109.

3- Ajibefun, I. A., and Abdulkadri, A. (1999). An Investigation of Technical Inefficiency of Production of Farmers under the National Directorate of Employment in Ondo State, Nigeria. APP Economics Letters, 6, 111-114.

4- Battese, G. E., and Coelli. T.J. (1995). A Model for Technical Inefficiency Effects in a Stochastic Frontier Production Function. Empirical Economics, 20: 325-32.

5- Bhatia, V. K and Rai, S. C. (2008). Evaluation of Socio- Economic Development in Small Areas, A project Report on Indian society of agricultural statistics Iasri Campus, Library Avenue, Pusa New Delhi - 110 012 (2003 – 2004)

6- Fakayode, B. S., Omotheso, O. A., Isoho, A. B., and Ajayi P. D. (2008). An Economic Survey of Rural Infractructures and Agricultural Productivity Profiles in Nigeria European. Journal of Social Science, 7, (2): 158-170.

7- Fan, S., and Zhang, X. (2004). Infrastructure and Regional Economic Development in Rural China. China Economic Review, (15): 203-214.

8-Idachaba, F. S. and Olayide, S.O. (1980). Rural Infrastructure and the Small Farmers. Pp. 245- 256 in S.O. Olayide *et al,*. (editors), Nigerian Small Farmers: Problems and Prospects in Integrated Rural Development, Ibadan, C.A.R.D., University, Ibadan.

9- Muhammad-Lawal, A., Omotesho, O.A., and Falola, A. (2009). Technical Efficiency of Youth Participation in Agriculture: A Case Study of The Youth - In Agriculture Programme in Ondo State, South Western Nigeria Nigerian Journal of Agriculture, Food and Environment, 5(1): 20-26.

10-NBS (2007) National Bureau of Statistic Abuja Nigeria

11- Nkonya E., Phillip, D., Mogues, T ., Yahaya, M. K., Pender, J., Adebowale, G., Arokoyo, T., and Kato, E. (2008). The Impact of a Pro-Poor Community Driven Development Project in Nigeria. International Food Policy Reseach Institute (IFPRI)

12- Ojo, M. O. and Akanji, O. O. (1996 June). The Impact of Macroeconomic Policy Reforms on Nigeria Agriculture. CBN Economic and Financial Review. 34 (2): 549-579.

13- Peng, D. (2002). Rural Infrastructure Investment and Poverty Reduction, Jingji Xuejia (The Economist) (5): 79-82.

14-Wanmali, S. (1985). Rural Household Use of Services. A Study of Miryalguda Taluka India. Research Report 48. Washington, D.C. IFPRI.

15- World Bank. (2002). Inadequate Infrastructural

Facilities as a Measure of Poverty, World Bank Economic and Social Studies Review. 44 (1).

16- Zonghang, L., and Liu, X. (2009). The effects of Rural Infrastructure Development on Agricultural Production Technical Efficiency; Evidence from the Data of Second National Agricultural Census of China. Paper presented at the International Association of Agricultural Economists Conference, Beijing, China.

# Using Probabilistic-Risky Programming Models in Identifying Optimized Pattern of Cultivation under Risk Conditions (Case Study: Shoshtar Region)

*Mohammad Kavoosi Kelashemi[1]*, Mohammad Aghapour Sabbagi[2] and Hamid Reza Alipour[3]*

Abstract

Using Telser and Kataoka models of probabilistic-risky mathematical programming, the present research is to determine the optimized pattern of cultivating the agricultural products of Shoshtar region under risky conditions. In order to consider the risk in the mentioned models, time period of agricultural years 1996-1997 till 2004-2005 was taken into account. Results from Telser and Kataoka models showed that due to accepting the risk amounts, most of the optimized amounts suggest the tomato cultivation during the cultivation period of fall, and watermelon cultivation during the cultivation period of spring. On the basis of results, due to allocation of agricultural lands of Shoshtar to tomato and watermelon cultivation and specializing the farming activity in this province, gross profit of agricultural production system can be increased to 6116047000 and 727782272 thousand Rials, respectively. The results of understudy models were investigated under different income scenarios and probabilistic levels of risk acceptance. Correct policy making in order to offer the suitable equipments for adjusting the effects of lack of certainty and risks due to the climatic unwanted conditions in production process of agricultural products of Shoshtar region improve the life situation of farmers of the mentioned region.

***Keywords:***
***Probabilistic-risky programming model, Cultivation pattern, Agricultural products, Shoshtar region***

[1] *PhD student of Agricultural Economics, Tehran of University, Karaj, Iran.*
[2] *Assistant Professor of Islamic Azad University Shoshtar Branch.*
[3] *Assistant Professor of Islamic Azad University Rasht Branch.*
* *Corresponding email: mkavoosi@ut.ac.ir*

## INTRODUCTION

Although risk is common problem in agricultural activities amongst all parts of the world, its greatness in developing countries is more that that in other countries (Norak and et al. 1991). On the other hand, bearing the risk by small farmer is harder in comparison with great land owners (Irima and et al., 2004). In risky conditions, maximizing the profit of agricultural activities through maximizing the gross margin of agricultural risk from which the risk expense has been reduced is realized by considering the limitations due to production factors and risk (Hassan Shahi, 2006). With area about 3538 kilometer squares, Shoshtar region has been located at the center of Khuzestan province. Population of this province is 210108 people. The areas of aquatic and non irrigated lands of the mentioned province are 92047 and 49830 hectare, respectively. This region, with producing 2136229 tons aquatic agricultural products and 18409 tons non-irrigated agricultural product in agricultural year 2004-2005, has allocated a special position for itself in agriculture of Khuzestan province (Jahad-Keshavarzi, Khuzestan province, 2006). The main agricultural products of Shoshtar region include aquatic and non-irrigated wheat, aquatic and non-irrigated barley, paddy, aquatic cucumber, watermelon, grain maize and tomato. (Table 1)

Producing strategic products such as wheat and rice, in Shoshtar region makes it clear to make correct policies about stability of production system. On the other hand, unfavorable climatic conditions and limitations in availability to agricultural basic inputs cause frequent damages in agricultural lands of this region. In the event that only agricultural year 2004-2005, an area amounting 1142 hectares of aquatic lands of Shoshtar region has experienced damages and products have been destroyed. Regarding the mentioned problems, necessity for determining a suitable cultivation pattern that firstly creates stable and primary income for farmers of this city and then makes the ground ready for specializing the region's agriculture, is felt.

Linear programming models have frequent applications in agricultural decision makings (Nicholson and et al. 1994). On the other hand, together with development of programming models and compilation of multi-period decision making models the possibility for considering market risk and production was provided in agricultural decision makings. For example, dynamic algebraic linear programming model has been used in most studies about for determining the optimized pattern of water consuming regarding the previous agricultural periods (Backeberg, 1997; Haile and et al. 2003; Van Schalkwyk and Louw, 2004). Linear programming had been widely used in determining cultivation pattern under risky conditions. Peykani and et al. (2009) determined the best variety of rice for cultivation, using Kataoka and Telser programming methods. Results showed that at different levels of the gross margin of the aim, two varieties include Hashemi and Ali Kazemi were the best for achieving considered goals.

Intense dependence of Shoshtar region's farmers on income from production of these products as the only means for living causes these people to

Table 1: Acreages, production quantity and yield of major cultivation product of Shoshtar region in cultivation year of 2004-2005.

| Product | Acreages (ha) | Production quantity (ton) | Yield (kg per ha) |
|---|---|---|---|
| Aquatic Wheat | 25967 | 117253 | 4515 |
| Non-irrigated Wheat | 15019 | 15417 | 1026 |
| Paddy rice | 13273 | 53971 | 4066 |
| Aquatic Barely | 3540 | 5982 | 1689 |
| Non-irrigated Barely | 4463 | 2992 | 674 |
| Aquatic Cucumber | 608 | 11031 | 18143 |
| Watermelon | 3532 | 104849 | 29685 |
| Maize | 3053 | 19502 | 64 |
| Tomato | 2412 | 73031 | 30278 |

Resource: Jahad-Keshavarzi of Khuzestan province.

take safety-first conditions and price risk into account in selecting cultivation pattern. Using probabilistic-risky programming models makes it possible to consider safety-first conditions in order to realize the considered gross margin per hectare and taking the price risk into account in determining the cultivation pattern. More studies have been made to analyze the farmers' decisions regarding the risk effect (Rudel, 2000; Tauer, 1983). On the other hand, using the safety-first models in most studies has been useful in explaining the farmers' behaviors (Patrik and et al., 1985; Roy, 1952). Lack of ability in exactly foreseeing the price of products, price of production inputs, and amount of production harvest from one hand and atmospheric and climatic conditions on the other, have caused instability in incomes of Shoshtar region's farmers. Taking this point into account that this region's farmers, considering the risk, want to gain maximum amount of profit, the questions are: 'From among the cultivation options of agricultural products, which of them and with what amount of risk should they select in order to gain maximum profit?', 'which combination of the production factors should they select to reduce the production costs down to the minimum level and cause the productions to reach the desired level?'. Answering the aforesaid questions requires applying an exact programming. Using Telser and Kataoka probabilistic-risky mathematical programming models, the present research is to determine the optimized pattern of agricultural products cultivation in Shoshtar region under the risky conditions. In order to consider the risk in aforesaid models, time period of the agricultural years 1996-1997 till 2004-2005 was regarded.

## MATERIALS AND METHODS

Safety-first models have been used as standards for a long time to make decisions under uncertainty conditions. Using the two models of Telser and Kataoka, this research introduces the most suitable agricultural activities to realize the pre-determined goals. In this research, selection of sequence of optimized agricultural activities has taken place regarding the risk rows that include gross margin per hectare of each understudy products during the agricultural years 1996-1997 till 2004-2005, realizing the income goals of the farmer (under the first-safety conditions) and limitation of Shoshtar region's agricultural resources. In Telser's model, expected gross profit regarding the risk rows of the gross profit per hectare of each one of agricultural products during the understudy years, limitation of safety-first conditions and limitation of agricultural resources is maximized. In Kataoka's model, gross profit level per optimized hectare has been determined a goal such as T in which the probability of gross profit's falling down per hectare to the level lower than the amount of T, is less than the amount of probabilities that have previously been determined in the model. Both models of Kataoka and Telser include probabilistic limitations for considering the safety-first conditions that is as follows: (1)

$$\Pr(Z \pi g) \leq 1/L^{*} \qquad (1)$$

At the above equation, $\Pr(Z \pi g)$ is the probability of occurrence of $Z \pi g$ . Z is random variable of gross margin per hectare and g is the goal gross margin per hectare. $1/L^{*}$ is upper limit for $\Pr(Z \pi g)$.

Different approaches are used to enter the safety-first conditions regarding the hypothesis of programming model. In constrained chance programming, a common approach is changing the probabilistic limitations into deterministic ones (Charles and Cooper, 1959). Another approach requires the multi-normality and ability to make effective E-V series of the answers. Pyle and Turnovsly have used this approach to compare the results from the two methods of optimized expected utility and safety-first (Pyle and Turnovsly, 1970). The third approach is imposing the probabilistic limitation using the probabilistic inequality. For the first time, Berk and Hin used the nonlinear form of the Lower Partial Moment Inequality (Berk and Hin, 1982). Creating the conservative confidence limits is one of the difficulties regarding the aforesaid non-linear form usage. Atwood offered a special form of Chebychev's linear inequalities that created less conservative confidence limits (Atwood, 1985). This inequality enjoys continuous distribution and obeyed the Lower Partial Moment Inequality principles. This probabilistic inequality has been applies in this research to consider the safety-first conditions in both Kataoka's and Telser's risky programming models. Lower partial moment is defined as (Atwood, 1985): (2)

$$R(a,t) = \sum (t-z_i)^{a.f_i} \quad (2)$$

In the above relation, $R(a,t)$ is lower partial moment in which t is the amount of gross margin per hectare below which deviations are measured and zi shows the amount of z (expected gross margin) when the ith case occurs. a, (a>0) is the force that, taking it into account, deviations that are below t are created and fi is the amount of probability that, taking it into account, the ith case is occurred. In order to enter the probabilistic inequality in linear programming model the following relation can be used (Atwood and et al., 1988): (3)

$$Pr\ (Z \leq t - pQ\ (a,t)) \leq (1/P)^a \quad (3)$$

In which $Q\ (a,t) = [R(a,t)]^{1/a}$ and its amount is always greater than zero. p is fixed amount and greater than zero. If p is defined as P = (t-g)/Q(a,t) , and if t>g, equation 3 can be expressed as (Atwood and et al., 1988): (4)

$$Pr(Z \leq g) = Pr\ (Z \leq t - pQ(a,t))[Q(a,t)/(t-g)]^a \quad (4)$$

In order to use the aforesaid probabilistic inequality in linear programming model, the amount of should be assumed as 1, therefore in the above relation Q (t) can be replaced with Q(1,t) or R (1,t): (5)

$$Pr(Z \leq g) = Pr(Z \leq t - pQ(t)) \leq Q\ (t)/(t-g) \quad (5)$$

Finally, in order to consider the probabilistic limitation of safety-first condition Pr ($Z\pi$ g) ≤ $1/L^*$ in considered risky programming models, inequality t- LQ (t) ≥ g is used.

Regarding the risk rows of gross margin per hectare of each one of agricultural activities during understudy years, limitations in safety-first conditions, and limitations in resources of agricultural production, expected gross margin in Telser's model is optimized. Specialty of the mentioned model is using the probabilistic limitation in order to consider the safety-first conditions. Taking the safety-first conditions into account, probabilistic-risky programming model can be defined as follows (Atwood and et al., 1988): (6)

$$\begin{aligned} &\text{Max } E\ (Z) = E'\pi_{ri} X_i \\ &\text{Subject to:} \\ &\Sigma_i\, a_{ij} X_i \leq b_j \qquad j = 1,\ldots, m \\ &\Sigma_i \pi_{ri} X_i - t + Z_r \geq O \qquad r = 1,\ldots, s \\ &r' Z_r - Q\ (t) = O \\ &t - LQ\ (t) \geq T \\ &X_i, b_j, Z_r \geq O \qquad i = 1,\ldots, n \end{aligned} \quad (6)$$

In the above model $a_{ij}$ is technical coefficient of the under use input j in the $i^{th}$ activity, $X_i$ is the $i^{th}$ activity level, $b_j$ is the existent amount of the jth input, $\pi_{ri}$ the gross margin per hectare of i activity in the year r, T is the amount of goal income, E (z) is expected accumulative gross margin, $E'\pi_{ri}$ is the transpose vector of expected gross margin for each activity, and is the vector of activity levels. t is the resource level for gross margin, $X_i$ is the transpose vector of probabilistic levels and $r' Z_r$ is equivalent with Q (t) . t- LQ (t) ≥ T is Atwood relation that after determining the amounts of goal income T and L, expected accumulative gross margin reaches to its maximum amount with considering this limitation, risk rows and limitations in resources.

In probabilistic-risky programming system of Kataoka's, taking the safety-first conditions into account, the goal is to realize ideal of maximizing the amounts of goal gross margin of the (T). Considered limitations in this model is identical with those of Telser's system, and only the probability limitation related to considering safety-first conditions is changed into t- LQ (t) - T ≥ O. Results from Kataoka's system show the maximum amounts of the goal gross margin regarding the limitation in resources, limitation in price risk or matrix of gross margin amounts of each activity

$$\begin{aligned} &\text{Max} T \\ &\text{Subject to:} \\ &\Sigma_i\, a_{ij} X_i \leq b_j \qquad j = 1,\ldots, m \\ &\Sigma_i \pi_{ri} X_i - t + Z_r \geq O \qquad r = 1,\ldots, s \\ &r' Z_r - Q\ (t) = O \\ &t - LQ\ (t) \geq T \\ &X_i, b_j, Z_r \geq O \qquad i = 1,\ldots, n \end{aligned} \quad (7)$$

during the understudy years and limitations in safety-first conditions. Considering the Kataoka's safety-first conditions, the probabilistic-risky programming model can be defined as (Atwood and et al., 1988): (7)

In the present research the sequence of understudy activities is cultivation of aquatic and non-irrigated wheat, aquatic and non-irrigated barley, paddy, aquatic cucumber, watermelon, grain maize, and tomato in Shoshtar region. In order to determine the optimized pattern of cultivation using the Telser's and Kataoka's approaches, the limitations in production resources – including labor, land, chemical fertilizers of nitrogen, phosphor, potassium, fungicide chemical poisons, herbicide, pesticide and water– were considered in programming models. In order to consider the production and yield risk in the processes of determining the optimized pattern of cultivation, gross margin of each one of the sequence of understudy activities during the agricultural years 1996-1997 till 2004-2005 was regarded. Data needed for doing the present research was gathered from Jahad-Keshavarzi of Shoshtar region, Khuzestan province's agricultural supportive services company, Khuzestan province's regional water company, Khuzestan province's coordination office and economical affairs and central bank of Islamic Republic of Iran.

## CONCLUSIONS AND DISCUSSION

In the present research, in order to estimate Telser's and Kataoka's programming models, Lingo software packages were used. After calculating the amounts of gross margin for each one of agricultural activities of aquatic wheat, aquatic barley, paddy, non-irrigated wheat, non-irrigated barley, cucumber, aquatic maize, tomato and watermelon during the agricultural years 1996-1997 till 2004-2005, the aforesaid amounts were changed into 2005 basic year with the use of suitable price index. Consequently, besides enjoying homogeneous risk rows in programming models, all of the results can be interpreted on the basis of prices relating to the year 2005. In order to consider the capacity of the region's lands and take the most important advantages of agricultural production resources, three programming groups were taken into account for non-irrigated lands, agricultural products relating to the cultivation period of fall and agricultural products relating to the cultivation period of spring.

Applying Telser model which includes probabilistic inequality for considering safety-first conditions has provided the possibility that taking the living conditions of farmers into account and measuring the probabilistic limitations, the most suitable sequence of agricultural activities for cultivation

Table 2: Telser's model Results for fall cultivation period.

| Goal gross margin per hectar[1] | Probability limitation | Expected gross margin[2] | Acreage of optimized enterprises (ha) | |
|---|---|---|---|---|
| | | | Tomato | cucumber |
| **6347** | 0.1 | 6116047 | 416766.5 | 0 |
| | 0.1 | 2349603* | 10836.5* | 262124.8* |
| | 0.2 | 6116047 | 416766.5 | 0 |
| | 0.2 | 2349603* | 10836.5* | 262124.8* |
| **8347** | 0.1 | 6116047 | 416766.5 | 0 |
| | 0.1 | 2349603* | 10836.5* | 262124.8* |
| | 0.2 | 6116047 | 416766.5 | 0 |
| | 0.2 | 2349603* | 10836.5* | 262124.8* |
| **10347** | 0.1 | 6116047 | 416766.5 | 0 |
| | 0.1 | 2349603* | 10836.5* | 262124.8* |
| | 0.2 | 6116047 | 416766.5 | 0 |
| | 0.2 | 2349603* | 10836.5* | 262124.8* |

1 in thousand rials.
2 in million rials.
* Scenario of keeping tomato's acreage at present region's level.
Resource: research findings.

Table 3: Telser's model Results for spring cultivation period

| Goal gross margin per hectar[1] | Probability limitation | Expected gross margin[2] | Acreage of optimized enterprises (ha) | |
|---|---|---|---|---|
| | | | Paddy | watermelon |
| 6106 | 0.1 | 727782272 | 0 | 90139 |
| | 0.1 | 587966528* | 68070* | 22069* |
| | 0.2 | 727782272 | 0 | 90139 |
| | 0.2 | 587966528* | 68070* | 22069* |
| 7106 | 0.1 | 727782272 | 0 | 90139 |
| | 0.1 | 587966528* | 68070* | 22069* |
| | 0.2 | 727782272 | 0 | 90139 |
| | 0.2 | 587966528* | 68070* | 22069* |

1 and 2 in thousands rials.
* Scenario of keeping watermelon's acreage at present region's level.
Resource: research findings.

be selected regarding the goal income. In Telser model, the optimized levels of activities are determined by selecting the amount of goal income and lower partial moment amounts. In this direction, after selecting the amounts of goal income, basic probabilistic levels are considered for each amount of the goal income. Aforesaid probabilistic amounts are expressive of littleness of expected accumulative gross margin in comparison with the level of goal income. Results from Telser's model have been reported in the following tables. (Table 2)

Goal income amounts have been determined on the basis of adjusted average gross margin of understudy agricultural activities in each time period, and continue up to the maximum average of the existent adjusted gross margin. In case of the cultivation period of fall, three goal incomes were selected as 6347, 8347, and 10347 thousand rials per hectare. Taking this goal into account that the amount of gross margin of farming on the basis of the price of the year 2005 with probability of 0.1 not to be less than 6347 thousand rials, the amount of under cultivation level of 416766/5 hectares tomato was selected as optimized activity. In this case the average amount of expected gross margin with optimized product cultivation is 6116047 million thousand rials. In this level of the goal income changing the probabilistic amounts into 0.2 or accepting more risk doesn't cause any changes in model results. Changing the goal income into 8347 thousand rials at probabilistic level 0.1 causes the optimized level of tomato cultivation activity to become 416766/5 hectare. Regarding the defined risk rows in Telser's model and fluctuations in gross margin of production system in understudy years in lieu of placing other amounts of goal income in basic probabilistic levels, optimized answers don't change by Telser model. In other words, in Telser model, the ideal of maximizing the expected accumulative gross margin in comparison with other probabilistic amounts and goal income, have priority. Beside each probabilistic and goal income limitation a scenario has been considered on the basis of retaining the under cultivation level of tomato at the present amount. Regarding the high consumed amounts of fertilizer and chemical poisons and labor intensify cultivation of tomato, compilation of retaining the under cultivation level of tomato at present amount of the region was considered. After exercising the aforesaid scenario the amount of expected accumulative gross margin was reduced from 6116047000 to 2349603000 thousand rials and the agricultural product of cucumber is also placed in optimized cultivation mode. (Table 3)

The ideal of Telser's model is to maximize the expected gross margin. Regarding the condition of gross margin of mentioned agricultural activities during understudy period, at first adjusted average of gross margin of mentioned activities (6106 thousand rials) was selected as goal income. Then two probabilistic levels of 10 and 20 percent for probability of expected gross margin reduction from the level of goal gross margin was taken into account. In case of goal income of 6106 thousand rials and probabilistic limitation of 10 percent, the offered optimized cultivation model by Telser's model is generally peculiar to watermelon cultivation. Under the aforesaid conditions, maximum amount of expected gross margin that farmers of the region gain is 727782272 thousand rials. Regarding the strategically of paddy cultivation, and emphasis of economical planners on retaining the under cultivation level of this product, scenario of retaining the present under cultivation level of watermelon was regarded. By exerting aforesaid scenario at probabilistic level of 10 percent and goal income 6106 thousand rials 68070 hectares were allocated to paddy cultivation. In this case expected gross margin amount of the region's agriculture is reduced by 23.8%. Ideal of Telser's model is such that changes in probabilistic amounts and goal income don't cause changes in model results and always maximizing the expected gross margin is taken into model's account. Regarding the negative expected gross margin amounts for non-irrigated products during the under study years, the ideal of Telser's model is not realized on the basis of maximizing the expected accumulative gross margin amounts and the model is not able to offer feasible results for non-irrigated products. (Table 4)

In Kataoka model no goal income is inspired into the model and the model is always to offer cultivation model that maximizes the goal gross margin level in the region. Means for accepting higher risk levels by farmers in Kataoka's model is probabilistic limitation. As results show, by accepting higher probabilistic amounts maximum goal gross margin level is also increased. Increase in probability level that expected gross margin being less than the goal gross margin from 0.2 to 0.3 causes 1% increase in maximum goal gross margin. This increase is in accordance with theoretical principles, because by accepting more risk levels, the farmer expects higher expected gross margin to be realized. Finally, when the probability of expected gross margin being less than the goal gross margin is increased to 40%, the offered model by Kataoka will be very similar to the results of Telser's model and this is an expressive of results not being conservative, in this case, maximum goal gross margin that can be expected on the basis of events of agricultural years 1996-1997 till 2004-2005 and limitations in production resources is 4172275 million rials. Regarding the labor intensify cultivation of tomato (162.6) and high amount of used chemical fertilizers (789.8 kilograms) and chemical poisons (10.7 liter) in the aforesaid agriculture and consequently, creating environmental risks in the region, optimized cultivation pattern has been offered by retaining the same probabilistic conditions and limiting under cultivation level of tomato in its present amount in the region.(Table 5)

Results from Kataoka's model showed that regarding the 10 percent probability of reduction in expected gross margin level from the goal income level, watermelon agricultural products with 82865.66

Table 4: Kataoka's model Results for fall cultivation period.

| Probability limitation | maximum Goal gross margin | Acreage of optimized enterprises (ha) | | |
|---|---|---|---|---|
| | | Tomato | cucumber | Aquatic Wheat |
| | 4025430 | 355726.4 | 61040.05 | 0 |
| | 987775.5* | 10836.5* | 74112.7* | 294856* |
| | 4025430 | 355726.4 | 61040.05 | 0 |
| | 1007539* | 10836.5* | 98716.39* | 256270.5* |
| | 4066941 | 386372.47 | 30394.02 | 0 |
| | 1128476* | 10836.5* | 262124.86* | 0* |
| | 4172275 | 416766.22 | 0.27 | 0 |
| | 1373593* | 10836.5* | 262124.6* | 0* |

1 in million rials.
* Scenario of keeping tomato's acreage at present region's level.
Resource: research findings.

Table 5: Kataoka's model Results for spring cultivation period.

| Probability limitation | maximum Goal gross margin | Acreage of optimized enterprises (ha) | | |
|---|---|---|---|---|
| | | Paddy | Maize | Cucumber |
| | 575325824 | 7273.33 | 0 | 82865.66 |
| | 418269216* | 57603.9* | 10466* | 22069* |
| | 575325824 | 7273.33 | 0 | 82865.66 |
| | 418269216* | 57603.9* | 10466* | 22069* |
| | 591480512 | 19326.87 | 0 | 70812.12 |
| | 426256928* | 68070* | 0* | 22069* |
| | 609046720 | 0 | 0 | 90139 |
| | 438237792 | 68070* | 0* | 22069* |

1 in thousands rials.
* Scenario of keeping watermelon's acreage at present region's level.
Resource: research findings.

hectares and paddy with 7273 .33 hectares are offered as optimized cultivation pattern by Kataoka's model. By accepting more risk amounts (increasing the probabilistic levels) the amounts of maximum goal gross margin is increased and, on the other hand, watermelon allocates more under cultivation level to itself. Finally, in 40% probability, watermelon is introduced as the only optimized agricultural product by the model. Taking the probabilistic amounts into account, a scenario was offered to retain under cultivation level of watermelon at the present amount of the region. By exerting this scenario the amount of maximum goal gross margin, is reduced at the same probabilistic limitation. The mentioned scenario at probabilistic levels of 30 and 40 percent allocates the whole available under cultivation level to two products of paddy and watermelon.

Kataoka's model determined the amount of under cultivation level of non-irrigated products as 0 in all risk accepting levels. The mentioned results are due to the negative expected gross margin amounts of non-irrigated products in under study years.

## CONCLUSION

As results from Telser's and Kataoka's models showed that by accepting more risk amounts (increasing the goal income and probabilistic levels in Telser's model and increasing probabilistic levels in Kataoka's model), the optimized amounts advise tomato cultivation during the cultivation period of fall and watermelon cultivation during the cultivation period of spring. On the basis of results, by allocating agricultural lands of Shoshtar region to tomato and watermelon cultivation and specializing the farming in this city, gross margin of agricultural production system can be increased to 6116047000 and 727782272 thousand rials, respectively. At the present conditions of Shoshtar region, because of unsuitable living conditions of the farmers, changing the cultivation model and leading the farmers towards the mentioned products cultivations besides increasing the productivity of productive units, are among the most suitable techniques. In this way, production stability can be guaranteed.

Accepting aforesaid risky levels by farmers of Shoshtar region depends on offering suitable equipments for adjusting the effects of uncertainty and risks due to unwanted climatic conditions in agricultural production process. Beside the equipments such as different kinds of agricultural insurances and cheap financial dept, on time performing of agricultural activities and reducing the variable expenses of production by mechanization development are among those cases that can be effective in realizing the above said things. Granting cheap financial dept in order for providing production inputs, strengthening productive structures, giving power to rural cooperatives in order to increase the power of bargaining of farmers and it effect on market, private sector's more participating and commercializing the production companies and developing distributive services to know about the methods of correct applying of production inputs are some of important techniques of risk reducing in production system and encouraging the farmers to change the cultivation model and specialize the farming.

## REFERENCES

1- Atwood, J.A., 1985, Demonstration of the Use of Lower Partial Moments to Improve Safety-First Probability Limit. American Journal of Agricultural Economics, Vol. 6: 787-793.

2- Atwood J.A., Watts M.J., Helmers G.A. and L.J. Held, 1988, Incorporating Safety-First Constraints in Linear Programming Production Models, Western Agriculture Economics Association, Western Journal of Agriculture Economics, Vol.13: 29-36.

3- Backeberg, G.R., 1997, Water Institutions, markets and decentralized resource management: Prospect for innovative policy reforms in irrigated agriculture, Agrekon, Vol. 36:350-384.

4- Berck, P. and Hihn J. M., 1982, Using the Semi variance to Estimate Safety-First Rules. American Journal of Agricultural Economics, Vol. 64: 298-300.

5- Central bank of Islamic Republic of Iran, 1996-2005, Wholesale price index statistics.

6- Charnes, A., and Cooper W. W., 1959, Chance-Constrained Programming. Management Science, Vol. 6: 73-79.

7- Haile, B.O., Grove, B. and L.K. Oosthuizen, 2003, Impact of capital on the growth process of a sugarcane farm in Mpumalanga, 41st Annual Conference of the Agricultural Economics Association of South Africa.

8- Hassan Shahi, M., 2006, Agricultural decision making under risk case study Arsanjan region, Journal of Iranian Agricultural Economics and Development, Vol. 54: 161-178.

9- Hazell P.B.R. and R.D. Norton, 1986, Mathematical Programming for Economic Analysis in Agriculture, Macmillan, New York.

10- Irima, M., J. Novak and P. Duffy, 2004, optimal crop insurance options for Alabama cotton-peanut producers: A Target MOTAD Analysis, Agricultural Economics, Agecon.

11- Jahad-Keshavarzi of Shoshtar region, 1996-2005, production cost statistics of cultivation products.

12- Khuzestan province's Agricultural Supportive Services Company, 2005, Allocation statistics of agricultural inputs.

13- Khuzestan province's Regional Water Corporation, 2005, Allocation statistics of agricultural water.

14- Khuzestan's Economical Affairs and Coordination Office of Governor's Office, 2005, Report of Khuzestan agricultural condition.

15- Nicholson, C.F., R.W. Black, C.I. Urbina, D.R. Lee, D.G. Fox, and P.J. Van Soet, 1994, Economic comparison of nutritional management strategies for Venezuelan dual-purpose cattle system, Journal of Animal Science, Vol. 72: 1680-1696.

16- Norak, J.L. and et al., 1991, Risk and Sustainable Agriculture: A Target MOTAD analysis of the 92 years, Old Rotection, Southern Journal of Agricultural Economics, Vol. 22: 145-154.

17- Patrick, G.R., Wilson P.N., Barry P.J., Boggess W.G. and D.L. Young, 1985, Risk Perception and Management Responses: Producer-Generated Hypothesis for Risk Modeling. Southern Journal of Agricultural Economics, Vol. 2: 231-238.

18- Peykani, Gh.R., Kavoosi Kelashemi, M. Sadat Barikani, S.H. and A. Azari, 2009, Determining the Best Variety of Rice for Cultivation Using Kataoka's and Telser's Risky-linear Programming Methods-Case Study of Gilan Province 2000-2006, J. Agric. Sci. Technol., Vol. 11: 1-8.

19- Pyle, D.H., and S.J. Turovsky, 1970, Safety-First and Expected Utility Maximization in Mean-Standard Deviation Portfolio Analysis, Rev. Econ. and Statist. Vol. 52: 75-81.

20- Roy, A.D., 1952, Safety-First and Holding of Assets. Econometrica, Vol. 20: 431-449.

21- Rudel, R. 2000, Target MOTAD for risk lovers: An Alternative Version, Southern Journal of Agricultural Economics, Vol. 18: 175-185.

22- Tauer, L.M., 1983, Target MOTAD. American Journal of Agricultural Economics, Vol. 65: 601-606.

23- Van Schalkwyk, H.D. and D.B. Louw, 2004, Market risk, water management and the multiplier effect of irrigation agriculture with reference to Northern Cape. Research Report, Water Research Commission.

# Size Distribution of Income Among Rice-Based Farming Households in South Eastern States of Nigeria

*Chikezie, C[1], Ibekwe U.C[2], Ohajianya D.O[2], Orebiyi, J.S, OguomaN.N[2], Obasi, P.C, Henri-Ukoha, A[2], Emenyonu, C.A[2] and Nwaiwu, I.U[2]*

Abstract

The study was designed to investigate the income distribution among rice-based farming households in Okigwe Agricultural Zone of Imo State, Nigeria. A sample of 120 rice-based farmers was selected by multistage sampling technique. Data were collected with a well structured questionnaire administered to 120 randomly selected rice farmers. Data collected were analyzed using descriptive statistics and inferential statistics. Gini-coefficient model was also employed. The socio-economic characteristic of the farmers revealed that majority of the farmers were educated. Besides, majority of them also had appreciable experience in rice farming which makes them better rice farmers. It was revealed that income share percentage of the richest households was 17.65% followed by the second richest group with income share percentage of 13.27%. Those in the twelfth decile represented the poorest group with a cumulative share percentage 2.82%. The Gini-coefficient of distribution of 0.32 showed that incomes were not highly concentrated but varied around the low per capita of N1442,859 per household. It becomes imperative therefore to enhance farmers' income through a sustainable improved technology in rice farming.

***Keywords:***
***Income distribution, Rice-based farmers, Gini coefficient, Sustainable.***

[1] *Abia Agricultural Development Programme, Abia State.*
[2] *Federal University of Technology, Owerri, Imo State.*
[*] *Corresponding Email: chykecoms@yahoo.com, haukoha@yahoo.com*

## INTRODUCTION

Rice is a very important primary food source and this is drawn from the understanding that rice-based systems are essential for food security, poverty alleviation and improved livelihoods by enhancing the status of the farmer (FAO, 2004). It is a staple food and a good source of income for the growers. Rice production occurs in all agro-ecological zones in Nigeria. Hawk Worth (1985) estimated that half the world's population subsists wholly or partially on rice. In the early 60's in Nigeria, the local production of rice was meeting consumption demand but from 1990's till date the decline in local production made demand for rice to exceed supply. This led to an increase in the importation of rice in Nigeria. This situation between 60's and 90's created a serious drain on Nigeria foreign exchange reserve. This also led to an increase in rice production bill from a level of US $60 Million in 1994 to US $160 Million in 2003 (FAO, 2003). As a result of this, majority of rural dwellers have fallen below the poverty line and a greater gap had continued to exist between the rich and the poor. One striking issue was that income levels among rice farming households was considered to be one of the major factors for sustained improved rice production. Meanwhile, in recent years, the introduction of new packages like Fadama projects to Nigerians farmers was expected to bridge the gap between demand and supply of rice but local production has however not been enough to meet the consumption demand of the growing Nigerian populace. One of the reasons for this trend is the issue of rice farmers' return which is becoming very low (CBN, 1997). This however is expected to boost local production of rice for the farmers thereby improving their income sustainably. This therefore, will ultimately reduce the gap in the distribution of income between the rich and the resource poor farmers.

The distribution of a nation's income indicates the proportion of the income that goes to the rich and that which goes to the poor. It also shows that the rich can purchase many, rather than few goods and services. Hence, the more shares a person gets as his/her purchasing power and consumption pattern. The size distribution of income determines how the total income earned by a specified population is distributed among members of that population (Ibekwe and Eze, 2002). The trend of income distribution has been a big concern to economists for a long time (Clarke et. al., 2003). Specifically, the 1990's witnessed an increased empirical attention by economists in the study of income and wealth (Atkinson and Bourguignon, 2000). Their interest became high because it was observed that high incidence of income inequality was unfavorable for political and socio-economic growth. However, previous studies have shown that over the last two decades, income inequality has risen in many developing countries like Cameroun and Nigeria (Kanbour and Lesting, 1999). The rapid economic growth that occurred between 1965 and 1974 in Nigeria was followed by a serious income disparity which has continued to widen substantially (Aigbokan, 1997; Ipinnaye, 2001; Ibekwe, 2001). Despite the past policy intervention to correct the abnormality, the problem has continued to increase among the nation's farming zones (Ibekwe, 2001). One of the aims of agricultural policy is to increase the economic and social standard of living of the farming households. Standard of living implies a broad definition of quality of life of the agricultural population which includes their happiness, utility, income, purchasing power etc (Hills, 1989). Poverty in the form of low income affects households in terms of consumption, production and investment. Poverty however is the lack of income needed to acquire minimum necessities of life (Ibekwe, 2001). It is most damaging where it affects the entire farming zone. Poverty in the form of low income reduces the operation of safety nets for farm households and cuts off entire farming communities from access to benefits of modern developments (Ibekwe, 2001). The widening dimension of income and poverty arouses serious humanitarian concerns and fears of political instability. Despite considerable progress made in recent decades, the world still falls short of the goal of adequate food supply as well as other amenities. Food and agricultural organization (1992), estimated that more than

790 millions of people in the developing countries are chronically undernourished, consuming too little food to meet even minimal energy and protein needs. They therefore advocate equitable, participatory rural development as the key to eradication of poverty. This is the major cause of low income, under nutrition and food insecurity. It also recommends the promotion of rural development that helps to increase the purchasing power of the poor as well as improving the consumption pattern. Since the millennium declaration was adopted in September 2000, the goals and targets it contains have become essential benchmark for assessing development all over the world. Given the importance of meeting the goals, progress towards them must be consistently monitored and particularly aimed at identifying potential problems. This aim is not easily achieved due to uncertainties associated with agriculture especially among rice farmers.

Few researchers have worked on income inequality in the eastern part of the country, hence the relevance of the study. The paper seeks to address the following objectives: to analyze the socio-economic characteristics of the rice farming households and to determine the size distributions in income of rice-based households in the study area. This would lead to the formulation and implementation of policies that would enable them to improve on their performance and productivity such that their income will be enhanced sustainably.

## MATERIALS AND METHOD

The study was conducted in Okigwe Agricultural Zone of Imo State. This zone is located on the North Eastern part of Imo State and bounded on the South East by Abia State (Imo Diary, 1992). The topography is flat and gently undulating (MFO, 1999). Okigwe zone was purposively selected, being the main rice farming zone in Imo State. Multistage sampling technique was adopted in the study. First stage involved the purposive selection of two local government areas (based on the highest concentration of rice farmers). Second stage involved the random selection of four communities from the 2 local government areas from which a list of rice farmers was compiled. From this sampling frame, 5 female headed rice farmers and 10 male headed rice farmers were selected using proportionate sampling technique, giving a sample size of 120 respondents. Primary data collection was by a bi-weekly interview with the aid of a questionnaire for a period of 12 months (March 2004 to February 2005). Field workers were trained to keep detailed information on the following; farm labour, inputs, purchases of crops and livestock, transport costs, sales of rice outputs etc. The secondary data were sourced from websites, journal articles and other relevant literature.

Data collected were analyzed using descriptive statistics such as average, percentage, frequency distribution. Also, Gini Coefficient model as specified by Gini, Matlon (1979), Adelman and Morris (1991) as shown below were used to analyze farm income distribution. The income distribution model used is specified as:

A value equal to 0 corresponds with perfect

$$G = \frac{1}{2n^2u} \sum_{i=1}^{n} \sum_{j=1}^{n} \left| Yi - Yj \right|$$

Where,

n = number of observations

U = mean income

Yi = income observation i

Yj = income observation of all other observations

equality while 1 corresponds with perfect inequality (Matlon, 1979).

## RESULTS AND DISCUSSION

### *Socio-economic characteristics of the farmers*

Table 1 showed that 7.5% of male headed households and 25% of female headed households had no formal education, this are likely to be those who are advanced in age. Secondary education has the highest percentage for male headed households (53.75%) and female headed households had (42.50%). Tertiary institution had (21.25%) for males and 12.50 for female headed households. The implication is that farmers in the area are educated. This is likely to influence the adoption behavior of the farmers

Table 1: Socio-economic characteristics of the respondents

| Characteristics | Male-headed | | Female-headed | |
|---|---|---|---|---|
| | Frequency | Percentage | Frequency | Percentage |
| **Educational level** | | | | |
| 0 | 6 | 7.50 | 10 | 25.00 |
| 1-6 | 14 | 17.50 | 7 | 20.00 |
| 7-12 | 43 | 53.75 | 17 | 42.50 |
| Above 12 | 17 | 21.25 | 6 | 12.50 |
| **Age** | | | | |
| 25-39 | 18 | 22.50 | 12 | 30.00 |
| 40-54 | 37 | 46.25 | 18 | 45.00 |
| 55-69 | 15 | 18.75 | 7 | 17.50 |
| 70 and above | 10 | 12.50 | 3 | 7.50 |
| **Farming experience** | | | | |
| 1-5 | 19 | 23.75 | 13 | 32.50 |
| 6-11 | 16 | 20.00 | 11 | 27.50 |
| 12-17 | 9 | 11.25 | 12 | 30.00 |
| 18-23 | 21 | 26.25 | 3 | 7.50 |
| 24 and above | 15 | 18.75 | 1 | 2.50 |
| **Farm size (Ha)** | | | | |
| 0.20-1.59 | 26 | 32.50 | 23 | 57.50 |
| 1.6-2.90 | 35 | 43.75 | 8 | 20.00 |
| 3.0-4.39 | 10 | 12.50 | 5 | 10.00 |
| 4.40-5.79 | 7 | 8.75 | 4 | - |
| 5.8 and above | 2 | 2.50 | - | - |

positively which has a strong bearing on the income and investment of the farmers. Education contributes to increase earning from agriculture and non-agricultural wage employment (Escobal, 2001).

Table 1 also showed that the mean age of the farmers fell within the productive age range of 40 – 54 years indicating that majority of the respondents were middle aged farmers who were still active, vibrant and dynamic and are more likely to adopt innovations better and faster than their elderly counterparts. Age range of (31-50 years) is innovative, motivated and active individuals. The table further revealed that the respondents were reasonably experienced in rice production as mean years of experience in rice production were 19 years. It is expected that the longer a person is in a business, the more skilled and experienced he becomes in its management. Although the males had more years of experience, majority of the male headed households had farm size of (43.75%) and female (57.50%) fell within the range of 0.2 to 2.9ha. This is attributed to the land tenure system obtainable in the state. This has a serious effect on mechanized farming (Idiong 2007) and also has an indirect effect on their income.

### *The Size Distribution of Income*

The size distribution of income determines how the total income earned by a specified population is distributed among members of that population. This size distribution is normally shown by disaggregating the total income into deciles (Ahearn, et al., 1985). The individual units are first ranked in ascending or descending order of income and then put into deciles with each decile consisting often 10 households. Here the entire sample population is said to be ranked in order of degree of income concentration (Miguel and Centeno, 2003; Ibekwe, 2001). The size distribution of aggregate income by household population decile is presented in table 2.

Table 2 shows the average income of 10 households in each decile with the corresponding cumulative percentage of income. Each decile consist of 10 households with the richest households represented in the first decile and the poorest in the twelfth decile. The table shows that income share percentage of the richest households was 17.65% followed by the second richest group with income share percentage of 13.27%. Those in the twelfth decile represented the poorest group with a cumulative share percentage 2.82%. Comparatively speaking, when the structure of income distribution is inequitable, it takes different forms depending on the study in question. It is also worthy to note that that fact that the richest decile has a large share does not necessarily mean that the poorest household and the richest household was found to be1:6. The ratio between the average income of groups at the top and bottom of income distribution also served to illustrate the wide disparity in access to monetary resources in the study area. Incomes were not highly concentrated but varied around the low per capita of N1,442,859 per household. This finding is consistent with what is obtainable in a population whose mean income do not greatly exceed the subsistence level (Ibekwe, 2001). According to Miguel and Centeno, 2003, the relative income level percentile shows the relative income levels of households at certain percentiles or fraction (for instance two – third) compared to the median income. By taking a fraction of the average as a relative threshold, instead of the average itself, it is also possible to determine the relative poverty or the approximate proportion of the population whose income does not afford them access to goods regarded as essential in the society (Lipton, 2001). This cumulative proportion below the percentile of the median income provides a useful information on the incidence of low income (ECLAC, 2003). The percentage of people whose income falls short of a given relative threshold such as median income also helps to illustrate the pattern of income. The Gini-coefficient of distribution of 0.32 was determined. This showed a lower level of income distribution when compared to World Bank determination of income distribution in Nigeria which was 0.45 in 1996. This Gini value of 0.32 showed that income disparity was not wide; therefore it was closely more concentrated. This has implication for income redistribution policy. Income distribution values depend on the area under study as well as the scope. For instance, Matlon (1979) studied income distribution among farmers in Northern Nigeria and found a Gini-coefficient value of 0.28 which was much lower than World Bank (2006) estimated value while Ibekwe (2001) found a ratio of 0.488 for farm households in Orlu agricultural zone of Imo State. This finding is consistent with what is obtainable in a population whose mean income do not greatly exceed the substance level (Aigbokhan, 1997; Ipinnanye 2001; Ibekwe 2001).

Table 2: The distribution of aggregate income by household population deciles

| Decile | Income (N) | Income share (percentage) |
|---|---|---|
| | 726,211.6 | 17.65 |
| | 546,063.8 | 13.27 |
| | 454,963.0 | 11.06 |
| | 408,621.9 | 9.93 |
| | 372,638.6 | 9.05 |
| | 348,342.9 | 8.47 |
| | 299,699.1 | 7.29 |
| | 255,809.7 | 6.22 |
| | 230,699.1 | 5.61 |
| | 192,136.9 | 4.67 |
| | 163,320.8 | 3.96 |
| | 115,852.6 | 2.82 |

## CONCLUSION

The result from the study showed that there were different levels of income unevenly distributed in the study area. Income levels varied among the entire Okigwe Agricultural zone creating a gap between the rich and the poor. Though the distribution was not very wide, yet it had some policy implication in the area. The distribution of incomes in the study area is important for farm credit support programmes and farm credit recovery. This is relevant for effective poverty alleviation programmes of the Federal and State governments in Nigeria. The socio-economic characteristic of the farmers revealed that majority of the farmers were young, educated had appreciable experience in rice farming which makes them better rice farmers.

## RECOMMENDATIONS

In order to improve the income of rice farmers, State Ministry of Agriculture in conjunction with Federal government should encourage the use of new and improved technology in farming to boost yield and income of the farming populace. This way the disparity in income distribution will be reduced. The effect of such policy intervention would depend on the structure of relevant institutions (Research Institutes, State Agricultural Development Programmes). There is also the need for farmers to have increased access to improved and subsidized inputs and credit facilities.

## REFERENCES

1- Adelman I. and C.T Morris (1971); "The Anatomy of income Distribution in Developing Nations" Mimeo (USAID) Washington D.C.

2- Ahearn M; Johnson J; and R. Strickland (1985); "The Distribution of Income and wealth of farm operator Households" Economic Research Society, United State Department of Agriculture, PP 1087-1094

3- Aigbokhan B.E (1997), "Poverty Alleviation in Nigeria: Some Macroeconomic Issues" NES Annual Conference Proceeding PP 181- 209.

4- Atkinson A. B and F. Bourguignon (2000); "Introduction to income Distribution and Economics in A.B

5-Atkinson and F. Bourguinon (eds) Handbook of income distribution, vol. 1 North Holland, Amsterdam.

6- Central Bank of Nigeria (1997). Annual Report and Statement of account. Lagos, Nigeria

7- Clarke G. L; Collin X and H. Zou (2003). "Finance and income inequality: Test of alternative theories" World Bank Policy Research Working Paper, 2984, Washington D.C.

8- Economic Commission for Latin America and the Carribean (ECLAC) (2003). "Social Panorama of Latin America 2004".(Santiago de Chile:ECLAC).

9- Escobal, J. (2001). "The determinants of Non-farm income diversification in rural Peru". World Development Series 29(3): 496-508.

10- Food and Agricultural Organization (1992): Africa Recovery. Vol. 6.No.3, Rome Italy.

11- Food and Agricultural Organization (2003): Production year book, Rome, Italy.

12- Food and Agricultural Organization (2004): Concept paper- The International Year of Rice 2004, Rome, Italy

13- Hawkworth D. L. (1985), Rice Diseases CMI Slough, UK. PP330

14- Hills, B (1989). Farm Income , Wealth and Agricultural Policy Amesbury Gower Publishing Company Limited. Aldeshot, Great Britain

15- Ibekwe U.C. (2001), "Income Distribution among farm Households in Orlu Agricultural zone of Imo State, Nigeria", Unpublished Ph.D, Thesis, University of Nigeria, Nsukka. PP.9-20,33-40

16- Ibekwe, U.C and C.C. Eze (2002). "Income and consumption pattern among inhabitants of Rural Communities of Ideato Area of Imo State, Nigeria" Jotten, Vol. 9, No.2, PP 25.

17- Idiong I. C.(2007), "Estimation of Farm Level Technical Efficiency in Small-scale Swamp Rice production in Cross River State of Nigeria: A Stochastic Frontier Approach", World Journal of Agricultural Sciences 3(5): 653-658, 2007 ISSN 1817-3047 Imo State of Nigeria, Diary 1992

18- Imo State of Nigeria, Diary 1992. Imo State Government.

19- Ipinnaiye, A. O. (2001), "A Decomposition Analysis of the sources of Income inequality in Ibadan" Unpublished B.Sc, project, Department of Agricultural Economics, University of Ibadan.

20- Kanbur, R and N. Lesting (1999). "Why is inequality back on the agenda" Paper prepared for the annual Bank Conference Development Economics, World Bank, Washington D.C

21- Lipton M.(2001). Reviving global poverty reduction: what is the role of genetically modified plants, Journal of International Development, 13:7, 823.

22- Matlon P. J. (1979), Income Distribution among Farmers in Northern Nigeria; Empirical Results and Policy Implication, Department of Agricultural Economics, Michigan State University. New York, U.S.A.

23-Miguel H.K. and A. Centeno (2003); Inequality in Latin America, Annual Review of Sociology", 29: 1,363.

24- Ministry of Forestry, Imo State (1999), Revised forest management Plan Prepared for formecy, by Beak/Geomatics International, Nigeria.

25- Ogunpola A. and O. Oladeji (1975); "Housing as an indicator of urban poverty: The case of metropolitan Lagos"The proceedings of the Nigerian economic Society, Ibadan, Nigeria.

26- West African Rice Development Association (2000). The Nigerian Rice Economy in a competitive World; Constraints, Opportunities and Stategies choices (Concept note submitted to USA).

# PERMISSIONS

All chapters in this book were first published in IJAMAD, by Islamic Azad University; hereby published with permission under the Creative Commons Attribution License or equivalent. Every chapter published in this book has been scrutinized by our experts. Their significance has been extensively debated. The topics covered herein carry significant findings which will fuel the growth of the discipline. They may even be implemented as practical applications or may be referred to as a beginning point for another development.

The contributors of this book come from diverse backgrounds, making this book a truly international effort. This book will bring forth new frontiers with its revolutionizing research information and detailed analysis of the nascent developments around the world.

We would like to thank all the contributing authors for lending their expertise to make the book truly unique. They have played a crucial role in the development of this book. Without their invaluable contributions this book wouldn't have been possible. They have made vital efforts to compile up to date information on the varied aspects of this subject to make this book a valuable addition to the collection of many professionals and students.

This book was conceptualized with the vision of imparting up-to-date information and advanced data in this field. To ensure the same, a matchless editorial board was set up. Every individual on the board went through rigorous rounds of assessment to prove their worth. After which they invested a large part of their time researching and compiling the most relevant data for our readers.

The editorial board has been involved in producing this book since its inception. They have spent rigorous hours researching and exploring the diverse topics which have resulted in the successful publishing of this book. They have passed on their knowledge of decades through this book. To expedite this challenging task, the publisher supported the team at every step. A small team of assistant editors was also appointed to further simplify the editing procedure and attain best results for the readers.

Apart from the editorial board, the designing team has also invested a significant amount of their time in understanding the subject and creating the most relevant covers. They scrutinized every image to scout for the most suitable representation of the subject and create an appropriate cover for the book.

The publishing team has been an ardent support to the editorial, designing and production team. Their endless efforts to recruit the best for this project, has resulted in the accomplishment of this book. They are a veteran in the field of academics and their pool of knowledge is as vast as their experience in printing. Their expertise and guidance has proved useful at every step. Their uncompromising quality standards have made this book an exceptional effort. Their encouragement from time to time has been an inspiration for everyone.

The publisher and the editorial board hope that this book will prove to be a valuable piece of knowledge for researchers, students, practitioners and scholars across the globe.

# LIST OF CONTRIBUTORS

**Tahir Munir Butt**
Department of Agri. Extension Education, University of Agriculture, Faisalabad

**Muhammad Zakaria Yousuf Hassan**
Agricultural Department, Punjab-Pakistan

**Shahbaz Talib Sahi**
Department of Plant Pathology, University of Agriculture, Faisalabad

**Khalid Mehmood**
Department of Agricultural Extension in University of Agriculture, Faisalabad-Pakistan

**Mashood-ul-Subtain**
University of Agriculture, Faisalabad, Depalpur Campus, Okara-Pakistan

**Naeem-ul Hassan**
University of Agriculture, Faisalabad, Depalpur Campus, Okara-Pakistan

**Soleiman Rasouliazar and Saeid Fealy**
Department of Agricultural Management, Mahabad Branch, Islamic Azad University, Mahabad, Iran

**Fatemeh Razzaghi Borkhani, Ahmad Rezvanfar, Hossein Shabanali Fami and Mahtab Pouratashi**
Department of Agricultural Extension and Education, Faculty of Agricultural Economics and Development, University of Tehran, Karaj, Iran

**Mojtaba Sookhtanlo, Hesamedin Gholami and Seyyed Reza Es'haghi**
Department of Agricultural Extension and Education, University of Tehran, Karaj, Iran

**Lounis Semara**
Ph.D Student of Animal Production, Department of Agriculture and Animal Science, Setif 1 University, Algeria

**Charefeddine Mouffok**
Assistant Professor, Department of Agriculture and Animal Science, Setif 1 University, Algeria

**Toufik Madani**
Professor, Department of Agriculture and Animal Science, Setif 1 University, Algeria

**Ademoye Emmanuel Akinboboye, Fadipe, A.A., Adigun Jubril Olayinka and Animashaun**
Department of Agricultural Economics and Farm Management, University of Ilorin, PMB 1515, Ilorin, Nigeria

**Ahmad Reza Ommani**
Associate professor, Department of Agricultural Management, Shoushtar Branch, Islamic Azad University, Shoushtar, Iran

**Laleh Salehi**
PhD of Agricultural Extension, Department of Agricultural Extension and Education, Faculty of Agricultural Economics and Development, College of Agriculture & Natural Resources, University of Tehran

**Ahmad Rezvanfar and Seyed Mahmood Hosseini**
Professor of Department of Agricultural Extension and Education, Faculty of Agricultural Economics and Development, College of Agriculture & Natural Resources, University of Tehran

**Seyed Hamid Movahed Mohammadi**
Associate Professor of Department of Agricultural Extension and Education, Faculty of Agricultural Economics and Development, College of Agriculture & Natural Resources, University of Tehran

**Maryam Bagheri**
Graduate Student of Department of Agricultural Extension and Education, Garmsar Branch, Islamic Azad University, Garmsar, Iran

**Mehrdad Nikanami**
Department of Agricultural Extension and Education, Garmsar Branch, Islamic Azad University, Tehran, Iran

**Hossein Shabanali Fami**
Department of Agricultural Development & Management, Faculty of Agricultural Economics & Development, College of Agriculture & Natural Resources, University of Tehran, Karaj, Iran

**Mehdi Charmchian Langerodi**
Assistant professor of Agricultural Extension and Education Department, Sari Branch, Islamic Azad University, Sari, Iran

**Zainab Mohd Karim and Nalini Arumugam**
School of Agriculture & Biotechnology, Faculty of Bioresource and Food Industry, Universiti Sultan Zainal Abidin (UniSZA)

**Boniface Bonaventure**
Centre for the Promotion of Knowledge and Language, Universiti Malaysia Sabah (UMS), Agricultural Campus

**Mehdi Alikhani Dadoukolaei**
Young Researchers and Elite Club, Sari Branch, Islamic Azad University, Sari, Iran

**Amir Zameni**
MSc. Graduate in Agricultural Extension and Education, Sari Branch, Islamic Azad University, Sari, Iran

**Javad Shahraki**
Associate Professor, Department of Economics, University of Sistan and Baluchestan, Zahedan, Iran

**Mosayeb Pahlavani**
Associate Professor, Department of Economics, University of Sistan and Baluchestan, Zahedan, Iran

**Nasrin Ohadi**
Ph.D. Candidate of Agricultural Economics, Department of Economics, University of Sistan and Baluchestan, Zahedan, Iran

**Mohammad Sadegh Sabouri and Siavash Noori**
Faculty of Agriculture, Department of Agricultural Extension and Education, Garmsar Branch, Islamic Azad University, Garmsar, Iran

**Mostafa Karbasioun**
Faculty of Agriculture, Department of Agricultural Development, Shahrekord University, Shahrekord, Iran

**Ghader Dashti, Babollah Hayati, Noushin Bakhshy and Mohammad Ghahremanzadeh**
University of Tabriz, Faculty of Agriculture, Department of Agricultural Economics

**Fatemeh Askary Bozayeh**
Ph.D. Student of Geography and Rural Planning at Kharazmi University, Master of Research in Guilan Agricultural and Natural Resources Research and Education Center

**Fathollah Keshavarz Shal**
Ph.D. of Human Geography, Director of Planning and Development Department of Agriculture in Gilan Agricultural and natural Resources Research and Education Center

**Zahra Mojib Haghghadam**
Ph.D. Studentof Entomology at Guilan University, Master of Research in Guilan Agricultural and Natural Resources Research and Education Center

**Parisa Shahinrokhsar**
Ph.D. Student of Irrigation and Drainage at Ferdosi University, Scientific Member of Guilan Agricultural and Natural Resources Research and Education Center

**Amolegbe Khadijat Busola, Usman Mutiat Abimbola and Animashaun Jubril Olayinka**
Department of Agricultural Economics and Farm Management P.M.B 1515, University of Ilorin, Ilorin, Nigeria

**Hossein Bevrani**
Professor, Department of Statistics, Faculty of Mathematical Sciences, University of Tabriz, Tabriz, Iran

**Simin Naseri**
PhD Graduate in Agricultural Extension from UAS, Bangalore, India

**Ehsan Gholifar**
Ph.D. Candidate, Agricultural Extension and Education Department, Tarbiat Modares University, Iran

**Enayat Abbasi**
Assistant Professor, Agricultural Extension and Education Department, Tarbiat Modares University, Iran

**Gholamreza Pezeshki Rad**
Professor, Agricultural Extension and Education Department, Tarbiat Modares University, Iran

**Hassan Salehi**
Associate Professor, Iran Fisheries Organization, Tehran, Iran

**Abdolmotalleb Rezaei**
Assistant Professor, Agricultural Extension and Education Department, University of Tehran, Iran

**Adigun James Kehinde, Ashimolowo Olubunmi R and Fabusoro Eniola**
Department of Agricultural Extension and Rural Development, Federal University of Agriculture Abeokuta Nigeria

**Adeogun Stephen Oluseun**
Department of Agricultural Administration, Federal University of Agriculture Abeokuta Nigeria

**Mesele Samuel Ayodele**
Department of Soil Science and Land Management, Federal University of Agriculture Abeokuta Nigeria

**Ejaz Ashraf and Abu Bakar Muhammad Raza**
Assistant Professor, Agriculture. Extension Education. University College of Agriculture, University of Sargodha, Punjab-Pakistan

**Samiullah**
Lecturer in Statistics, University College of Agriculture, University of Sargodha, Punjab-Pakistan

**Muhammad Younis**
University College of Agriculture, University of Sargodha, Punjab-Pakistan

**Agom, Damian Ila, Susan Ben Ohen and Kingsley Okoi Itam**
Deptartment of Agric Economics, University of Calabar, Calabar, Nigeria

**Nyambi N. Inyang**
Commercial Agric Development Project (CADP), Cross River State, Nigeria

**Abura Odilla Gilbert and Barchok Kipngeno Hillary**
Chuka University College, Chuka, Kenya

**Onyango Christopher Asher**
Egerton University, Egerton, Kenya

**O. L Balogun, A. Adeoye, S. A Yusuf and R. J Akinlade**
Department of Agricultural Economics, University of Ibadan, Nigeria

**A. Carim-Sanni**
National Institute of Social and Economic Research Ibadan, Nigeria

**Mohammad Kavoosi Kelashemi**
PhD student of Agricultural Economics, Tehran of University, Karaj, Iran

**Mohammad Aghapour Sabbagi**
Assistant Professor of Islamic Azad University Shoshtar Branch

**Hamid Reza Alipour**
Assistant Professor of Islamic Azad University Rasht Branch

**Chikezie**
Abia Agricultural Development Programme, Abia State

**Ibekwe U.C, Ohajianya D.O, Orebiyi, J.S, Oguoma N.N, Obasi, P.C, Henri-Ukoha, A, Emenyonu, C.A and Nwaiwu, I.U**
Federal University of Technology, Owerri, Imo State

**Michael G. Kanyi, Tom Vandenbosch, Dr. Fredrick U. Ngesa and Dr. Joash K. Kibett**
Global Coordinator, Farmers of the Future Initiative, World Agroforestry Centre (ICRAF), Nairobi, Kenya
Senior lecturer, Department of Agricultural Education & Extension, Egerton University, Kenya
Senior lecturer, Department of Agrcultural Education & Extension, Egerton University, Kenya

**Kaimba George Kinyua**
Chuka University College, Kenya

**Guliye Abdi Yakub**
Egerton University, Keny

**Njehia Bernard Kamau**
Kenyatta University, Kenya

**Hillary Kiplangat Bett**
Humboldt University of Berlin, Germany

# Index